Multiple Perspectives on Difficulties in Learning Literacy and Numeracy

Claire Wyatt-Smith · John Elkins · Stephanie Gunn
Editors

Multiple Perspectives on Difficulties in Learning Literacy and Numeracy

Springer

Editors
Prof. Claire Wyatt-Smith
Griffith University
Fac. Education
Kessels Road 170
4111 Nathan Queensland
Mt Gravatt Campus
Australia
c.wyatt-smith@griffith.edu.au

Prof. John Elkins
Faculty of Education
Griffith University
Mt Gravatt Campus
176 Messines Ridge Road
Mt Gravatt Queensland 4122
Australia
j.elkins@griffith.edu.au

Stephanie Gunn
Griffith University
Fac. Education
Kessels Road 170
4111 Nathan Queensland
Mt Gravatt Campus
Australia
s.gunn@me.com

ISBN 978-1-4020-8863-6 e-ISBN 978-1-4020-8864-3
DOI 10.1007/978-1-4020-8864-3
Springer Dordrecht Heidelberg London New York

Library of Congress Control Number: 2010938618

Printed on acid-free paper

Springer is part of Springer Science+Business Media (www.springer.com)

Acknowledgements

Any book is the sum of the efforts of many individuals, and this is especially true for an edited book. We'd first like to extend our appreciation to our chapter authors. We approached you all, valuing your insights on issues in learning difficulties and your high professional standing in your respective fields. Thank you so much for agreeing to contribute to this book, despite your busy professional lives. Together, your chapters provide yet another contribution to the goal of improving the educational opportunities for students with difficulties in learning.

We also would like to offer special thanks to Peta Colbert, who has assisted with chapter editing and the book's creation from the beginning. You are a valued long-term colleague, and it is a pleasure to have the opportunity to work with you. We appreciate your skills and willingness to engage with the creative process of book production and with the dialogues contained within.

Our editorial consultant, Renée Otmar, has been invaluable for her editorial eye on each chapter as well as for ensuring the authors met timelines, as much as was within her control! Renée, our authors have commented on how pleasant their interactions with you have been in the development of this book and how smooth the process has been.

The development of this book is a part of our academic work. However, it also reflects the deep beliefs we hold of the importance of research into learning, theoretical perspectives and methodologies. Finally, we would like to say that, as co-editors, we have enjoyed once more the opportunity to collaborate with each other and with colleagues around the world.

Brisbane, QLD — Claire Wyatt-Smith
Brisbane, QLD — John Elkins
Brisbane, QLD — Stephanie Gunn

Contents

Contributors

Robyn Beaman Macquarie University, Sydney, NSW, Australia, robyn.beaman@speced.sed.mq.edu.au

Mere Berryman University of Waikato, Hamilton, New Zealand, mere@waikato.ac.nz

Laura Black University of Manchester, Manchester, UK, laura.black@manchester.ac.uk

Raymond Brown Griffith University, Brisbane, QLD, Australia, ray.brown@griffith.edu.au

Maria Lucia Castanheira Federal University of Minas Gerais, Belo Horizonte, Brazil, lalucia@gmail.com

Jill Castek University of California, Berkeley, CA, USA, jcastek@berkeley.edu

Peta Colbert Griffith University, Brisbane, QLD, Australia, p.colbert@griffith.edu.au

John Elkins Griffith University, Brisbane, QLD, Australia, j.elkins@griffith.edu.au

Patricia A. Gallant University of Michigan-Flint, Flint, MI 48502, USA, pgallant@umflint.edu

Judith Green University of California, Santa Barbara, CA, USA, green@education.ucsb.edu

Stephanie Gunn Griffith University, Brisbane, QLD, Australia, s.gunn@me.com

Colin Harrison University of Nottingham, Nottingham, UK, colin.harrison@nottingham.ac.uk

Anne Jobling The University of Queensland, Brisbane, QLD, Australia, a.jobling@uq.edu.au

Robyn Jorgensen (Zevenbergen) Griffith Institute for Educational Research, Griffith University, Brisbane, QLD, Australia, r.jorgensen@griffith.edu.au

David Koppenhaver Appalachian State University, Boone, NC, USA, koppenhaverd@appstate.edu

Anneliese Kramer-Dahl National Institute of Education, Nanyang Technological University, Singapore, Singapore, anneliese.kramer@nie.edu.sg

Dennis Kwek National Institute of Education, Nanyang Technological University, Singapore, Singapore, dennis.kwek@nie.edu.sg

Donald J. Leu University of Connecticut, Storrs, CT, USA, donald.leu@uconn.edu

J. Greg McVerry University of Connecticut, Storrs, CT, USA, jgregmcverry@gmail.com

Robyn Miller The University of Queensland, Brisbane, QLD, Australia, robynmiller07@gmail.com

Karen B. Moni The University of Queensland, Brisbane, QLD, Australia, k.moni@uq.edu.au

Marjorie Montague University of Miami, Coral Gables, FL, USA, MMontague@aol.com

John Munro The University of Melbourne, Melbourne, VIC, Australia, jkmunro@unimelb.edu.au

W. Ian O'Byrne University of Connecticut, Storrs, CT, USA, ianobyrne@uconn.edu

Robert M. Schwartz Oakland University, Rochester, MI 48309, USA, rschwart@oakland.edu

Christina E. van Kraayenoord The University of Queensland, Brisbane, QLD, Australia, c.vankraayenoord@uq.edu.au

Janice Wearmouth University of Bedfordshire, Bedford, UK, janice.wearmouth@beds.ac.uk

Kevin Wheldall Macquarie University, Sydney, NSW, Australia, kevin.wheldall@mq.edu.au

Claire Wyatt-Smith Griffith University, Brisbane, QLD, Australia, c.wyatt-smith@griffith.edu.au

Beth Yeager University of California, Santa Barbara, CA, USA, eyeager@education.ucsb.edu

Lisa Zawilinski University of Connecticut, Storrs, CT, USA, lisa.zawilinski@uconn.edu

About the Authors

Robyn Beaman is a research fellow at Macquarie University Special Education Centre (MUSEC) and deputy director of the MULTILIT ('Making Up Lost Time In Literacy') Research Unit, based in Sydney, Australia. Robyn has extensive experience in the implementation of remedial literacy programs in community settings, including the co-direction, for the Exodus Foundation in Sydney, of the Schoolwise Programs for disaffected, older low-progress readers since 1996 and the establishment of MULTILIT tutorial centres for indigenous low-progress readers in Cape York, since 2005. In addition to effective literacy instruction, Robyn also researches troublesome classroom behaviour. In 2005, she was awarded a Macquarie University Community Outreach Award for her MULTILIT work.

Mere Berryman's work in special education began after more than 20 years as a classroom practitioner. This led to the proposal of a research centre focused on working collaboratively with families and educators in order to raise student participation and achievement. This research centre is now part of the New Zealand Ministry of Education Special Education. As the manager of this centre, Mere and her team have found that approaches that acknowledge the importance of first developing relationships with students and their families, in ways that maintain respect for each other, are most successful. This involves culturally responsive and collaborative approaches to understanding and resolving problems.

Laura Black is a lecturer at the University of Manchester. Her research interests focus on pedagogic processes in the classroom and the construction of learner identities, particularly in relation to mathematics. Laura is currently working on several projects on widening participation in mathematically demanding programs within post-compulsory education.

Raymond Brown is a senior lecturer in the School of Education and Professional Studies at Griffith University, Gold Coast Campus, Australia. He came to this position after 31 years as a classroom teacher, where he implemented and studied inclusive approaches to teaching and learning. Over the past 10 years, Raymond's research has been concerned with providing insights into how inclusive classroom practices emerge and are sustained within classrooms; how students construct and display certain identity positions within classroom learning environments; and how learning processes interact at the personal and social levels to motivate and guide people to 'speak' and 'act' as members of learning communities.

Maria Lucia Castanheira is professor of Education at the Federal University of Minas Gerais, Brazil. She is particularly interested in examining the social construction of opportunities for learning in diverse classrooms (elementary school and adult education) and literacy practices in and out of school. As a member of CEALE/UFMG, a literacy centre, Maria works in national projects supported by the Brazilian Ministry of Education (Brazilian Textbook Evaluation and Teacher Developmental Programs). She has published on the discursive nature of classroom interaction, access of learning disabled and hearing-impaired students to literate practices, textbook analysis and research on epistemological issues.

Jill Castek is a post-doctoral scholar with the Seeds of Science/Roots of Reading project at the University of California, Berkeley, where she develops and researches integrated science and literacy curricula. Her classroom-based research explores the opportunities for reading, writing and learning on the Internet, and the contexts and conditions through which new literacies are acquired. She investigates ways in which to use a wide variety of instructional strategies to support the skills and strategies needed to read, write and collaborate online. Jill earned her doctorate in educational psychology, with an emphasis on cognition and instruction, from the University of Connecticut, United States.

Peta Colbert is a research fellow in the Faculty of Education at Griffith University in Queensland, Australia. She has co-authored several reports on large government-funded projects and evaluations focused on students with disabilities and learning difficulties, assessment reform and literacy professional development. Peta has also worked on national, competitive research projects, including an Australian Research Council-funded project exploring standards-driven reform in assessment, with a focus on teacher judgment and moderation.

John Elkins John Elkins is Emeritus Professor at The University of Queensland and Adjunct Professor of Literacy at Griffith University in Brisbane, Australia. He served on the Board of Directors of the International Reading Association, was president of the Australian Reading Association and edited the *Journal of Adolescent and Adult Literacy* and the *International Journal of Disability, Development and Education*. His research interests include reading difficulties and inclusive education.

Patricia A. Gallant is an associate professor of education in the School of Education and Human Services at the University of Michigan—Flint in the United States. In addition to teaching literacy methods courses, she is the coordinator for the K-12 Literacy Specialization masters program. Patricia's research interests include early literacy instruction (especially kindergarten), teacher change, the effects of educational policies on teachers and classrooms and field-based teacher education program reform. She currently serves as the president of the Michigan Reading

Association and also served for 3 years as co-editor of the *Michigan Reading Journal*.

Judith Green is professor of education at the University of California, Santa Barbara, United States. She engages in research on discourse and the social construction of disciplinary knowledge in linguistically and culturally diverse classrooms, from preschool through higher education. Judith also writes on issues of epistemology related to collecting, archiving, searching and analysing video records within ethnographic archives. She served as editor of the *Handbook of Complementary Methods in Education Research* (2006) and is editor of the *Review of Research in Education* (2006, 2008, 2010). Judith is a fellow of the American Anthropology Association and the National Conference for Research in Language and Literacy.

Stephanie Gunn is a research fellow at Griffith University in Queensland, Australia. She has been involved in educational research for nearly 20 years, including as National Coordinator, Literacy and Numeracy Projects for the Australian government, where she assisted in the development of national policies on literacy and numeracy. Stephanie has been co-author on several large, government-funded projects. Her current research includes work on a large Australian Research Council-funded project that is exploring standards-driven reform in assessment, with a focus on teacher judgment and moderation.

Colin Harrison's main field of research is literacy, with a focus on the use of information and communication technologies and on teachers' professional development. His research typically has been carried out in classrooms in primary and secondary schools. Colin has written reviews of research into literacy learning for the Scottish and English governments that have been distributed to every school in the country and has been a member of the government's Post-Rose Advisory Group, which advised on the professional development needs of literacy teachers. He directed the Interactive Classroom Explorer project, which aimed to enhance the knowledge and confidence of literacy teachers, using specially developed online video discussion software.

Anne Jobling is an adjunct senior lecturer and honorary researcher in the School of Education at the University of Queensland, Australia. She is a well-respected teacher and researcher in the area of intellectual disability. For many years Anne worked on and directed the Down Syndrome Research Program. She became the executive director of the Latch-On program, developed with Dr. Karen Moni. Anne's research focuses in education are on health-related issues and the literacy development of individuals with intellectual disabilities. She has published several books and book chapters and has over 50 peer-reviewed articles, some with international researchers. She has been successful with several Australian Research Council grants.

Robyn Jorgensen (Zevenbergen) is professor of education at Griffith University and is currently the chief executive officer of the Nyangatjatjara Aboriginal

Corporation in central Australia. She is the chief investigator on eight Australian Research Council Grants. Her research work focuses primarily on issues of equity and exclusion and spans early childhood through to post-compulsory education and into workplaces. Robyn draws on sociological theories to understand the ways in which taken-for-granted practices work to the exclusion of some children and students from learning, particularly mathematics. Her work has been predominantly focused on students from working-class families, indigenous families and from families living in rural and remote areas.

David Koppenhaver is professor of language, reading, and exceptionalities at Appalachian State University in Boone, North Carolina, United States, where he teaches graduate and undergraduate classes on literacy and technology. He is the co-founder and former director of the Center for Literacy and Disability Studies at the University of North Carolina—Chapel Hill and was a Fulbright Scholar at the University of Queensland in 2002. David's research focuses on literacy instruction and assessment of children and adults with developmental disabilities, including cerebral palsy, autism and intellectual disabilities.

Anneliese Kramer-Dahl is an associate professor in English language and literature at the National Institute of Education in Singapore, where she teaches discourse and genre analysis, academic literacy, language across the curriculum and literacy pedagogy. Anneliese has published in all these areas, in journals such as *Linguistics and Education*, *Language and Education*, and *Discourse*. Most recently she has researched issues of inequality and differential literacy instruction in Singapore classrooms and has also served as the principal investigator for two major intervention studies that seek to broaden teachers' textual repertoires in secondary classrooms.

Dennis Kwek is a senior research associate at the Centre for Research in Pedagogy and Practice, National Institute of Education, Singapore. He is currently involved in a study on whole-school curriculum reform in Singapore, and a large-scale, classroom observation research project utilising multidimensional qualitative and quantitative approaches. A doctoral candidate working on the notion of 'weaving', a pedagogical model of connected learning across different types and forms of knowledge, Dennis is also interested in discourse analysis, temporal and conceptual analysis of classroom interactions, whole-school change, professional development and alternative models of pedagogical routines, from East and West.

Donald J. Leu occupies the John and Maria Neag Endowed Chair in Literacy and Technology at the University of Connecticut, United States, and is the Director of the New Literacies Research Lab. He has degrees from Michigan State University (B.A.), Harvard (Ed.M.) and Berkeley (Ph.D.). Donald is a member of the Board of Directors of the International Reading Association and was elected to the Reading Hall of Fame. He is an international authority on the new literacies of online reading comprehension and how best to prepare students for reading and learning in an online age of information.

J. Greg McVerry is a Neag fellow, doctoral student and researcher with the New Literacies Research Lab at the University of Connecticut, United States. He was awarded the Masters in Elementary Education from the University of Hartford. Greg currently teaches in the University of Connecticut Two-Summers Educational Technology Masters program and provides professional development to schools in the areas of literacy, technology, e-portfolio assessment and online instruction. His research interests examine the skills and dispositions students need in order to critically evaluate online texts.

Robyn Miller is a speech language pathologist and psychologist. She has worked as a research assistant and project manager on several research projects in the School of Education at The University of Queensland, Australia. Robyn's research interests are centred on literacy development, with an emphasis on students with learning difficulties and developmental disabilities, inclusive education, teacher education and professional development for teachers.

Karen B. Moni is a senior lecturer in the School of Education at the University of Queensland, Australia. She coordinates English curriculum, literacy and designing effective teaching courses in undergraduate and postgraduate programs and is the Executive Director of Latch-On, a post-school literacy research and teaching program for young adults with intellectual disabilities, developed at the University of Queensland. Karen's research interests include literacy and young adults with intellectual disabilities, and teaching, learning and assessment in higher education. She is a past president of The English Teachers' Association of Queensland and is the current Editor of *English in Australia.*

Marjorie Montague Marjorie Montague, Ph.D., Professor at the University of Miami in Coral Gables, Florida, United States, specialises in learning disabilities, emotional and behaviour disorders, and attention deficit hyperactivity disorder. Her areas of research include behavioural and attentional problems in children and adolescents and interventions to improve mathematics achievement for students with learning disabilities. Marjorie has overseen numerous federally funded research and training projects.

John Munro is an associate professor and head of studies in exceptional learning and gifted in the Graduate School of Education at The University of Melbourne, Australia. He is a trained primary and secondary teacher and a psychologist. John's research interests, teaching and publications are in literacy learning and learning difficulties, maths learning disabilities, learning internationally, gifted learning and learning disabilities, gifted learning in African and Asian cultures, gifted mathematics learning and instructional leadership and school improvement. He is a consultant to several school improvement projects in Victoria, Australia. John was Chair of the College of Educational and Developmental Psychologists (Aus) and is a Life Member of Learning Disabilities Australia.

W. Ian O'Byrne is a Neag fellow, doctoral student and researcher with the New Literacies Research Lab at the University of Connecticut, United States. He was awarded the Masters of Education from the University of Massachusetts in the 180 Days Program. Ian has been involved in initiatives ranging from online coursework and integrating technology in the classroom and in school, to career pathways. He has also been involved in department leadership. Ian's research interests include examining the reading and content-creation practices employed by adolescents in online spaces.

Robert M. Schwartz is a professor in Department of Reading and Language Arts at Oakland University, Rochester, Michigan, United States. He is a reading recovery university trainer and a past president of the Reading Recovery Council of North America. Robert's research interests include self-monitoring in beginning reading, early literacy intervention, research design and professional development for literacy teachers. He has recently published on these topics in *Reading Research Quarterly*, *Journal of Educational Psychology, International Journal of Disability, Development and Education* and *The Reading Teacher.* Robert is particularly interested in the role of one-to-one tutoring in supporting the literacy development of the lowest-performing beginning readers and the professional learning of their teachers.

Christina E. van Kraayenoord is an associate professor in the School of Education at The University of Queensland, Australia. She teaches courses related to literacy development and literacy education, diversity and inclusive education, and the teaching and learning of students with learning difficulties and disabilities. Christa's research is in the areas of learning difficulties, literacy, metacognition and motivation, whole-school improvement and teacher change as it relates to literacy and pedagogical practices. She is a vice president of the International Academy for Research in Learning Disabilities and the Editor of the *International Journal of Disability, Development and Education.*

Janice Wearmouth is currently a professor of Education at the University of Bedfordshire in England. She has held chairs in education in New Zealand and the United Kingdom. In her work, Wearmouth brings together a concern for the learner whose educational experience is problematic with a concern for professionals who have to deal with, and mitigate, the problems that are experienced and facilitate opportunities for learning. Her particular interests lie in the areas of pedagogies and curriculum effective in responding to the needs of all students, particularly those who experience difficulties in learning, literacy learning in schools, approaches to understanding, assessing and addressing behavioural issues in schools and teacher professional development.

Kevin Wheldall is Director of Macquarie University Special Education Centre and the MULTILIT ('Making Up Lost Time in Literacy') Research Unit, based in Sydney, Australia. He is a fellow of the Academy of Social Sciences in Australia and the International Academy for Research in Learning Disabilities. In 2008, Kevin received the Mona Tobias Award of Learning Difficulties Australia and in 2009 was inducted into Macquarie University Innovators' Hall of Fame for his research on MULTILIT. He has researched and written extensively on learning and behavioural difficulties, is the author of over 200 academic publications and edits the international journal *Educational Psychology* and (jointly) the *Australian Journal of Learning Difficulties*.

Claire Wyatt-Smith Claire Wyatt-Smith is Professor of Education and Dean of the Faculty of Education at Griffith University in Queensland, Australia. She has been chief investigator on a number of Australian Research Council (ARC) projects and other funded research studies focusing on teacher judgment, standards and evaluative frameworks as these apply in policy and practice. Claire's current research includes an investigation of the dependability of teacher assessment for statewide reporting in Queensland. A current ARC project includes a longitudinal study of standards-driven assessment reform in middle schooling, the focus being on teacher judgment and moderation, both face-to-face and ICT-mediated.

Elizabeth (Beth) Yeager is assistant researcher in Education and Executive Director, Center for Literacy and Inquiry in Networking Communities at the University of California, Santa Barbara, United States. She is interested in research on discourse and social construction of disciplinary knowledge and academic identities in linguistically and culturally diverse settings, and on developing technology enhanced models that support construction of disciplinary knowledge and practice, and professional development. Elizabeth has published on her work as a teacher-researcher, and on the discursive nature of classroom interaction, access of diverse students to disciplinary knowledge and academic identities, constructing networking learning communities using broadband digital 'highways', and engaging students as ethnographers in classroom settings.

Lisa Zawilinski is a Neag fellow, doctoral student and researcher with the New Literacies Research Lab at the University of Connecticut, United States. She earned both her masters and bachelor degrees from the University of Connecticut. Lisa has taught and provided professional development in education for more than 15 years. She is a trained reading consultant and former Co-director/Tech Liaison for the Connecticut Writing Project. Lisa is currently investigating the ways in which elementary students read, write and communicate on the Internet.

List of Figures

List of Tables

Chapter 1
Theoretical Frameworks and *Ways of Seeing*: Operating at the Intersection—Literacy, Numeracy and Learning Difficulties

Claire Wyatt-Smith and John Elkins

This book is concerned with learning difficulties in literacy and numeracy and the need for greater appreciation of varied approaches to research that have been used in efforts to improve learning. Literacy and numeracy education is core in quality learning across all phases of schooling and all curriculum areas. While these ideas will be familiar to many readers, there has been limited dialogue across the domains of literacy and numeracy research, particularly as they relate to learning difficulties. In effect, each has tended to operate in a separate entity, likened to 'silos'.

This chapter serves to introduce the challenge of breaking down barriers to illuminate the points at which literacy, numeracy and learning difficulties intersect. It also attends to the theoretical and methodological diversity and isolation that characterise research in these domains. These considerations are critical if teaching is to be made more effective for struggling students. Further, given that policy priorities in many countries relate to equity of opportunity, it is timely to review how different theoretical frameworks and methodologies provide different lenses through which to study students' learning needs and, more importantly, through which to understand different approaches to improving students' learning. This chapter lays out the motivation for this approach, presenting the case that harnessing a range of theoretical orientations and methodologies will generate evidence-based insights not otherwise possible. The book seeks to enable readers to engage with the chapters in ways that correspond to their own contexts and research–practice interests, while also encouraging them to seek insights from domains they do not commonly explore.

One dimension of importance is that of contexts, not only differences such as urban/suburban/rural, but also those between states and nations, and between in-school and out-of-school literacy and numeracy practices. For example, emphases on student achievement in literacy and, to a lesser extent, numeracy, are common to various countries, including the United States, United Kingdom, Finland, Canada, Australia and New Zealand. However, they may be different from those countries in which the challenge is to provide basic education to all children and young

C. Wyatt-Smith (✉)
Griffith University, Brisbane, QLD, Australia
e-mail: c.wyatt-smith@griffith.edu.au

C. Wyatt-Smith et al. (eds.), *Multiple Perspectives on Difficulties in Learning Literacy and Numeracy*, DOI 10.1007/978-1-4020-8864-3_1,

people. Within a country, there may be major differences between what is entailed in addressing student achievement in different regions, based on factors such as population density, socioeconomic status, ethnicity and educational level of parents. In Australia, the education of Indigenous students continues to defy government improvement initiatives, especially where these are related to literacy and numeracy testing (Ministerial Council on Education, Employment, Training and Youth Affairs (MCEETYA[1]), 2008).

Another dimension of interest is the basic skill area traditionally known as 'the three Rs'. Today, reading and writing have morphed into literacy,[2] which itself has been enriched by the incorporation of digital skills (Leu, Coiro, Knobel, & Lankshear, 2008), recognition of multimodalilty (Kress, 2009; Unsworth, 2009) and the addition of a critical perspective (Freebody & Luke, 1990; Willinsky, 2008). Often neglected in the past—in comparison to literacy—numeracy is receiving increased attention, though confusion as to the distinction between numeracy[3] and mathematics is common (Department of Education & Science and Training, 2005). Indeed, the term numeracy may not be familiar in some countries.

There are many approaches to researching the difficulties in learning that students experience in the key areas of literacy and numeracy. In some cases, the theories and research methods for studying these difficulties are those that apply in the fields of literacy and numeracy, while some have been developed within the Learning Disability or Special Education areas. This book seeks to advance understanding of these difficulties and the interventions that have been used to improve outcomes. By including authors drawn from several countries and with expertise in a variety of research traditions, the book illustrates the sometimes complementary and sometimes contradictory results of research, and suggests new approaches to understanding and serving students experiencing difficulties in learning literacy and numeracy.

The editors, together with colleagues, have recently completed a large-scale research program (Wyatt-Smith, Elkins, Colbert, Gunn, & Muspratt, 2007), funded by the Australian government, that examined actual intervention practices for supporting students experiencing difficulties in literacy and/or numeracy and drawing upon diverse theoretical and methodological orientations to literacy and numeracy difficulties and to interventions. The context for the research was the Australian government's *Effective Teaching and Learning Practices for Students with Learning Difficulties Initiative*, under the Adelaide Declaration on National Goals for Schooling in the 21st century (MCEETYA, 1999). This concern parallels the *No*

---

[1]MCEETYA is now known as the Ministerial Council for Education, Early Childhood Development and Youth Affairs (MCEECDYA).

[2]Literacy is defined as 'the flexible and sustainable mastery of a repertoire of practices with the texts of traditional and new communication technologies via spoken language, print and multimedia' (The State of Queensland, 2000, p. 9).

[3]'To be numerate is to use mathematics effectively to meet the general demands of life at home, in paid work, and for participation in community and civic life' (Australian Association for Mathematics' Teachers, 1997, p. 15).

Child Left Behind Act of 2001 in the United States and similar emphases in other countries, including the *Primary National Strategy* 2003 in the United Kingdom.

In our research program[4] we used qualitative and quantitative methodologies to generate new insights into the nature of what counts as effective teaching and learning practices for students with learning difficulties in literacy and numeracy. Following completion of the research, we saw a need to bring together an international group of scholars to engage with difficulties in student learning of literacy and numeracy as they manifest in their particular contexts. Our approach in the research and in this book is sympathetic to that used by Beach, Green, Kamil, and Shanahan (2005) and Green, Camilli, and Elmore (2006) in addressing multidisciplinary perspectives. Motivating our approach is our understanding that harnessing a range of theoretical orientations and methodologies across literacy, numeracy and learning difficulties will generate insights not otherwise possible. Several authors (for example Green et al., 2006; McCardle & Chhabra, 2004) suggest that a sound basis for action is converging evidence from multiple sources and different perspectives. Evidence-based convergence lends strength to findings as no single study, methodology, finding or view is considered, in and of itself, a sufficient basis for action.

As mentioned earlier, a distinctive feature of this book is its focus on the intersection of literacy education and numeracy education, with particular concern for the students who experience learning difficulties. Traditionally, these have tended to be addressed separately by researchers and policy makers, leading to compartmentalised thinking. Thus, in the United States, there are distinct communities of teachers and researchers who focus on literacy, such as the International

---

[4]Our research study that provided the stimulus for this book attempted to understand how schools in Queensland, Australia, identified and supported students who experienced difficulties in literacy and numeracy (Wyatt-Smith et al., 2007). Some schools had relatively homogeneous student characteristics, while others were varied in ethnicity, home language and socioeconomic factors. Where schools recognised that most students were performing at low levels in literacy and numeracy, classroom programs tried to address this situation directly. Other schools saw difficulties only among a small proportion, and thus targeted support to these students. The first element of our study was to administer questionnaires to school principals and support teachers. This generated much useful information on what schools were doing, and emphasised the high degree of local decision-making that pertained. The second element was to analyse statewide, standardised achievement test data obtained at years 3, 5 and 7. We were able to show that average and low-achieving students made similar rates of growth across the 4-year-period. However, the large spread in achievement that existed at year 3 was maintained at each subsequent testing. Students who had been identified prior to year 3 (by *Reading Recovery*, Clay, 1993 or Year 2 Diagnostic Net, Education Queensland, 1995) also showed similar rates of learning through subsequent primary school years, and continued to be low achievers. The third element comprised detailed case studies of schools identified by school systems as providing effective support in literacy or numeracy. The case reports showed a wide range of practices for identifying struggling students and a similarly wide range of intervention practices. Common across the cases were collaborative planning and careful monitoring of student outcomes. However, longitudinal tracking and analysis of the effectiveness of specific interventions were generally lacking. These three methodological approaches were used against the backdrop of an extensive review of research literature, reported selectively in Chapter 2 of this book.

Reading Association, on numeracy, such as the National Council of Teachers of Mathematics, or on learning disabilities, such as the Division for Learning Disabilities and the International Academy for Research in Learning Disabilities. The ability of such organisations to contain the diversity of views, particularly of researchers, has in some cases led to fracture and result in the formation of new groups with narrower theoretical perspectives. One consequence of separation and narrow affiliation may be that researchers often demonstrate lack of awareness of developments in the other domains that might enrich understanding. In short, to date there has been limited exchange across these fields, but rather a defensive posture designed to secure a claim as the 'true' scholarship.

Such narrow affiliation has a role, of course, in the specialisation that is a necessary condition for scientific advancement. It also may acquire a political dimension, as has happened in literacy education with the interpretation of what constitutes evidence-based research being a significant area of contestation (Department of Education & Employment and Workplace Relations (DEEWR), 2005; Snow, Griffin, & Burns, 2005). Interestingly, it may be that the quest for evidence-based practice has set up the conditions for restricting evidence to studies that purport to be the educational equivalent of double-blind clinical trials as used in the biomedical sciences. However, it is doubtful that an educational equivalent of the Cochrane Collaboration[5] can be developed without expanding the types of educational research beyond controlled experimentation. We take this issue further in the final chapter.

Most scholars and graduate students draw upon ideas from only one of the three domains focal in the book, and usually from one single or dominant theoretical frame in their chapters. Typically, readers affiliate with reading education, mathematics education or learning disabilities, and belong to one of the corresponding professional associations. This book's scope opens a scholarly forum for engaging readers who are familiar with one of these domains with research currents in others.

The plan of the book

The first three chapters serve as a foundation for the specific discussions in Chapters 4–17. In Chapter 2, Gunn and Wyatt-Smith present a wide-ranging review of literature that may assist readers to obtain an overview of research in the three domains of literacy, numeracy and learning difficulties. They consider the ways in which these domains are defined and the various theoretical frameworks that have been influential on research in each. They also examine the accounts of effective provision across the domains, drawing attention to generic considerations. These include instructional approaches, transition and continuity across phases of schooling, time allocation, leadership, student motivation, monitoring and assessment,

[5]The Cochrane Collaboration is a peer-reviewed system of meta-analyses of the effectiveness of various medical treatments. It aims to establish evidence-based practices.

classroom talk, new technologies, community partnership, student diversity and teacher education.

Sourced predominantly from the literacy domain, the notion of three waves of provision (classroom instruction, early interventions and long-term support) has been influential in the field. A main issue associated with these three waves is the allocation of resources among them. Gunn and Wyatt-Smith note the resourcing challenges of providing continuing student support, of maximising time for learning, of improving teaching competence and maximising student engagement. Further, they highlight the pedagogical change needed for teachers to be able to incorporate new technologies in effective classroom interventions.

In Chapter 2, Green, Castanheira and Yeager present a compelling *telling case* in which they explore different dimensions of learning experiences for a bilingual student, Sergio, in a linguistically and culturally diverse Grade 5 class. This is a new study constructed from a continuing ethnographic research project that focuses on theorising Sergio's opportunities for learning to be literate. In this exciting study, the authors centre on individual–collective relationships and address two key questions that were unresolved in earlier work: How can the interdependence of collective and individual development be explored and made visible? And, how can individual students' developing understandings be documented? In addressing these questions, they provide readers with what they refer to as 'a means for making visible the complementary basis of the theoretical traditions guiding [the] ethnographic telling case' (p. 51) they present. Crucial in this chapter is the issue of the nature of *evidence*—*what* different theoretical perspectives count as evidence. The achievement of this chapter is how it carefully lays open for scrutiny the theoretical and methodological decisions that informed the analyses. On offer to readers is a rich discussion of the telling case and its particular focus on individual–collective relationships. Also, canvassed are the consequences of particular theoretical and methodological decisions for what comes to be uncovered as a result of their application.

In Chapters 4–16, the authors identify their theoretical and research orientations, focusing on ways in which difficulties in learning may be overcome. In this way, the research–practice connection is featured. Additionally, the authors follow a general framework for their chapter to ensure continuity and coherence throughout the book. The general framework includes the following:

1. chosen theoretical and methodological approach
2. identification of other perspectives that have been used by researchers in related investigations
3. application of the theoretical/methodological choices to some empirical data
4. strengths and limitations of these choices for what they make available about learning difficulties, literacy and/or numeracy
5. positing of Essential Next Questions for research, practice and policy.

In many countries, improving the ability to read and comprehend appears as a pressing educational priority for governments and school communities. This is especially the case for students who struggle with reading and for whom education systems are challenged to make provision that is recognisably in addition to mainstream or classroom instruction. Castek, Zawilinski, McVerry, O'Byrne and Leu (Chapter 4) and Harrison (Chapter 5) take up these issues and challenge readers to rethink reading and comprehension at a time when the Internet and developments in information and communication technologies have become routine in daily life. Common to both chapters is the view that any approach seeking to improve students' literacy learning, and their reading abilities more specifically, must recognise that the nature of writing and reading, including reading comprehension, has expanded in the 21st century (International Reading Association, 2009).

Castek et al. draw directly on New Literacies theory to address the opportunities and challenges involved in attempts at increasing reading achievement, making clear the differences between online reading and offline reading (reading traditional print texts). The central issue confronted in the chapter is whether adolescent readers can benefit from instruction with online reading comprehension skills before they are fully proficient with offline reading comprehension. Drawing on empirical data, Castek et al. present evidence suggesting that 'many struggling readers appear to benefit in important ways from online reading experiences and instruction in the new literacies of online reading comprehension' (p. 105). The compelling insights in the chapter call for a rethinking of what is valued as reading achievement, and include the observation that traditional evaluations of reading (and the priority given to offline reading) may well cause students skilled in the use of online information to remain undetected.

In Chapter 5, Harrison adopts an approach that is grounded in classroom use of the Internet and other digital technologies. He offers valuable insights into how digital technology and the Internet can be used by teachers in promoting literacy learning, especially for those students who are likely to have difficulty in traditional classrooms. He shows that students in the 21st century need all the traditional components of literacy, together with their extension to multimedia situations and the non-linear navigation around such multimedia texts, and the continuing challenge of learning to use social-networking technologies such as blogs, Facebook, YouTube, Twitter and others not yet invented. Harrison describes computer software and the Internet applications that engage and challenge students to give them a sense of power as readers and actually develop their critical faculties.

In Chapter 6, Colbert draws on the case studies of the Australian project (Wyatt-Smith et al., 2007) mentioned earlier, to identify and discuss key features of practice found to be effective in supporting the literacy and numeracy development of students with learning difficulties. The features are: school leadership; the direct involvement of consultative committees, including parents, to inform decisions about support provision; personnel networking (teachers and other health professionals) with active involvement in planning, delivery and assessment of support; a range of programs and strategies tailored to individuals and the phase of schooling; effective evidence-based approaches that enable teachers and school systems to monitor the impact of support; and strategic decision-making at the local level to attend to both the built environment and the social contexts for learning.

In Chapter 7, Kramer-Dahl and Kwek start with theories of deficit thinking in relation to social class and ethnicity, drawing on sociological accounts of youth, family and schooling. The authors use a multi-layered analysis of teachers' deficit constructions to examine classroom interaction (layer 1), teacher beliefs (layer 2) and alignment of beliefs with policy discourses (layer 3). They consider home and school issues in the unique context of Singapore, where English is the required language, yet students have other home languages, such as a Chinese dialect, Malay or Tamil. The school system in Singapore is highly meritocratic and is seen as having a key role in the economic success of the nation. According to these authors, teachers in secondary English classes use a reduced syllabus with students who are from ethnic minority homes. They describe the pedagogies used by teachers and identify what the teachers valued. In their discussion they refer to interviews with teachers about the learning difficulties and the family backgrounds of the students. Singapore teachers accept as normal that students should be streamed and taught at different rates, that achievement depends upon innate abilities and that curriculum should be adjusted to match students' future roles in society. Kramer-Dahl and Kwek identify areas for future study as how teachers' deficit beliefs are perpetuated and how teachers can interrupt the cycle of deficit thinking and its realisation in classrooms and society.

In Chapter 8, Wearmouth and Berryman consider literacy as a social practice and provide an account of how parents, families and the community can assist literacy learning. They adopt an interactive model of reading and situate the student learner in a social context in which culture is powerful, but not necessarily aligned with the classroom view of being literate. Using New Zealand and the United Kingdom experiences, they show how parental involvement can take many forms and that power may be located differently in these models. Wearmouth and Berryman delineate the implications for teachers' professional development, especially where teachers belong to social communities different from those of the families in the school. The authors report that most teachers appreciated the value of families and communities participating in school-sanctioned literacy activities such as story reading, though they still needed support in encouraging this in sensitive ways. A limitation, however, is that many teachers have little idea of how community practices can be incorporated into school teaching and learning situations.

In Chapter 9, Munro adopts a metacognitive approach to supporting how readers comprehend. He describes studies that involve the explicit teaching of comprehending strategies and looks at the problems in understanding text experienced by students who have progressed beyond the early oral reading of narrative that is the focus of Schwartz and Gallant (Chapter 11). Munro shares their attention to the mental activities of children as they are reading (such as paraphrasing, visualising and predicting). He also considers how struggling students can be helped to use better comprehending strategies within mixed-ability classes, and goes on to recognise that effective instruction will need to support transfer of successful strategies to independent reading. Further, Munro recognises that teachers of students in the middle years often overlook the extent of students' difficulties with decoding text, and may fail to implement appropriate intervention in parallel with comprehension instruction.

While this volume demonstrates the imbalance between literacy and numeracy that characterises the literature, within literacy research, writing is given very much less attention than reading. In Chapter 10, van Kraayenoord, Moni, Jobling, Elkins, Koppenhaver and Miller use a sociocultural framework. They present an intervention with middle school students who exhibited literacy-learning difficulties. The authors discuss how they tried to develop effective instruction that has the goal of making students more independent writers who are able to critique their own output and engage in productive editing of earlier drafts. These authors provide a model of teaching writing that can be used in classrooms to support those students who experience difficulties. While Munro presents metacognitive approaches for reading, these authors focus on how metacognitive strategies for writing should be used alongside explicit teaching and carefully planned feedback. A feature of the chapter is how it raises issues associated with changing teacher practices through in-service professional development and the duration of support that may be needed, particularly with older students.

In Chapter 11, Schwartz and Gallant work within a framework of developmental psychology, drawing heavily on the work of Marie Clay (2001). They argue that children who find early narrative reading difficult need to heighten their monitoring of success. They cast the problem of assisting such children as entailing more-than-better instruction. Specifically, they focus on children's own problem-solving as they gain fluency and accuracy. The authors claim that the central task is to help children cope with the complexities of reading, and argue that individual help from teachers, such as provided within the Reading Recovery service, is essential for a substantial minority of 5–6-year olds. Rather than seeing word recognition and reading comprehension as alternative emphases in the early years, they regard both as crucial and constantly changing elements in the developing skills of children, and explain many failures as consequences of insufficiently sensitive observation (and consequent intervention) by teachers. Schwartz and Gallant suggest that struggling readers need to be helped to develop their own self-improvement systems, and teachers can use modelling and searching supportive questioning to prevent these children from 'learning to be learning disabled' (Clay, 1987, p. 155).

In Chapter 12, Wheldall and Beaman adopt a view of literacy as reading and writing, and centre on the relationship between print and sound, in which difficulties are viewed as stemming from limited phonological knowledge. The authors describe an intervention to support students who have marked difficulties in reading. They applied the intervention to socially disadvantaged students, in particular Indigenous students in the remote Cape York region of northern Australia, where students were taught through a program known as MULTILIT in a tutorial centre. In a second project, a main goal was to ensure the use of MULTILIT approaches within classrooms, with components being delivered by teachers, assistants and aides. The authors show how the program worked to reverse the downward trajectory of literacy achievement that characterised most Australian Indigenous students. They note that circumstances that adversely affect education in remote Indigenous communities, such as inexperienced teachers and high turnover of staff, also presented challenges for the implementation of MULTILIT. Continuing research in a larger number of sites is designed to improve the effectiveness of implementation and, in particular, the reading comprehension of the students.

Drawing on a sociocultural framework (Vygotsky, 1987), in Chapter 13 Brown challenges the notion of 'learning difficulty' and explores the development of what he refers to as mathematical literacy in the every classroom. Brown's starting proposition is the right of every child to pursue the goal of accessing the dispositions and practices of a mathematically literate person. His focus is on the social dimension of classroom practice and, from here, he draws readers into a picture of how sociocultural approaches to teaching and learning work to support the development of mathematical literacy by all students, including those identified as displaying learning difficulties. Brown's achievement in this chapter is to challenge a traditional notion that concern for all students to master basic mathematics facts and algorithms should be used to justify students being denied access to challenging tasks that 'span the different strands of a general mathematics curriculum' (p. 276). To this end, Brown invites readers into his own classroom to meet three students as they present their solution to a part of a teacher-set problem on the whiteboard. Through analysis of talk and interaction, Brown challenges readers to engage with 'learning difficulty' as being constructed in each curriculum encounter, depending on what has gone before and what beliefs and dispositions teachers and students bring to these encounters. He also brings to light how the teacher enacts the role of mathematician in the classroom in a way that promotes participation in the literate practice of mathematics. For Brown, access to being literate—cracking the code—involves 'making explicit the mathematical language that scaffolds mature thinking within the context of completing a task' (p. 286).

In Chapter 14, Montague reviews the literature on mathematical learning difficulties and describes the characteristics of students that may impede learning success. She notes that several theoretical approaches have been used in studying students who have difficulties in school mathematics. She recognises that, depending on the needs of each student, behaviourist, information-processing, metacognitive, motivational, sociocultural and other perspectives may be needed to plan effective instruction and support. Sometimes these or other theoretical frameworks lead to similar interventions, and sometimes they may conflict. The challenge of delivering best practice is not just to determine which method delivers the largest effect size, in part because the research base is limited, and in part because students exhibit a variety of difficulties in mathematics. She identifies two instructional approaches (direct instruction and cognitive strategy instruction) that have been demonstrated to be effective in addressing these difficulties. A feature of her chapter is how, in response to policy driven accountability pressures, researchers in children's mathematical learning difficulties are seeking to establish evidence-based criteria for choosing effective interventions.

In Chapter 15, Jorgensen adopts a framework different from those considered by Montague. Focusing on language and culture, she draws on Bourdieu (1991) to elucidate the difficulties in learning mathematics that are experienced by many

Indigenous students in remote Australia. She centres her analysis on the intersection of literacy, numeracy, culture and context. While students in remote parts of Australia have lower scores on achievement tests, Jorgensen argues that it is not appropriate to explain this as inherent in the learners. Rather, Jorgensen shows that, partly through language differences and partly through different cultural meanings for mathematics, Australian Indigenous students are constrained in their learning of the mathematics expected by standard curricula. She explores the role of language in mediating the learning of mathematics, and that of cultural context in placing obstacles in the path of Indigenous learners.

In Chapter 16, Laura Black draws on sociocultural theory (Lave & Wenger, 1991; Wenger, 1998) to consider the communicative behaviours of teachers and students, and how this impacts on student participation (and non-participation) in learning and whole-class discussions. She focuses attention on the nature of whole-class teaching and the role of classroom discussions for supporting learning for all students. She hones in on 'how teachers are to orchestrate such discussions in an environment where the construct of "ability" and the drive for individualised notions of success (through performance in examinations) dominate' (p. 332). While Black refers directly to the educational policy context of England, her interest in whole-class provision for students with learning difficulties is one shared by educators who face the inevitable tensions associated with a focus on learning and learning improvement on the one hand, and accountability and measurement, on the other. Black argues the case for whole-class discussions to be regarded and valued as 'a key site for the transformation of students into "successful" learners of the relevant knowledge domain' (p. 333). She makes the argument for class discussions to be understood as building for teachers and students alike a shared foundation of experience from which both parties negotiate various concepts, representations and methods. Through a study using participant observation, Black presents a re-storying of a student, Jason, attending a school in a large town in the north-west of England. Readers may wish to consider how Black's account of classroom communicative behaviour and learning opportunities in England relates to Brown's portrayal in an Australian classroom. For both writers, the message is clear: it is through interaction and classroom talk that teachers and students co-construct knowledge and student identities, with the potential for what Black refers to as 'long lasting effects on future participation in educational practices' (p. 345).

In Chapter 17, we identify and discuss the overarching themes that emerge from the chapters and consider questions for further research posed by the authors. Against this backdrop, we revisit the dual catalysts for the book. First is the urgent need to break down the longstanding, traditional 'silos' that exist across literacy, numeracy and learning difficulties. Second is the issue of how evidence-based practice could utilise the findings from a wide range of research methodologies and theoretical perspectives. As Green, Castanheira and Yeager (Chapter 3) point out, diversity of perspectives is both a resource and a challenge. We propose that in order to work productively within and across perspectives, we need to attend to the governing assumptions that have, to the present, framed the questions that we have asked about learning difficulties and, even more fundamentally, the relational dimension of learning, teaching and assessment. This shift will no doubt be risky. It may well mean that hitherto taken-for-granted understandings about learning disabilities and effective literacy and numeracy teaching come to be challenged when other ways of 'seeing' (available through other theoretical perspectives) are explored.

Reading this book

Readers will no doubt choose their own pathways through the chapters. However, we suggest there are some guiding questions that may be kept in mind and used to create threads and patterns across the chapters. Different chapters have different sorts of implicit answers to the questions. More essentially, different chapters offer different 'ways of seeing' literacy, numeracy, learning difficulties, and even what it means to be 'student' and 'teacher'.

Who are the students? Some researchers focus on students who score below some desired or mandated level on large-scale tests of literacy and/or numeracy, and their research does not ask questions about how sub-groups of these low achievers differ in their needs or responses to different types of intervention. Other researchers are concerned primarily with students who share certain characteristics such as ethnicity, low socioeconomic status, language background or some disability. These scholars may focus on a narrower set of variables as these relate to learning and achievement in the sub-group of interest. Others regard the classroom as the matrix in which support is extended to some students through the nature of the interactions and relationships that are fostered. For these researchers, the focus is on how the classroom operates, not the labels or categories that others might apply.

The previous question connects to *Who delivers learning support?* Some efforts to improve literacy and numeracy have general application in classrooms to most or all children, and the classroom teacher is the key. In contrast, other practices are specific to students with particular needs and may require personnel trained in specific techniques. In addition, complex collaboration may be needed among various professionals, paraprofessionals and parents in supporting student learning.

What is special about literacy and numeracy? Are literacy and numeracy narrow ends in themselves, or are they best understood as the means through which learning occurs in all areas of the curriculum? Answers to these questions relate directly to understandings about teaching and, more specifically, teaching likely to lead to improved student outcomes. If there is a starting proposition that the terms literacy and numeracy refer to a fixed body of skills, then teaching literacy and numeracy can be understood as the execution of an accepted set of classroom procedures. If, however, the proposition is that literacy and numeracy relate to learning in the curriculum, then teaching for struggling students can and should address literacy and numeracy demands as they relate to specific curriculum areas.

What are demands for literacy and numeracy learning in the 21st century?

Researchers appear to operate in one or other of the two domains of print-dependent and digitally mediated learning. In this book, for example, other than the two chapters that specifically address screen-based activities, researchers give emphasis to traditional paper-and-pencil and chalkboard teaching and learning activities. Researchers also tend to assume that teaching is the crucial element, though some, particularly those with interest in digital technologies, recognise the agency of students in learning and overcoming difficulties through problem-solving, which is arguably crucial for living and working in the 21st century. While it is clear in the published research that some special educators have studied the applications of technologies in improving student learning, it is also clear that much intervention remains print-dependent. Where print is the sole or dominant mode, students may be doubly disadvantaged in terms of their preparation for the complexities of modern communication practices and citizenship more generally.

What classrooms offer the best opportunities to learn? If we accept the usefulness of the three-wave framework for supporting students who experience difficulties in literacy and numeracy learning, then the classroom is the key to the first phase that is initial instruction and opportunities for students to become autonomous learners. However, the classroom may also provide early intervention, which is the second wave, though often early intervention is delivered outside classrooms and by personnel other than the class teacher. If we accept the idea that it is through talk and other interactions, including student and teacher modelling, that learning is best supported, then classrooms rich in talk and supportive relationships are likely to facilitate effective learning. The third wave is the provision of continuing support for students for whom early intervention was insufficient. Again, the classroom is the environment in which such continuing support will need to be delivered and sustained, which may require curriculum, and instructional adaptations may be crucial. In such ideal classrooms, assessment should not only establish what learning has occurred, but should illuminate the difficulties that students experience, and guide future learning and teaching practice. The potential of digital technologies for delivering such support in new ways remains unrealised, with little known currently about how changed classroom interaction patterns through the use of technologies can change the nature of learning itself, as well as growth over time. It is, however, recognised that developments in this area must be given high priority with regard to resourcing and training of teachers, with educators working together with digital communication carriers for the best results. If this were realised, effective practice in classrooms could be complemented by literate and numerate social practices enacted outside classrooms, in homes and in wider local and global communities (both actual and virtual).

How are students experiencing difficulties identified? Throughout the world, education authorities are using large-scale, standardised testing of necessarily limited scope, along with arbitrary standards setting. Currently, these tests are confined to print technology and fail to identify the mastery (or lack thereof) of students in negotiating the digital world. Despite this limitation, even cross-national testing regimes are exerting substantial influence on the school experiences of students. A rarely asked question is whether the majority of students who do not experience difficulties in learning literacy and numeracy are being adversely impacted by this emphasis on performance in standardised testing. Furthermore, while large-scale testing yields data about students at class, school and education system levels, it does not have the sensitivity needed for diagnostic purposes (Cumming, Wyatt-Smith, Elkins, & Neville, 2006). More specifically, its role in informing the selection of interventions for students with learning difficulties is necessarily limited. Generally speaking, the testing regimes are developed for reporting and accountability purposes. The focus on 'fitness for purpose' is therefore essential when considering how classroom assessment in the hands of teachers can complement the evidence that large-scale testing data generates, especially for students with learning difficulties. If parents and communities generally are to have a rich picture of student learning, it is essential that testing practices be nuanced to allow their use for diagnostic purposes.

What are the resources needed to improve learning? While most researchers are interested in questions of effectiveness, few consider the costs and practicalities of the intervention approaches they studied. A constant tension exists between expenditure directed at a limited number of students and the general improvement of all students. This may be exacerbated by funding being earmarked for students who meet particular criteria, such as English as a second language or learning disabilities, or for schools serving particular demographics, such as inner-city location, areas of disadvantage or rural and remote schools. A further matter in relation to resources, or more specifically resourcing, relates to the categories used to diagnose students, with those same categories sometimes tied to funded learning support. Even within a country or state, there can be regional variations in how funding allocations for learning support are determined. From a research perspective, the interest then lies in equity matters and how schools manage limited resource allocations: which categories of students are singled out for support (over others)?

Finally, while we recognise that there are different pathways through this book, we encourage each reader to take up the questions above. Our intention is for readers to reveal to themselves how different theoretical orientations and research approaches open up (and close down) different ways of seeing and knowing learning difficulties, and even the very nature of what counts as quality literacy and numeracy learning.

References

Australian Association for Mathematics Teachers. (1997). *Numeracy = everyone's business*, Report of the Numeracy Education Strategy Development Conference. Adelaide: Author.

Beach, R., Green, J. L., Kamil, M., & Shanahan, T. (Eds.). (2005). *Multidisciplinary perspectives on literacy research* (2nd ed.). Cresskill, NJ: Hampton Press.

Bourdieu, P. (1991). *Language and symbolic power* (G. A. M. Raymond, Trans.). Cambridge: Polity Press.

Clay, M. M. (1987). Learning to be learning disabled. *New Zealand Journal of Educational Studies*, *22*(2), 155–173.

Clay, M. M. (1993). *Reading recovery: A guidebook for teachers in training*. Wellington: Heinemann.

Clay, M. M. (2001). *Change over time in children's literacy development*. Portsmouth, NH: Heinemann.

Cumming, J. J., Wyatt-Smith, C. M., Elkins, J., & Neville, M. (2006). *Teacher judgment: Building an evidentiary base for quality literacy and numeracy education*. Brisbane: Queensland Studies Authority. Available at: www.qsa.qld.edu.au/downloads/publications/research_qsa_teacher_judgment.pdf.

Department of Education, Employment and Workplace Relations (DEEWR). (2005). *National inquiry into the teaching of literacy*. Retrieved from DEEWR: http://www.dest.gov.au/nitl/report.htm

Department of Education, Science and Training. (2005). *Numeracy research and development initiative 2001–2004: An overview of the numeracy projects*. Canberra: Author.

Education Queensland. (1995). *Year two diagnostic net procedures and validation guidelines*. Brisbane: Author.

Freebody, P., & Luke, A. (1990). 'Literacies' programs: Debates and demands in cultural context. *Prospect*, *5*(3), 7–16.

Green, J. L., Cammili, G., & Elmore, P. B. (2006). *Handbook of complementary methods in education research*. Mahwah, NJ: Lawrence Erlbaum Associates for the American Educational Research Association.

International Reading Association. (2009). *New literacies and 21st century technologies: A position statement of the International Reading Association*. Newark, DE: Author.

Kress, G. R. (2009). *Multimodality: A social semiotic approach to communication*. London: Routledge/Falmer.

Lave, J., & Wenger, E. (1991). *Situated learning: Legitimate peripheral participation*. Cambridge: Cambridge University Press.

Leu, D. J., Coiro, J., Knobel, M., & Lankshear, C. (Eds.). (2008). *Handbook of research on new literacies* (pp. 899–940). Hillsdale, NJ: Lawrence Erlbaum Associates.

McCardle, P., & Chhabra, V. (2004). *The voice of evidence in reading research*. Baltimore: Brookes Publishing Co.

Ministerial Council of Employment, Education, Training and Youth Affairs (MCEETYA). (1999). *The Adelaide declaration on national goals for schooling in the 21st century*. Retrieved June 21, 2009, from MCEECDYA: http://www.mceecdya.edu.au/mceecdya/default.asp?id=11576

Ministerial Council on Education, Employment, Training and Youth Affairs (MCEETYA). (2008). *National report: Achievement in reading, writing, language convention and numeracy*. Retrieved June 21, 2009, from MCEETYA: www.naplan.edu.au/naplan_2008_reporting/naplan_2008_reporting.html

Snow, C. E., Griffin, P., & Burns, S. M. (Eds.). (2005). *Knowledge to support the teaching of reading: Preparing teachers for a changing world*. San Francisco: Jossey-Bass.

The State of Queensland. (2000). *Literate futures report of the literacy review for Queensland state schools*. Brisbane: Education Queensland.

Unsworth, L. (Ed.). (2009). *Multimodal semiotics: Functional analysis in contexts of education*. London: Continuum International Publishing Group.

Vygotsky, L. S. (1987). *The collected works of L. S. Vygotsky, volume 1: Problems of general psychology*. New York: Plenum Press.

Wenger, E. (1998). *Communities of practice: Learning, meaning and identity*. Cambridge: Cambridge University Press.

Willinsky, J. (2008). *Critical literacy in action: Social perspectives and teaching practices*. Rotterdam: Sense Publishers.

Wyatt-Smith, C., Elkins, J., Colbert, P., Gunn, S., & Muspratt, S. (2007). *Changing the nature of support provision: Students with learning difficulties: Interventions in literacy and numeracy project (InLaN).* Canberra: Department of Education, Science and Training, Australian Government. Available from http://www.dest.gov.au/sectors/school_education/publications_resources/profiles/effective_teaching_learning_practices_stud_learn_difficult.htm#authors.

Chapter 2
Learning Difficulties, Literacy and Numeracy: Conversations Across the Fields

Stephanie Gunn and Claire Wyatt-Smith

Introduction

The published literature on effective teaching and learning practices for struggling students spans at least three fields: *learning difficulties, literacy* and *numeracy*. The literature in each of these fields is already well developed, with any single topic within a field easily able to provide material for a substantial book. This chapter draws on the key findings from a wide-ranging review of literature across these fields. The review was undertaken by Gunn (2007) as part of an Australian, large-scale research program investigating the effectiveness of school interventions in literacy and numeracy for students with learning difficulties (Wyatt-Smith, Elkins, Colbert, Gunn, & Muspratt, 2007).[1] The program was mentioned previously in Chapter 1 and is discussed further in Chapter 6. The literature review aimed to present a range of views from the three fields without privileging any particular perspective. A starting proposition was that there was merit in taking a *multiperspective* approach, providing an opportunity to benefit from the wealth of different, even contradictory, perspectives to further our understanding of the complexities of these fields. A related proposition was that a single theoretical or research perspective could not provide all the answers demanded in today's complex times. Traditionally, the three fields have been addressed separately by researchers and policy makers, often resulting in compartmentalised thinking and a lack of awareness of developments in the other fields. The chapter provides a space for a useful colloquy of perspectives across fields to inform and frame effective provision for struggling students.

The chapter starts with a discussion of how the terms 'learning difficulty', 'literacy' and 'numeracy' are variously defined. This is followed by an investigation

---

S. Gunn (✉)
Griffith University, Brisbane, QLD, Australia
e-mail: s.gunn@me.com

[1] A chapter of this length cannot provide the detailed discussion presented in the review on the theoretical perspectives and literature on effective practices for these three intersecting fields. See the full review at: <http://www.dest.gov.au/sectors/school_education/publications_resources/profiles/Effective_Teaching_Learning_Practices_Stud_Learn_Difficult.htm>.

C. Wyatt-Smith et al. (eds.), *Multiple Perspectives on Difficulties in Learning Literacy and Numeracy*, DOI 10.1007/978-1-4020-8864-3_2,

of the various theoretical perspectives and debates from the three fields of interest. Next is an overall commentary on some of the main findings across fields, based on an examination of recent major studies, reviews and meta-analyses into literacy and numeracy provision and practices.

Defining learning difficulties, literacy and numeracy

Numerous attempts have been made to find a single definition for each of these three fields. However, a consensus on a single definition may not be feasible, and possibly not even desirable, given our rapidly changing times. This is particularly relevant when considering the advent of new technologies that inevitably impact upon our communication practices and even our understandings about the nature of knowledge and learning, in and out of school. It has been argued that new and emerging technologies generate new literacies (for example Leu, 2000; Leu, Kinizer, Coiro, & Cammack, 2004; Unsworth, 2002, 2008) and numeracies (for example Cumming, 2000a; Zevenbergen, 2004) as well as new understandings of learning difficulties (for example Berninger, 2001). Such developments, therefore, require revisiting of previously accepted definitions to ensure they address contemporary demands faced by students. Definitional issues for each of these fields are now briefly examined.

The field of *learning difficulties* has been replete with different terms and labels (for example learning disabilities, dyslexia, word-blindness, minimal brain dysfunction, special needs), generated according to the particular discipline and philosophy of the researcher or practitioner. Most of the definitional debate has occurred in North America, where the processes of classifying students according to specific needs and characteristics are inevitably linked to funding of services for the educational needs of students and research. The precise measurement and labelling required for funding purposes have proven problematic, particularly for encouraging a deficit construction of difficulties in learning, where difficulties are constructed as reflecting a deficit in the ability (internal capacity) of individual students. In practice, the heterogeneous nature of those experiencing learning difficulties and the varying assumptions on the nature and causes of learning difficulties have impacted on resultant definitions (Elkins, 2002; Lerner, 2003). A continuum of causality ranges from those difficulties perceived as almost exclusively biological (small in number) in nature and only marginally responsive to environmental factors, to those that appear to be more socially determined and shaped. Inevitably, most students fall within these extremes, with those presenting with difficulties in learning literacy and numeracy as varied as the interactional factors that shape them.

Consequently, in recent years, the literature reflects a move away from precise measurement and labelling of 'in-child' deficits to a closer focus on the success or otherwise of an intervention. For example, in the United States there has been a move away from measures such as IQ tests for identifying students with learning difficulties towards an examination of responsiveness to intervention. In the latter approach, students who are identified as not achieving at the same level and rate as their peers are provided with specified interventions, and the response to the intervention is monitored (Fletcher, Coulter, Reschly, & Vaughn, 2004). Hence, in many respects the actual term used to describe an individual student becomes secondary to the practices that support the child.

In Australia, the definitional debate has been less categorical in focus. The term 'learning difficulty' has been used to describe the significant minority of students who do not seem to respond to their classroom programs, with the term 'learning disability' reserved for those who have or are expected to have persistent problems in learning over an extended period of time (Elkins, 2002). Again, the focus is on supporting students who are not responding to the classroom program as expected, with attention given to the classroom program or intervention and students' responsiveness to that program.

In the past decade, *literacy* education has generated various accounts of the nature of literacy and different views of the features of quality literacy education in schooling and beyond. This is most evident in different definitions of literacy. Typically, these have ranged from skills-based conceptions of functional literacy through to broad and all-encompassing definitions that identify repertoires of literate practice and integrate social, cultural and, in some cases, political empowerment. Coiro, Knobel, Lankshear, and Leu (2008) argue that a precise definition may not be possible because the key characteristic of literacy today is that it regularly changes, so that as new technologies appear, new literacies continually emerge at a pace faster than we are able to evaluate or define them. Ultimately, most definitions encompass the need for a set of knowledge and skills, or repertoire of practices, that enable students to contribute to social, civic and economic spheres of activity.

Historically, *numeracy* is a recent term. First coined in the United Kingdom (Crowther Report, 1959) and later discussed in the Cockcroft Report (1982), the term numeracy was seen to emphasise components of mathematics in the contexts of everyday lives, with a broad interpretation relating to social practices and the social context of numeracy use (Baker & Street, 1993; Brown, Askew, Baker, Denvir, & Millet, 1998). Elkins (2005) noted that 'numeracy has become a term for those aspects of mathematics that are related to functioning in society' (p. 217). However, while the term numeracy is now common in the United Kingdom and in countries such as Australia and New Zealand, other terms (for example quantitative literacy, critical numeracy, mathematical skills, statistical literacy and critical mathematics) are also used internationally. Despite some differences in terminology and definition, common elements are evident, with a focus on the importance of context, whereby a person can be more or less numerate in relation to a particular situation (Australian Association for Mathematics Teachers (AAMT), 1997; Organisation for Economic Co-operation and Development (OECD), 1999; Willis, 1998).

Further, regardless of the definition or label applied, the social, political and economic contexts in which a definition is developed and used must be considered. Also essential is consideration of the often-unstated assumptions about knowledge and learning, especially in schooling, that motivate definitions. It may no longer be useful to think in terms of definitions for fields in isolation (learning difficulties, literacy and numeracy), given the wide recognition of emergent forms of knowledge and understandings about learning. Instead, as discussed next, there is a need to focus on the particular practices adopted in the respective fields and how they might converge or achieve coherence in new ways.

Learning difficulties—changing perspectives

The field of learning difficulties (henceforth LD) has been characterised by 'change' over the past four decades, with the focus moving from a medical orientation to a more psychological and educational one. In that time, the field has witnessed the emergence of several theoretical perspectives and paradigm shifts. The development of theory in the LD field has been continuing, reflecting diverse, even contradictory, views that can be confusing for those working with children who are experiencing learning difficulties and those working in the field itself. In an attempt to clarify these varying views, several authors (Heshusius, 1991; Iano, 1986; Poplin, 1988a, 1995) proposed two fundamental worldviews in the LD field—*reductionism* and *constructivism*. The first paradigm, variously named 'reductionism' (Poplin, 1988a), 'positivism' (Warner, 1993), 'the natural-science technical view' (Iano, 1986) and 'mechanistic paradigm' (Heshusius, 1991), assumed that nature can be observed from a detached, objective point of view. Within a *reductionistic paradigm,* researchers attempt to reduce complex phenomena into their component parts. Thus, theories about a phenomenon and the relationships among its component parts are developed as a series of studies progressing through a deductive process (O'Shea, O'Shea, & Algozzine, 1998). A number of theoretical models represent this framework—biological/medical, psychological process, behavioural and cognitive/metacognitive.

The medical model has been highly prominent, with the LD field founded on the assumption that neurological factors were the basis of learning difficulties. However, recent techniques in neuroscience, brain imaging and genetics have led to considerable progress in LD theory at the medical/biological level. Forness and Kavale (2001) noted, what they called the 'new medical model' not only acknowledges the role of contemporary medicine, but also recognises environmental considerations. In recent years, a key journal in the LD field devoted a special issue (*Learning Disability Quarterly*, Summer 2001) to articles written by representatives of five biologically orientated research approaches,[2] providing a tutorial on their main research tools and a succinct summary of current research. Berninger (2001) provided a useful summary and noted that 'although each contributor was given the charge of writing about his or her biological research, each addressed, without any solicitation, the issue of interactions between biological and environmental factors' (p. 139). The biggest challenge for the medical model has been to explore ways

[2]The five approaches are: genetics—Raskind; event-related potentials (ERP)—Molfese & Molfese; brain neuroanatomy—Leonard; functional magnetic resonance imaging (fMRI)—Richards; cognitive neuroscience—Booth.

in which this medically based research might impact on how teachers work with students in their classrooms (Shaywitz & Shaywitz, 2004; Sousa, 2001). From a definitional perspective, the sample of students referred to in these studies is pertinent. In Australia, for example, the term 'learning disability' (sometimes referred to as 'dyslexia' in other countries) is reserved to refer to a more specific group of students who experience persistent problems over an extended period of time. It is likely that this body of work refers to this group, and reminds us that although some students within the broader definition of LD may demonstrate unusual and different brain activity during educational tasks, many children with LD do not. It is also possible that many children who do not experience learning problems may well demonstrate unusual brain activity patterns (Bender, 2004). Despite this, it is likely that medical/biological theories will continue to excite controversy and may exert increasing influence over educational practice for some years to come.

The second model under the reductionist banner is the psychological-processing model. This approach is based on the idea that the mind contains certain basic learning processes whose efficient functioning is prerequisite for learning. These processes include the auditory, visual, tactile, motoric, vocal, attention, sequencing and memory processes. Ultimately, it is fair to say that research evidence has been kinder to the medical/biological theorists previously discussed than the psychological processing theorists, where according to Hammill and Larsen (1974) 'efficacy of training psycholinguistic functionings has not been conclusively demonstrated' (p. 12). Despite this evidence, the debate remains alive, with Torgesen (2004) arguing that the psychological process model is an idea ahead of its time and may look very different in the future.

Theorists' failure to support the assumption of an internal physical or psychological-processing deficit led to the application of behaviour theory, the third model within the reductionist paradigm. The application of behaviourism caused the focus to shift to investigating the interactions between the learner and the learning environment (O'Shea et al., 1998). Basically, in a behavioural model, learning problems are seen as reflecting ineffective interactions between learners and various instructional variables—that is inadequate teaching. Behaviour is believed to be shaped and maintained by its consequences, with behaviour, including academic behaviour, learned from environmental feedback. Numerous forms of instruction and assessment have been developed that reflect the concentration on skill, including criterion-referenced testing, curriculum-based assessment, precision teaching, direct and daily measurement, direct instruction and skill assessment. While some researchers (for example Heshusius, 1994; Poplin, 1988a) have questioned the theoretical model's ability to explain and predict developments in the learning process, there has been widespread application of behavioural models, with basic teaching methods developed from these models still currently being used by teachers.

Finally, in response to some concern about the behavioural perspective, particularly in terms of generalisation and maintenance of skills, a number of cognitive theorists believed that children with LD, when presented with some academic tasks, did not think or attempt to use their cognitive processes in planning, carrying out or monitoring their own progress (Jacobs & Paris, 1987; Wong, 1986), nor did they self-instruct with the same frequency and degree of accuracy as other children. That is, they were often disengaged or inactive in their learning efforts. Based on this theoretical perspective, several instructional approaches were developed to use with children who had learning difficulties. These approaches focused on the research conceptualisation of metacognition (thinking about thinking), which suggests that children need to think and plan out their thinking and their learning activities in order to complete a complex educational task (Bender, 2004). Despite some dissenting voices (for example Meichenbaum, 1980), particularly around generalisation of learning, many aspects of this approach have found acceptance in schools and are in use today (van Kraayenoord & Goos, 2003). In fact, there are still advocates of each of the four theories discussed within the reductionist paradigm, with traces and mixes of these models currently found in classrooms.

A shift to more holistic principles was seen as a transformation of basic assumptions of previous beliefs and, therefore, was viewed as a change in paradigmatic beliefs themselves. This paradigm, variously termed 'holism', 'constructivism' or 'social constructivism', worked from a number of key principles and assumptions. The first was that reality is dependent upon our construction of it (Harris & Graham, 1996; Poplin, 1988b; Reid, Robinson, & Bunsen, 1995). Here, knowledge is viewed as being shaped by contextual conditions and meanings. Education and educational inquiry are seen as value-impregnated ideological activities. Heshusius (1991) explained that 'the observer and observed are inextricably connected' (p. 441). Second, the 'whole is both more than and different from the sum of its parts' (Heshusius, 1991, p. 442), with learners viewed as active agents in their learning who construct new knowledge in complex, challenging learning environments with authentic tasks. That is, knowledge of the parts does not lead to knowing the whole. A prominent example of an educational approach emanating from constructivist theory relates to learning difficulties in reading—the whole-language approach. Here, the principles of holism are applied to reading instruction by having students engage in activities that are purposeful to them and where the unit of meaning is the whole word within a whole story, rather than some smaller part (for example letter sounds or isolated words).

Further, within the *social* constructivist framework, the emphasis is placed on the context in which learning occurs, and social activities are considered context-dependent. Poplin (1995) explains some differences in the terms used to describe this paradigm, emphasising that constructivism comes from a largely cognitive orientation drawing on a Piagetian framework, while social constructivism emphasises the role that sociocultural contexts play in the construction of meaning drawing on Vygotsky's (1978) theories. Recently there has been increasing attention to the sociopolitical and sociocultural analysis of the field. Reid and Valle (2004), for example, interrogated the process of meaning-making in the LD field based on the ideological underpinnings of Foucault's (1972) discourse analysis. They asserted that students with LD are not an objective fact; instead, they are historically and culturally determined. They explained that within this lens (as distinct from deficit views), difference is seen as just difference, shifting the focus on to 'redesigning the context, not on "curing" or "remediating" individuals' impairments' (p. 468). Thus, while conceding that there is a 'place at the table' for empirical, medical, legal or other kinds of theory and research, the emphasis is placed on human variation rather than pathology, and on a reframing of what is considered 'normal'.

While the debate over the merits of the two worldviews continues, as mentioned earlier, it is argued that one view or approach cannot address the complex nature of learning difficulties for this heterogeneous group. By examining the various theoretical perspectives for their underpinning assumptions about the learner and learning, educators can move beyond the divide that has the potential to impede progress and damage the credibility of the field (Andrews et al., 2000). Dixon and Carnine (1994) best explain this position

> Every ideology – not just constructivism or behaviourism – has its fringe elements, variously described as radical constructivists or fanatical behavourists. When the educational fringe groups declare war, the rest of us expend precious resources responding to charges that are, when stripped of rhetorical trimmings, often groundless.
>
> We suspect that when ideological rhetoric is set aside, mainstream educational researchers' (and other educators') best hope for advancing the field might be realized through our commitment to develop and verify the best possible curricular and instructional practices – best in the sense of effectively resulting in well-understood knowledge for all learners, and doing so efficiently, particularly for those children for whom efficiency is no luxury. (p. 364)

In order to move towards some complementarity for the field, some (for example O'Shea et al., 1998) suggest that researchers and practitioners work towards a meta-theory that moves beyond the confines of a single theoretical model, in search of the connections and disconnections between competing views.

The above discussion of the LD theoretical models has presented the basis for educational approaches in this area. The preceding theoretical work demonstrates how, for over more than 40 years, scholars investigating learning difficulties in children have contributed to education in general, and particularly to the areas of literacy (for example reading acquisition) and numeracy. Against this backdrop, attention turns to the work of scholars in the literacy and numeracy domains to better understand the impact upon and possible connections with the LD field.

Literacy today—competing views

Since the 1950s there has been increasing polarisation of views on the very nature of literacy itself. Concurrently, there have been competing views about what counts as effective approaches to the teaching of literacy, including as literacy relates to learning in curriculum areas. McMeniman (1997) noted that the savagely competing paradigms had led to a tyranny of false dichotomies and called for complementarity of these bipoles, not their dichotomisation. This section of the chapter looks briefly at these dichotomies and competing paradigms and moves on to explore some frameworks developed in the literacy field in Australia that represent a movement towards bringing these competing views together to ensure all students have opportunity to develop a full repertoire of skills and competencies.

Mills (2005) offered some clarity on the complex set of opposing views permeating the literacy field by proposing, as an organiser, three binary oppositions that exemplify the continuing major debates: the skills-based versus whole-language debate; the exclusively print-based approach versus multiliteracies; and the opposition between the cultural-heritage model of English and the sociocultural models of language and literacy education.

Skills-based versus whole-language approaches

One of the most contentious debates in literacy pedagogy is the skills versus whole-language debate (Mills, 2005), with each side of the debate bringing with it not only a particular view of what literacy is, but also a particular worldview. This dichotomy represents a clash of paradigms: the skills-based approach reflects a more compartmentalised view that focuses on a generic set of portable skills through direct instruction; the other draws on constructivist principles whereby learning is said to be shaped by contextual conditions and meanings, and advocates a whole-language approach to instruction. Essentially, one emphasises identified units and individual skills in isolation, while the other stresses use in context and meaning, though both focus on the individual child (Rassool, 2002). Though the debate is broadly referred to as being about the features of effective literacy education, the focus has been primarily on reading, with the skills approach building knowledge of words from part to whole, resulting in 'an emphasis on phonics, phonological awareness, common letter-strings and initial sound blendings in order to decode and write text' (Soler, 2002, p. 5). The other side of the debate has witnessed a movement away from viewing literacy—and reading in particular—as the neutral decoding of print to a view of literacy as a 'range of meanings produced at the interface of person and text, and the linguistic strategies and cultural knowledges used to "cue" into meanings embedded in the text' (Rassool, 1999, p. 28).

There have been advocates for and against both approaches, resulting in what has sometimes been described as 'the reading wars'. Recently, in an attempt to provide evidence-based advice to governments on this debate, several major studies and reviews of research have been undertaken in Australia (Department of Education, Science and Training, 2005), New Zealand (Education and Science Committee, 2001), United Kingdom (Rose, 2006) and United States (National Reading Panel, 2000), and these have reached similar conclusions as follows:

- Systematic phonics instruction is highly effective in preventing reading difficulties (National Reading Panel, 2000).
- The use of phonetic, word-level decoding skills is an important element in a balanced reading program (Education and Science Committee, 2001).
- Systematic phonics instruction is critical if children are to be taught well, although teachers must draw on an integrated approach to reading that includes phonics, fluency, vocabulary knowledge and comprehension (Department of Education, Science and Training, 2005).

- High-quality, systematic, synthetic phonic work taught discretely and consistently should be the primary approach to establishing word recognition, but it should be set within a broad and rich language curriculum that takes into account speaking, listening, reading and writing (Rose, 2006).

While these more recent studies are strong advocates for a more skills-based approach, with a focus on phonics instruction, there is recognition that the reading process involves more than word knowledge. Most importantly, some have argued that ultimately the opposition between these two views is unhelpful. Stanovich (2000), for example, an advocate for phonological awareness training and proficient decoding, argued that there were more points of agreement than disagreement between the opposing positions. He suggested both sides look at the defining differences, which are probably few, in order to decide whether they are worth the cost of 'war'. Often the point of departure lies solely in the importance the two camps attach to explicit and systematic instruction on how to decode words. Coles (2003) questioned how much and to what extent phonics instruction (including phonemic awareness) should be prioritised over other skills and strategies, and when it should be part of the reading instruction. Hence, while it is generally acknowledged that explicit teaching of word skills is important, some (Davis, 2002; Soler, 2002) raise concerns about policy makers turning to narrowly conceived short-term interventions (for example, legislated phonics instruction) and measures in the face of a perceived literacy decline. They argue that there must be acknowledgment of the complexities and issues surrounding literacy teaching. As Davis (2002) notes, the majority of teachers 'continue to use both of the major contested approaches—and others—as they seek to help children with different talents and backgrounds to learn to read' (p. 85). Mills (2005) similarly argues that 'the debate should no longer be framed as "either or" but "when" and "for which students"' (p. 71).

Print-based approach versus multiliteracies

A more recent debate that has emerged in literacy research 'concerns exclusively print-based literacy practice versus multiliteracies practice' (Mills, 2005, p. 71). Some (Gee, 1990; New London Group, 1996, 2000) have argued that students must acquire *multiple literacies* to be able to fully participate in the new global community, which has witnessed the emergence of mass digital computer and online communications (Coiro et al., 2008; Leu, Mallette, Karchmer, & Kara-Soteriou, 2005). The New London Group (1996, p. 60) coined the term *multiliteracies* to account for what they considered to be two principal aspects of the multiplicity of these new literacies or the multidimensional nature of literacy: (1) the burgeoning variety of text forms associated with information and multimedia technologies; and (2) the context of our culturally and linguistically diverse and increasingly globalised societies, for the multifarious cultures that interrelate and the plurality of texts that circulate. Leu et al. (2005) add a third aspect, which is the 'fact that new technologies will appear repeatedly in our future, generating even newer literacies on a regular basis' (p. 2). Working from this position, a key issue is how best to prepare students for new and continually changing literate futures, including work, public and private lives, where print-dependent literacy will inevitably be insufficient.

Moves in this direction require a radical rethinking of literacy pedagogy, focusing on how technologies shape communication practices and meaning-making possibilities in local and global contexts (Wyatt-Smith & Elkins, 2008). As each new technology and associated social practice changes, teachers are challenged to find new ways to talk about new literate practices that, despite their sharing certain elements (Bearne, 2009; Burke, 2009), are different from earlier generations of literacies (Leu et al., 2004; Unsworth, 2008). Further, past conceptions of exclusively print-based literacy 'need to be reconceptualised to account for the increasing range of textual practice that now counts as literacy' (Mills, 2005, p. 72), with literacy and technology integrally related. While this does not suggest a need to replace print-based literacy, there is a 'need to acknowledge that conventional, hard-copy forms of "linear" texts will continue to coexist with electronic hypertext for some time, and that old and new literacy technologies will frequently have complementary roles in a range of contexts' (Unsworth, 2002, p. 73). Given this, teacher knowledge will need to incorporate and make the connections among written, visual, oral and digital contexts.

Cultural heritage versus critical literacy: the final area of debate in language and literacy education is between cultural heritage and sociocultural perspectives, the latter taken as extending to what has come to be known as 'critical literacy'. The cultural-heritage model was identified by Dixon (1969) and 'dates back to the Greek view of literature as moral and spiritual influence . . . [which] emphasised the transmission of culture through the study of literature' (Cumming, Wyatt-Smith, Ryan, & Doig, 1998, p. 13). This perspective considers that the most important outcome of language education is 'access to the cultural and linguistic heritage of a culture, expressed most richly in its canon of valued literary works' (Freebody, Ludwig, & Gunn, 1995, p. 42). Here, there is belief in the unchanging merit and meaning in historically ratified texts, usually taken to be high literary texts. There is also the implicit affirmation of the conservative systems of belief represented in these texts (Hollindale, 1995, cited in Mills, 2005).

Conversely, sociocultural perspectives on language and literacy, including critical literacy, require 'a fundamental shift to viewing language as social practice, which is institutionally and culturally located in sites which are neither benign nor neutral' (Kamler & Comber, 1996, p. 1). Such perspectives draw upon a number of theoretical frameworks and are interdisciplinary. However, there are a number of 'shared assumptions: that literacy is a social and cultural construction, that its functions and uses are never neutral or innocent, that the meanings constructed in text are ideological and involved in producing, reproducing and maintaining arrangements of power which are unequal' (Kamler & Comber, 1996, p. 1). As with the preceding debates, the various views of advocates for both the cultural heritage and critical literacy (or literacies) stances have been questioned. On one side it has been argued that 'cultural heritage advocates need to acknowledge that their criteria for judging quality of literature reflects the dominant cultural interests and ideologies' (Mills, 2005, p. 76) with a need to consider the interests of marginalised groups and the diverse purposes of literacy in today's society (Hollindale, 1995; West, 1992).

Similarly, the critical literacy perspectives have also been subject to critique. One of the claims of critical literacy is that it has the potential to oppose and make evident the prevailing structures that limit access, entitlement and empowerment to those groups marginalised in society (Mills, 2005). However, some (for example Comber & Hill, 2000; Hodgens, 1996; Mellor & Patterson, 2005) have cautioned against such lofty claims for critical literacy alone, suggesting a need to recognise the multiple factors that influence marginalisation in society. For example, genre theorists, Christie and Misson (2002, p. 57) remind us that, 'while excellent work has been done on teaching against discrimination ... it is worth noting that this, like anything else in the classroom, can become a rather empty routine ... [where] the students can produce the expected answer and mouth the appropriate sentiments without any notable impact on their actual attitudes'. Ultimately, explicit knowledge of how language works and the mastery of critical literacy do not automatically ensure that social class and power structures are overcome (Mills, 2005).

Towards a complementarity of views: Mirroring the LD literature, the field of literacy research, including the term 'literacy' itself, reveals a complex and rich phenomenon that can be viewed from multiple theoretical and disciplinary perspectives. Again, like the LD field, considerable debate has occurred regarding this phenomenon. What becomes clear is that a focus on a unitary approach to literacy pedagogy may not be helpful, with none of the aforementioned views sufficient for literacy education in today's world. To date, in Australia, no particular perspective or method of literacy education has been legislated. Rather, several literacy scholars have provided frameworks that attempt to capture the multiple perspectives and dimensions of literacy to make available to students the full repertoire of skills and competencies required in contemporary society. A summary table of these frameworks is provided (Table 2.1), based on Unsworth (2002, p. 70).

Essentially, none of the above-named frameworks advocates a particular hierarchy or order for working with the different dimensions of literacy. Instead, the authors view them as providing a useful template for coordinating and addressing different dimensions where 'literate practice is ideally an integrated expression of all the roles and dimensions in question' (Durrant & Green, 2000, p. 102).

Numeracy today—varying perspectives

Since the term 'numeracy' was coined more than 40 years ago, several reports and research studies have focused on this area. However, while most recognise the contribution made by the two fields—numeracy and literacy—some (for example Milton, 2000) consider that one of the barriers to improved focus for children experiencing learning difficulties with numeracy is the conflation of numeracy and literacy. Most numeracy researchers insist that numeracy be viewed as distinct, since

Table 2.1 Literacy frameworks

Dimensions of literate practice (Unsworth, 2002)	Three dimensions (Durrant & Green, 2000)	Four roles of the literacy learner (Freebody & Luke, 1990)
Recognition: involves learning to recognise and produce the verbal, visual and electronic codes that are used to construct and communicate meanings	Operational: involves being able to read and write within a range of contexts in an adequate and appropriate manner, employing conventional print and electronic media	Code breaker: the practices required to 'crack' the codes and systems of written and spoken language and visual images
Reproduction: involves understanding and producing conventional visual and verbal text forms that construct and communicate the established systematic knowledge of cultural institutions	Cultural: involves understanding texts and information in relation to the contexts—real-life practices—in which they are produced, received and used. Here, literacy acts are not only context-specific, but also entail a specific content. Rather than being literate in and of itself, but of being literate with regard to something, some aspect of knowledge or experience	Text participant: the practices required to build and construct cultural meanings from texts. That is, how do the ideas represented in the text string together? What cultural resources can be brought to bear on the text? Text user: the practices required to use texts effectively in everyday, face-to-face situations. That is, how do the uses of this text shape its composition? What do I and others do with this text?
Reflection: necessitates an understanding that all social practices, and hence all literacies are socially constructed. Because of this, literacies are selective in including certain values and understandings and excluding others. This entails interrogating the visual and verbal codes to make explicit how other choices of visual and verbal resources construct alternative views	Critical: it is based on the understanding that social practices and their meaning systems are always selective and sectional; they represent particular interpretations and classifications. It involves being able to innovate, transform, improve and add value to social practices and the literacies associated with them	Text analyst: the practice required to analyse, critique and second-guess text. That is, what kind of person, with what interests and values, could both write and read this naively and unproblematically? What is this text trying to do to me? In whose interests? Which positions, voices and interests are at play? Which are silent and absent?

it is underpinned by a fundamentally different area of learning: mathematics. This, in turn, raises another issue regarding the relationship between mathematics and numeracy. Steen (1999a, cited in Board of Teacher Registration & Queensland, 2005) suggested that 'nearly everyone seems to agree that numeracy is both broader than and different from mathematics—at least as mathematics has traditionally been viewed by schools and society' (p. 13). Willis (1998) reminded us that numeracy is much more than learning mathematics in school and talks about lifting it out of school to be applied in real-world situations.

In addition to these considerations, there has been increasing interest in numeracy research over recent years. This has raised many of the issues evident in literacy research, but without some of the strident debates and dichotomies. Issues related to basic number skills versus more constructivist approaches to numeracy, the complex and multiple versions of numeracy, akin to multiliteracies, and discussions regarding critical numeracy and sociocultural frameworks are evident.

Theoretical approaches to numeracy: Traditionally in mathematics education, teaching of basic maths skills by explicit, teacher-directed instruction was deemed necessary (Grobecker, 1999) before moving to higher-level skills. Recently, this more reductionistic view has been expanded, with attention directed to fostering higher cognitive functioning in mathematics. This has included: big ideas; linkage between operations (that is adding, dividing, multiplying, etc.); depth of understanding and problem-solving (Carnine, 1993). Big ideas represent the central ideas within mathematics that will make learning other concepts easier and more meaningful: 'keys that unlock a content area for a broad range of diverse learners' (Kameenui & Carnine, 1998, p. 8). As in the LD and literacy literature, the approaches that reflect reductionistic theoretical views, have been criticised as decontextualised, mathematical concepts taught as an abstract body of knowledge, which is sometimes connected to arbitrary contexts by embedding tasks in 'real-world' word problems (Zevenbergen, 1995a).

More recently in Australia, there has been a significant shift in approaches to numeracy education. One of the major changes has been the dominance of constructivist approaches with an emphasis on concepts rather than procedures (Elkins, 2005). Within these approaches, numeracy is viewed as an ever-changing set of ideas that alter and expand as a consequence of experiences so that students are seen as being actively involved in the construction of meaning. Dialogue and communication are integral components as students and teachers work through problems, and 'mathematical ideas are learnt by building on or reconstructing what is already known' (Zevenbergen, 1995a, p. 78). Teachers provide students with scaffolding by asking questions that help students examine their ideas and solve problem more effectively. In addition, Renshaw (1996) provides a useful sociocultural view of numeracy education based on Davydov's (1975) interpretation of Vykotsky's work in relation to the numeracy curriculum. Here, the necessity 'for building mathematical concepts on the fabric of the children's existing forms of speaking, representing, and conceptualizing' (Renshaw, 1996, p. 63) is emphasised. While the debate of these varying approaches and theoretical standpoints may not be as vitriolic as witnessed in the literacy arena, some numeracy researchers have documented parallel criticisms to those seen in literacy research. For example, when commenting on the constructivist approaches, Gersten and Chard (1999) argued that students who lacked knowledge of the most basic numerical information to solve problems would be disadvantaged if only exposed to such approaches. Similarly, Westwood (2000, p. vii, cited in Elkins, 2005) considered that 'there is a real danger that the educational pendulum will swing so far in the opposite direction that teachers will feel they must abandon all forms of direct teaching'. He argued for the value of a combination of explicit teaching and constructivist learning opportunities for numeracy teaching.

Multiple numeracies: As with literacy, numeracy in the singular can be reconceptualised at numeracies that are complex and multiple (Cumming, 2000a). Here too, new skills and dispositions are required in our technologised society. Zevenbergen (2004) explains, 'new times' with increased technological and scientific innovations, increased global trade, less secure workplaces and a society formed about knowledge economies (for example via the web) require a retheorising of school mathematics. The current emphasis on numeracy is seen as central to this retheorising.

Critical numeracy: Here, the myth of objectivity of mathematics is challenged with a call for using socially critical numeracy as a 'tool for unmasking the politics and hidden assumptions being built into claims which effectively marginalize a significant proportion of the population' (Zevenbergen, 1995b, p. 100). Steen (1999b) argues that a key skill for students today is to develop the capacity to comprehend the nuances of quantitative inferences, while Frankenstein (1990) describes a process by which students come to understand the statistics and other numerical representations that are used in particular contexts and use them for empowerment. However, as with literacy, while a notion of critical numeracy has much to offer, claims that critical numeracy can assist in ameliorating oppression need to be treated with caution. Rather, the ability to sift through, understand and, most importantly, question quantitative information is more pertinent and realistic.

Towards a complementarity of views in numeracy

While the numeracy field has not witnessed the debates evident in the other fields, as mentioned earlier, it is clear from the competing views that a single approach is not seen by the field as facilitating effective learning outcomes when dealing with the complexity of knowledge that is numeracy in new times. As with literacy, numeracy education may be best served by bringing together a network of competing views to inform practice. As Cumming (2000b) explains, 'it is critical to draw on theoretical models and research from a range of perspectives to inform decision-making in numeracy teaching and learning, and to have a good sense of the complementarity of different theoretical models and research outcomes' (p. 43). In an attempt to capture some of the complexity of numeracy education in Australia, Willis (1998) and Hogan (2000) together provided a useful framework for consideration of the multiple elements of numeracy teaching and learning (Table 2.2). First, Willis identifies three aspects of numeracy and the types of know-how associated, with each arguing that to develop numeracy as practical knowledge a blending of these three elements is required, along with an element of what is colloquially described as 'nous'. Second, Hogan extends this by arguing that the blend of these three types of numeracy 'know-how' needed for particular contexts is determined, in part, by a student's capacity to take up three corresponding roles.

Table 2.2 Multiple elements of numeracy teaching and learning

Aspects of numeracy (Willis, 1998, pp. 33–37)	Roles of numeracy learner (Hogan, 2000, cited in Board of Teacher Registration & Queensland, 2005, p. 2)
Numeracy as mathematics: here, the numeracy is used more or less synonymously with 'mathematics' and is described in terms of mathematical concepts, procedures and skills students need to know—a view that dominates schooling and is more a 'basic skills' notion of numeracy. The focus is on enhancing what and how much mathematics	The fluent operator: showing fluency of use of mathematical knowledge and skill in familiar contexts. This is comfortable and efficient use of mathematical knowledge and being 'at home' with the everyday uses of mathematics
Numeracy as communicative competence: here, skills and knowledge are not independent of the contexts—mental, physical and social—in which they are used. Numeracy is seen as quite context-specific, with individuals more or less numerate with respect to particular settings or circumstances. The focus is on increasing the repertoire of situations with which students can deal mathematically	The learner: using mathematics to make sense of something new or to cope with unfamiliar situations
Numeracy as strategic mathematics: here, numeracy is described in terms of general or strategic mathematical processes, appreciations and dispositions needed to apply mathematics to familiar and unfamiliar situations and problems, and is about how well individuals choose and use mathematical skills they have in service of things other than mathematics. The focus is on increasing the choosing and using skills that students can access; that is, their strategic repertoire	The critic: being critical of the mathematics chosen and used in order to judge and question the appropriateness of its use

This section of the chapter points to a need to move away from false dichotomies and competition between theoretical perspectives towards a 'connective web' that bonds various theories, and moves beyond the confines of a single theoretical model that searches for the connections and disconnections between competing views (O'Shea et al., 1998). Some of the frameworks developed in the literacy and numeracy fields represent a movement towards bringing a multiperspective lens to literacy and numeracy education, ensuring that all students, including those with learning difficulties, have opportunity to develop a full repertoire of skills and competencies. By endeavouring to encapsulate the multiple theoretical and disciplinary perspectives that inform the three fields—learning difficulties, literacy and numeracy—this section of the chapter has provided a foundation for the following discussion, as well as a reference point for other chapters within the book.

Effective provision for students with learning difficulties: what the major studies tell us

Much has been written about teaching and learning practices that underlie effective provision in the fields of learning difficulties, literacy and numeracy. In this section we provide a brief overview[3] of key messages that have been distilled from an examination of major studies, reviews and meta-analyses that have significantly informed the current state of knowledge in these fields. In recent years, several governments (for example Australia: DEST, 2005; Vincent, Stephens, & Steinle, 2005; Canada: Expert Panel on Mathematics, 2004; New Zealand: Education & Science Committee, 2001; Ministry of Education, 2005; United Kingdom: Askew, Brown, Rhodes, & Johnson, 1997a, b; House of Commons Education and Skills Committee, 2005; United States: Kilpatrick, Swafford, & Findell, 2001; National Reading Panel, 2000; RAND Mathematics Study Panel, 2003) have commissioned major studies that drew on evidence-based research and have presented a number of recommendations for approaches to improve literacy and numeracy outcomes for all students. These studies often have particular relevance for those students experiencing difficulties in the areas of literacy or numeracy.

The studies have drawn on a variety of methodological (quantitative and qualitative) and disciplinary perspectives, providing different lenses through which to view learning and improvement. The greatest challenge—and particularly for teachers—is to bring together this wide variety of information into a coherent framework. The key messages discussed below have been synthesised in an attempt to distil the recurring ideas in this body of work on effective provision for learning literacy and numeracy.

Selection of approaches: One consistent finding from the research literature was that no single approach or blend of programs has been deemed the definitive answer for effective literacy and numeracy education. While the power seems to be in a combination of strategies, unprincipled eclecticism with a confusing mix of methods or blend of practices may be equally detrimental. Rather, what is required is a thoughtful, carefully integrated selection of validated instructional approaches based on contemporary research and theoretical understandings by informed teachers. Hence, while full investigations of the complex interactions among instructional approaches have been rare (Guthrie & Wigfield, 2000), when synthesising the outcomes of the major studies, reviews and other informing work, it is clear that we do have some consensus on 'what works', as is evident by the key messages below. What we have not fully chartered is the fine detail on how best to orchestrate a combination of approaches *in situ*. This stance recognises the teacher, and more specifically, the craft of teaching, as central to efforts to improve learning for cohorts

---

[3]As mentioned earlier, a chapter of this length is not able to provide the detailed discussion presented in the literature review that forms the basis for this section of the chapter. The reader is directed to Sections 3 and 4 of the full review for more detailed discussion: <http://www.dest.gov.au/sectors/school_education/publications_resources/profiles/Effective_Teaching_Learning_Practices_Stud_Learn_Difficult.htm>.

and for individual students. In what follows, readers are asked to consider how approaches could be adopted and combined in particular contexts with optimum effect.

What works

Three waves of provision: Several stages of schooling have been suggested that emphasise the need for instructional approaches to be systematically organised over time in response to diagnosed student need. These stages have been variously described as three waves or tiers, particularly in the case of literacy. While there are some minor inter-country differences, the first wave usually refers to initial whole-class teaching and the importance of quality teaching to minimise the risk of children falling behind. For some, quality first-teaching means a greater emphasis on 'basic skills' and, in the case of literacy, focused phonics instruction (for example see DEST, 2005a; Education and Science Committee, 2001; House of Commons Education and Skills Committee, 2005; National Reading Panel, 2000; Snow, Burns, & Griffin, 1998). While most would not argue against quality classroom teaching by well-informed teachers, some disputed the degree of focus on phonics in literacy or basic computational skills in numeracy, and preferred to leave the particular emphasis of any approach to teachers in response to individual need. The second wave refers to *early interventions* for those students falling behind; these being taken to include individual and group interventions. The emphasis is on a wide range and balance of literacy and numeracy approaches. In literacy, for example, this may include explicit teaching of sound–letter associations and phonological awareness using direct and strategy instruction approaches (for example Ellis, 2005; Louden et al., 2000). Clearly, care must be taken to link any intervention taking place outside the classroom (say, in withdrawal or pull-out settings) with the regular classroom through close communication between the teacher providing the intervention and the classroom teacher.

Finally, the third wave of provision acknowledges the continuing nature of support required by some students. Students who experience learning difficulties may present a chronic rather than an acute difficulty, with continuing support necessary beyond the early years of schooling (for example D'Agostino & Murphy, 2004; Snow et al., 1998). This minority of students require highly qualified specialist assistance (for example DEST, 2005a; Snow et al., 1998). Some argue for structured, explicit word-level instruction (Wheldall & Beaman, 2000; for example the MULTILIT program) and others focus on authentic texts with real-world connections, strategy instruction to foster metacognitive skills, student autonomy and choice in lesson design (for example Luke, Woods, Land, Bahr, & McFarland, 2002)—these are not mutually exclusive. The timing and duration of interventions are also vital: optimal engagement and motivation with support are to be provided before persistent failure occurs and with any support continually monitored and adjusted based on assessment (Alloway, Freebody, Gilbert, & Muspratt, 2002).

Transition and continuity: The literature reminds us that students experience many transitions during schooling, and consequently, there is a need to plan for continuity within school and across the various stages of schooling (for example primary to secondary) to ensure seamless transitions over time (for example Cumming et al., 1998; DEST, 2005a; Delina & van Kraayenoord, 1996; Hill, Comber, Louden, Rivalland, & Reid, 1998; Lankshear et al., 1997). This includes dialogue and sharing of information among teachers throughout the various stages of schooling and possibly a highly trained specialist teacher responsible for linking the whole-school planning process (for example, DEST, 2005a).

Additionally, literacy and numeracy teaching and learning were viewed as continuing throughout the years of schooling in all areas of the curriculum, and hence were the responsibility of all teachers. This was particularly the case as literacy and numeracy demands change as students progress through the middle and secondary years of schooling. Here teachers need to be aware of the interface between a specific curriculum and its literacies—or 'curriculum literacies' and the various numeracies that student encounter. Wyatt-Smith and Cumming (2003) argued for the need for teachers to take into account as they plan and design curricular tasks, the specific literacy demands of the particular curriculum area, as well as the literacy demands of assessments students are required to undertake.

Time for literacy and numeracy: The literature indicates that focused time with minimal disruptions is an essential element for both literacy and numeracy activities (for example DEST, 2005b; Rose, 2006). Some studies designated particular time slots, such as 20 minutes of systematic phonic work daily (Rose, 2006), while others placed greater emphasis on a lack of disruption and need for carefully planned and structured lessons (for example Snow et al., 1998). Time was also a factor in terms of the careful pacing of lessons to allow students time to ask questions, share their work and make necessary links to previous learning (for example Cumming et al., 1998).

Supportive leadership: Effective provision for teaching and learning required staff and school leadership to work in a coordinated manner, with opportunities for regular professional exchanges and collaboration. It was suggested that school leaders need to ensure there is an infrastructure for necessary resources and, most importantly, support (including time for literacy and numeracy) for continuing professional development and creation of learning clusters with other schools (DEST, 2005b; Education and Science Committee, 2001; House of Commons Education and Skills Committee, 2005; Luke et al., 2002; Rose, 2006). Some studies also mentioned the importance of specialist leadership to be available in each school (for example, DEST, 2005a; Snow et al., 1998) with, as mentioned earlier, the employment of, for example, a literacy specialist to coordinate a whole-school literacy program to ensure continuity, identify students at risk of failure, offer informal and formal professional development to colleagues, organise support, maintain and analyse a database on performance outcomes, and monitor progress. The professional development offered to colleagues may take the form of a coaching model, where teachers are provided with opportunities to observe, critique and reflect on good practice.

Student motivation and engagement: Another finding from the investigation of the literature includes the need to pay attention to student motivation and engagement, including: providing a range of choices and interesting tasks in literacy and numeracy activities; encouraging collaboration; learning goals co-developed by the teacher and student; real-world connections, and praise and reward for successful engagement in literacy and numeracy practices (for example Luke et al., 2002). For literacy, issues around student motivation, engagement and self-efficacy are considered important for improved outcomes (Alloway et al., 2002; Alvermann, 2001; Guthrie & Wigfield, 2000; Snow et al., 1998). For numeracy, it was noted that attitudes and motivations to numeracy were developed early, with teacher actions, attitudes and beliefs seen as key elements in the development of a positive attitude to mathematics/numeracy learning (for example Elkins, 2005; Hill, 2000).

Monitoring and assessment: Nearly all the studies and reviews identified earlier included monitoring and assessment as essential elements for effective provision. More specifically, several made recommendations for using continuous and varied means of monitoring and assessment to build up detailed profiles at both class and individual student levels to inform planning and teaching, and to permit timely responses when difficulty or delay is apparent (for example DEST, 2005; Hill et al., 1998; Hill, Comber, Louden, Rivalland, & Reid, 2002; Louden & Wildy, 2001; Louden et al., 2005; Snow et al., 1998). The need for multiple sources of evidence and continuing assessment strategies was also a recurring concern. Several of the studies and reviews proposed move towards new forms of assessment, away from current paper-and-pencil tests (for example Cumming, 2000a). Given this, several of the studies called for improved school-based diagnostic capacity and continuing assessment in order to respond to the diverse range of students. This raises a particular note of caution. The multidimensional aspects of both literacy and numeracy and the heterogeneous nature of those experiencing difficulties in learning point to the limitations of current moves to develop evidentiary bases for measuring school effectiveness, through, for example, large-scale cohort testing. While such measures are important, there is a need to avoid an approach that shuts down assessment to that which is more easily measured (for example decoding and basic written comprehension and computational skills), ignoring the complexities and multidimensional nature of teaching literacy and numeracy, and the diversity of students in most classrooms.

Recent work by Wyatt-Smith and Bridges (2008) demonstrated how classroom assessment can be used to improve the learning experiences and outcomes for all students, including those experiencing difficulties in learning literacy and numeracy. This work was based on the belief that 'being explicit about assessment expectations has a focusing effect on pedagogy and facilitates deeper learning' (p. 44). Here, 'front-ending assessment was a process whereby the planned, culminating tasks for assessment were critically analysed to identify the explicit knowledges that needed to be built into the unit planning and learning opportunities' (Wyatt-Smith & Gunn, 2009, p. 91). It was found that significant changes in learning and teaching occurred when reflection on assessment evidence began before teaching started. Additionally, with such a focus on assessment, Wyatt-Smith and Bridges (2008) reported that teachers gained confidence in providing feedback and students were able to take ownership of the learning process and work more independently when introduced to the assessment expectations prior to learning.

Classroom talk: A finding that received attention in several of the studies was the critical nature of the oral medium of communication, with a focus on listening, speaking and classroom talk (Cormack & Wignall, 1998; Department for Education and Children's Services (DECS), 1995; Freebody et al., 1995; Hill et al., 1998; Louden et al., 2000; Luke et al., 2002). It was argued that talk should be at the heart of the curriculum because effective learning is a socially interactive process that is conducted primarily through talk (DECS, 1995). Talk was also found to assist teachers to gain a greater depth of knowledge about students' learning and provided improved opportunities for immediate follow-up and reteaching compared to written forms of assessment (Cormack & Wignall, 1998). In consideration of the importance classroom talk has as a teaching and learning medium, clarity (that is purposefulness) of talk becomes essential. It was considered important that the teacher makes clear what is to be learnt, is mindful of the extended periods of listening required of students and listens purposefully to gain a richer awareness of individual student understandings (for example Cumming et al., 1998). Further, Freebody et al. (1995) suggested that teachers not only need to ensure they make clear the particular focus or goal of any literacy activity but, at times, hear answers students provide as an analysis of teacher questions (talk), rather than as a lack of student understanding or knowledge.

Emergence of new technologies: Considerations of the impact of new technologies ranged from the use of technology as an instructional tool (for example hypertext, word processors—National Reading Panel, 2000) for assisting with teaching basic word skills to a blending of traditional literacy with mastery of new technologies, where new literacies and new ways of shaping and communicating meaning are available (Lankshear et al., 1997; Leu, 2002; Leu et al., 2005; The State of Queensland, 2000). Also, it was argued that New Literacies (for example the Internet searching) have also become important, with the advent of open networks and free publication, but perhaps most important is the capacity of the teacher in this changing environment (for example Kimber & Wyatt-Smith, 2009; Leu et al., 2005). In an earlier research study, Lankshear et al. (1997) wrote about the principle of 'teachers first', with policy makers reminded to ensure teachers feel comfortable with emerging technologies and are able to adequately measure the value that any new technology may add to the teaching and learning cycle. The other element in the equation is the student and the observation that contemporary young people are often characterised as having a strong proclivity for, and competence with, new technologies. Interestingly, such competence is reported as having the potential to build alternative pathways to literacy and assist with those experiencing learning difficulties (Luke et al., 2002).

Community partnerships: There currently is greater appreciation that literacy and numeracy education is a shared concern. The studies and reviews examined point to the importance of effective home–school partnerships as a contributing factor in effective literacy and numeracy provision. Research indicates persuasive reasons for establishing such partnerships, including better attendance, higher academic performance, positive student attitudes, increased parental skills and leaderships, connecting families with others in the school community, and greater support for schools and teachers (Epstein, 1995; Goos et al., 2004; Marcon, 1999). A range of views on how these partnerships may be established was evident. These included one-way partnerships, whereby schools informed and updated parents on particular learning strategies and provided literacy activities to do at home (for example House of Commons Education and Skills Committee, 2005; Snow et al., 1998) and two-way partnerships, whereby home and school are viewed as equal sources of expertise and parents take greater responsibility for educational outcomes (for example Cairney & Ruge, 1998; Hill et al., 1998; Louden et al., 2000). Here it was suggested that parents be viewed as central to the involvement of educational provision. Schools faced the challenge of developing productive partnerships on a developmental basis, moving through several levels, from schools as transmitters of expertise, to schools as sharers of expertise and, finally, to school and home as equal (if different) sources of expertise (Leler, 1983, cited in Louden et al., 2000). Hill (2000, p. 25) similarly emphasised partnerships, claiming that 'when parents, teachers and students view one another as partners in education, a caring community forms around students in a way that supports learning'.

Student diversity: It is inevitable with globalisation that the diversity of the student population is increasing, with teachers often working with children from different cultural, economic and social backgrounds, in addition to those children experiencing difficulties with learning. This diversity requires an acknowledgment of, and response to, student diversity, including recognition of community knowledges, students' home backgrounds and the impact of gender and learning difficulties. A response to this diversity requires teachers who have the necessary theoretical and pedagogical knowledge to combine essential elements of multiple approaches to accommodate this diversity, along with an attitude that all students can learn (Ellis, 2005; Louden et al., 2000). Given this, the need for pre-service and continuing professional development on teacher beliefs and views regarding diverse student populations is considered essential, with particular regard to professional vocabulary and high expectations for all students.

Teacher education: The most clamorous message evident in most of the reviews and studies was the need for quality teacher education, including both pre-service and inservice. A growing body of research (for example, Barber & Mourshed, 2007; Darling-Hammond, 2000; Hattie, 2003; Hill & Rowe, 1998; OECD, 2005) indicated that a substantial proportion of school effectiveness can be attributed to teachers, with teacher effects being cumulative and additive (Louden et al., 2005). Given this, teacher knowledge about the history of a field, theoretical perspectives, and varying approaches and beliefs were seen to be at the heart of any effort to improve education. The importance of the teacher in the provision of high-quality instruction and the amelioration of learning difficulties was routinely mentioned in the major studies.

This chapter serves to highlight the importance of teachers' discipline and pedagogic knowledges. More specifically, it shows that quality teaching involves being knowledgeable about a multiplicity of perspectives and approaches in order to orchestrate a complex blend of practices in response to individual student attributes and to work with particular literacy and numeracy demands within the curriculum (curriculum literacies/numeracies), in the context of rapidly changing technologies. Clearly there is a need for a greater focus on teacher education in all three fields (learning difficulties, literacy and numeracy), both in pre-service and inservice programs. In addition, for teachers to be able to be at the forefront of education, optimum teacher education will require effective communication and strong connections across research, policy and practice. Here the emphasis needs to shift from the teacher-as-the-researched to teacher-as-researcher—with the teacher being a genuine agent in the research process.

Often theory development and research represent a top-down process from academic researchers to practitioners. Due to restraints on teacher time, robust mechanisms to allow practicing teachers to be genuine partners in the research process have not been developed. Ideally, collaboration should exist between researchers and teachers in order to advance the profession, with mechanisms routinely established to ensure active teacher participation. Several studies funded by the Australian government found that, when opportunities arose for teachers to be actively involved in the research process, enhanced teacher knowledge and confidence ensued (for example Cormack & Wignall, 1998; DEST, 2004; Wyatt-Smith & Bridges, 2008).

Currently there is not an extensive body of research on the nature of teacher preparation and the optimum processes for both inservice and pre-service education. A comprehensive investigation into the full range and scope of Australian provision of teacher education (Commonwealth of Australia, 2007) made several key recommendations for improving the quality of teacher education, including further research, university–school partnerships and support for early career teachers. Similar investigations were conducted into the diversity of national provision of teacher education in the United States (Levine, 2006). An earlier Australian study focused exclusively on pre-service education (Louden et al., 2005) and indicated that several practical issues needed to be addressed, including greater length and status of pre-service programs, more time devoted to preparing teachers to teach literacy and numeracy, and an improvement of professional experience components of programs in terms of duration, quality and structure. Such issues are equally relevant for inservice programs where short, one-off courses were deemed insufficient with teachers requiring continuing, coordinated approaches to professional development with time to work and collaborate with colleagues within their own school and within clusters of schools, to talk with expert teachers and have opportunities for reflection on their practice. As Hargreaves and Fullan (1991) explain, professional development needs to be intensive, sustained and theoretically based, yet, practically situated learning, with opportunities to observe good practice, to be involved in coaching and mentoring processes and to take time for reflection.

In addition to these practical issues is the central question of the exact nature of the content of teacher education. That is, what particular domains of knowledge and attributes do teachers in all levels of schooling require, particularly when working with those who are experiencing difficulties learning literacy and numeracy? Drawing on the work of Gunn, Wyatt-Smith and Cumming (2006), who considered the characteristics of 'masterful' teachers, there are several domains and attributes:

- *Personal competency in literacy and numeracy*—teachers need to have highly developed personal levels of literacy and numeracy competence, including ability to use information and communication technologies. While some of the research did not suggest that teachers needed to be highly qualified English or Mathematics specialists to teach literacy and numeracy, and in fact this may not always have a positive effect, teachers did need to have a rich understanding of the literacy and numeracy they taught.
- *Pedagogical knowledge*—includes knowledge of how students learn literacy and numeracy, understanding the different teaching approaches and theories of learning.
- *Discipline knowledge*—that is, program knowledge and understandings, including that for early childhood, primary/middle school and secondary English/Mathematics teachers around all aspects of literacy and numeracy, including multiliteracies and multiple numeracies.
- *Curriculum knowledge*—includes an understanding and ability to plan and design curricular tasks that take into account the specific literacy and numeracy demands of a curriculum area.
- *Assessment knowledge*—this involves knowledge of continuing monitoring and assessment *of* individual student learning, and program effectiveness, and *for* learning where continuing assessment and feedback to students are part of the learning process.
- *Metaknowledge of professional practice*—which involves critical reflection by teachers as they identify their assumptions of knowledge, and learning and learners. This includes an inquiry into teachers' own beliefs with an aim to moving beyond deficit views and having high expectations based on a belief that all children can learn and having the ability to communicate these expectations.
- *Knowledge of community contexts*—this involves understanding of the assessment and teaching of literacy and numeracy to a diverse range of students, and includes an ability to move beyond stereotypical deficit views of student and the capacity to build strong community partnerships. It was acknowledged that community contexts may also include students' literacy and numeracy practices outside school, in community and digital online contexts with local and global reach.

Conclusion

This review has brought to light the diversity of theories, methodologies and perspectives that characterise the respective fields of literacy, numeracy and learning difficulties. It has also given some insight into the challenges that educators face in seeking coherence across fields and perspectives. This is a key observation given the heterogeneous nature of struggling students and that no single approach or generic list of approaches is suitable for all students. For fundamental and effective change in support provision, we need to turn to the teacher, who must have a deep understanding (including the theoretical underpinnings) and knowledge of literacy and numeracy education and the efficacy of various teaching and learning approaches for those experiencing difficulties. This requires treating each child as an individual, bringing together a particular blend of theories and approaches. As Pressley (2002, p. 337) described, 'balanced teaching is the orchestration of many components' or approaches, and is about masterful teachers weaving together these approaches. Given this, one of the most frequently recurring messages from this analysis has been that a crucial element for effective provision, and the most valuable resource in any school, is the teacher. Effective teaching is not as simple as applying a generic list of what works; there can be no doubt that it requires highly developed professional skill and knowledge to bring together a connective web of theory and approaches to provide each child with quality learning experiences.

Essential next questions

How much and to what extent should direct, explicit instruction of phonics (including phonemic awareness) be prioritised over other skills and strategies, and when should it be part of the reading instruction?

As mentioned earlier, the point of departure between the skills-based and whole-language camps lies solely in the importance they attach to explicit and systematic instruction to decode. Considering the notion of literacy as social practice, the complexities of our technological age, the diversity of the student population and the many confusing messages currently being conveyed to teachers, this is a crucial question. While a definitive answer may not be possible, some guiding principles to address the continuing pendulum swing between the two camps would be helpful to educators.

How best to orchestrate a combination of approaches for effective literacy and numeracy learning?

As discussed, we have not fully chartered the fine detail on how best to bring together a connective web of theories and perspectives to provide effective learning for all children. While the highly heterogeneous and diverse population of students may not make this feasible or desirable, it would be helpful to provide teachers with some guiding principles or frameworks for guiding teachers' own case investigations into their practices, *in situ*.

What are the optimum processes for both inservice and pre-service education?

The preceding findings from a large body of research attest to the need for knowledgeable and masterful teachers (Pressley, 2005) in order to provide optimum support for students with difficulties in learning literacy and numeracy. Teachers need to be well prepared for the difficult task of bringing together a connective web of theory and approaches for diverse populations of students. Quality teacher education needs to be in place to ensure teachers do not run to the nearest packaged program, amidst a confusion of competing views and discordant paths. Currently, however, this is a reality for some teachers, reflecting a lack of concerted research in the area of teacher preparation and continuing teacher education and support.

References

Alloway, N., Freebody, P., Gilbert, P., & Muspratt, S. (2002). *Boys, literacy and schooling: Expanding the repertories of practice*. Canberra: Department of Education, Science and Training.

Alvermann, D. E. (2001). *Effective literacy instructions for adolescents*. Executive summary and paper commissioned by the National Reading Conference. Chicago: National Reading Conference.

Andrews, J. E., Carnine, D. W., Coutinho, M. J., Edgar, E. B., Forness, S. R., & Fuchs, L. S. (2000). Bridging the special education divide. *Remedial and Special Education*, *21*, 258–260.

Askew, M., Brown, M., Rhodes, V., & Johnson, D. (1997a). *Effective teachers of numeracy: Report of a study carried out for the Teacher Training Agency*. London: London: Kings' College, University of London.

Askew, M., Brown, M., Rhodes, V., & Johnson, D. (1997b). *Effective teachers of numeracy in primary schools: Teachers' beliefs, practices and pupils' learning*. Paper presented at the British Educational Research Association Annual Conference. Retrieved June 28, 2006, from www.leeds.ac.uk/educol/documents/000000385.htm

Australian Association for Mathematics Teachers Inc. (1997). *Numeracy = everyone's business*. Report of the numeracy education strategy development conference. Adelaide: Author.

Baker, D. A., & Street, B. (1993). Literacy and numeracy. In *International encyclopaedia of education* (pp. 3453–3459). Oxford: Pergamon.

Barber, M., & Mourshed, M. (2007). *How the best performing school systems come out on top*. McKinsey & Company. Retrieved July 23, 2009, from: http://www.mckinsey.com/clientservice/socialsector/resources/pdf/Worlds_School_Systems_Final.pdf

Bearne, E. (2009). Assessing multimodal texts. In A. Burke & R. F. Hammett, *Assessing new literacies: Perspectives from the classroom* (pp. 15–33). New York: Peter Lang Publishing, Inc.

Bender, W. N. (2004). *Learning disabilities: characteristics, identification, and teaching strategies* (5th ed.). Boston: Allyn and Bacon.

Berninger, V. W. (2001). Introduction to special issue: Understanding the nature-nurture interactions in language and learning difference. *Learning Disability Quarterly*, *24*, 139.

Board of Teacher Registration, Queensland. (2005). *Numeracy in teacher education: The way forward in the 21st century*. A report of the Numeracy in Preservice Teacher Education Working Party. Brisbane: Author.

Brown, M., Askew, M., Baker, D., Denvir, H., & Millet, A. (1998). Is the national numeracy strategy research based? *British Journal of Educational Studies*, *46*(4), 362–385.

Burke, A. (2009). Checkmarks on the screen: Questions of assessment and new literacies in the digital age. In A. Burke & R. F. Hammett, *Assessing new literacies: Perspectives from the classroom* (pp. 35–54). New York: Peter Lang Publishing, Inc.

Cairney, T. H., & Ruge, J. (1998). *Community literacy practices and schooling: Towards effective support for students*. Canberra: Department of Employment, Education, Training and Youth Affairs.

Carnine, E. W. (1993). Effective teaching for higher cognitive functioning. *Educational Technology, 33*(10), 29–33.

Christie, F., & Misson, R. (2002). Framing the issues in literacy education. In J. Solar, J. Wearmouth, & G. Reid (Eds.), *Contextualising difficulties in literacy development: Exploring politics, culture, ethnicity and ethics*. London: Routledge.

Cockcroft, W. (1982). *Mathematics counts*. Department of Education and Science. London: HMSO.

Coiro, J., Knobel, M., Lankshear, C., & Leu, D. J. (2008). Central issues in new literacies and new literacies research. In J. Coiro, M. Knobel, C. Lanksear, & D. J. Leu (Eds.), *Handbook of research on new literacies* (pp. 1–21). New York: Lawrence Erlbaum Associates.

Coles, G. (2003). *Reading the naked truth: Literacy, legislation, and lies*. Portsmouth, NH: Heinemann.

Comber, B., & Hill, S. (2000). Socio-economic disadvantage, literacy and social justice: Learning from longitudinal case study research. *The Australian Educational Researcher, 27*(3), 79–97.

Commonwealth of Australia. (2007). *Top of the class. Report on the inquiry into teacher education*. Canberra: House of Representatives Publishing Unit. Retrieved July 23, 2009, from http://www.aph.gov.au/house/committee/evt/teachereduc/report.htm

Cormack, P., & Wignall, P. (1998). *Classroom discourse project*. Canberra: Department of Employment, Education, Training and Youth Affairs.

Crowther Report. (1959). *15 to 18: A report of the Central Advisory Council for Education*. London: HMSO.

Cumming, J. J. (2000a). *Envisaging the future – Our changing technological society: Demands and links between numeracy performance and life outcomes for employment, education and training*. Contextual paper commissioned by Australian Association of Mathematics Teachers. Adelaide: Australian Association of Mathematics Teachers.

Cumming, J. J. (2000b). *Computational numeracy*. Melbourne: ACER Research Conference Report.

Cumming, J. J., Wyatt-Smith, C. M., Ryan, J., & Doig, S. M. (1998). *The literacy-curriculum interface: The literacy demands of the curriculum in post-compulsory schooling*. Canberra: DEETYA.

DECS/Department for Education and Children's Services. (1995). *Now you're talking*. Canberra: DEETYA.

DEST/Department of Education, Science and Training. (2004). *Developing computation*. Canberra: Author.

DEST/Department of Education, Science and Training. (2005a). *Teaching reading, national inquiry into the teaching of literacy: Report and recommendations*. Canberra: Author.

DEST/Department of Education, Science and Training. (2005b). *Teaching reading, national inquiry into the teaching of literacy: Literature review*. Canberra: Author.

Darling-Hammond, L. (2000). Teacher quality and student achievement: A review of state policy evidence. *Education Policy Analysis Archives, 8*(1), Available from http://epaa.asu.edu/epaa/v8n1

Davis, O. L. (2002). When will the phonics police come knocking? In J. Soler, J. Wearmouth, & G. Reid (Eds.), *Contextualising difficulties in literacy development: Exploring politics, ethnicity and ethics* (pp. 83–86). London: Routledge Falmer.

Davydov, V. V. (1975). Logical and psychological problems of elementary mathematics as an academic subject. In L. P. Steffe (Ed.), *Children's capacity for learning mathematics* (pp. 55–108). Chicago: University of Chicago Press.

Delina, M., & van Kraayenoord, C. (1996). *Whole school approaches to assessing and reporting literacy*. Canberra: DEETYA.

Dixon, J. (1969). *Growth through English*. London: National Association for the Teaching of English.

Dixon, R., & Carnine, D. (1994). Ideologies, practices, and their implications for special education. *Journal of Special Education, 28*, 356–367.

Durrant, C., & Green, B. (2000). Literacy and the new technologies in school education: Meeting the L(IT)eracy challenge? *Australian Journal of Language and Literacy, 23*(2), 89–108.

D'Agostino, J. V., & Murphy, J. A. (2004). A meta-analysis of reading recovery in United States schools. *Educational Evaluation and Policy Analysis, 26*(1), 23–38.

Education and Science Committee. (2001). *Me Panui Tatou Katoa – Let's All Read: Report of the Education and Science Committee on the inquiry into the teaching of reading in New Zealand*. Wellington, NZ: House of Representatives.

Elkins, J. (2002). Learning difficulties/disabilities in literacy. *Australian Journal of Language and Literacy, 25*(3), 11–18.

Elkins, J. (2005). Numeracy. In A. Ashman & J. Elkins (Eds.), *Educating children with diverse abilities* (2nd ed, pp. 216–245). Sydney: Prentice Hall.

Ellis, L. A. (2005). Balancing approaches: Revisiting the educational psychology research on teaching students with learning difficulties. In *Australian education review no. 48*. Camberwell: Australian Council for Educational Research.

Epstein, J. L. (1995). School/family/community partnerships. Caring for the children we share. *Phi Delta Kappan, 76*(9), 701–713.

Expert Panel on Mathematics. (2004). *Teaching and learning mathematics*. Ontario, ON: Ministry of Education. Available at: http://www.edu.gov.on.ca

Fletcher, J. M., Coulter, W. A., Reschly, D. J., & Vaughn, S. (2004). Alternative approaches to the definition and identification of learning disabilities: Some Questions and answers. *Annals of Dyslexia, 54*(2), 304–331.

Forness, S. R., & Kavale, K. A. (2001). Ignoring the odds: Hazards of not adding the new medical model to special education decisions. *Behavioral Disorders, 26*, 269–281.

Foucault, M. (1972). *The archeology of knowledge and the discourse on language*. New York: Pantheon.

Frankenstein, M. (1990). *Relearning mathematics. A different third R-radical maths*. London: Free Association Books.

Freebody, P., Ludwig, C. M., & Gunn, S. (1995). *Everyday literacy practices in and out of schools in low socio-economic urban communities*. Canberra: Commonwealth Department of Employment, Education and Training.

Freebody, P., & Luke, A. (1990). 'Literacies' programs: Debates and demands in cultural context. *Prospect, 5*(3), 7–16.

Gee, J. P. (1990). *Social linguistics: Ideology in discourse*. London: Falmer Press.

Gersten, R., & Chard, D. (1999). Number sense: Rethinking arithmetic instruction for students with mathematical disabilities. *Journal of Learning Disabilities, 33*, 1–18.

Goos, M., Lincoln, D., Coco, A., Frid, S., Galbraith, P., Horne, M., et al. (2004). *Home, school and community partnerships to support children's numeracy*. Canberra: Department of Education, Science and Training.

Grobecker, B. (1999). Mathematics reform and learning disabilities. *Learning Disability Quarterly, 22*(1), 43–58.

Gunn, S., Wyatt-Smith, C. M., & Cumming, J. J. (2006). *Evidence-based research to inform policy and practice in improving literacy outcomes for students across the stages of schooling*. Unpublished report for Department of Education and Training, Melbourne, Victoria.

Gunn, S. J. (2007). Literature review. In C. M. Wyatt-Smith, J. Elkins, P. Colbert, S. J. Gunn, & A. Muspratt, *Changing the nature of support provision – Students with learning difficulties: Interventions in literacy and numeracy project (InLaN)* (Vol. 2). Canberra:

Department of Education, Science and Training, Australian Government. Available from www.dest.gov.au/sectors/school_education/publications_resources/profiles/effective_teaching_learning_practices_stud_learn_difficult.htm#authors

Guthrie, J., & Wigfield, A. (2000). Engagement and motivation in reading. In M. Kamil, P. Mosenthal, P. Pearson, & R. Barr (Eds.), *Handbook of reading research* (Vol. 3, pp. 813–834). Mahwah, NJ: Erlbaum.

Hammill, D. D., & Larsen, S. (1974). The efficacy of psycholinguistic training. *Exceptional Children, 41*, 5–14.

Hargreaves, A., & Fullan, M. (1991). *Understanding teacher development*. London: Cassell.

Harris, K. R., & Graham, S. (1996). Constructivism and student with special needs: Issues in the classroom. *Learning Disabilities Research and Practice, 11*, 134–137.

Hattie, J. A. (2003). *Teachers make a difference: What is the research evidence?* Background paper to invited address presented at the 2003 ACER Research Conference, Carlton Crest Hotel, Melbourne, Australia, October 19–21 2003. Retrieved from www.acer.edu.au/workshops/documents/Teachers Make a Difference Hattie.pdf

Heshusius, L. (1991). Future perspectives. In D. K. Reid, W. P. Hresko, & H. L. Swanson, *A cognitive approach to learning disabilities* (2nd ed). Austin, TX: Pro-ED.

Heshusius, L. (1994). Freeing ourselves from objectivity: Managing subjectivity or turning toward a participatory mode of consciousness? *Educational Researcher, 23*(3), 15–22.

Hill, P. W. (2000). *Numeracy education: What do we know and what can we learn from the literacy experience*. Contextual paper commissioned by Australian Association of Mathematics Teachers (AAMT). Adelaide: AAMT.

Hill, S., Comber, B., Louden, B., Rivalland, J., & Reid, J. (1998). *100 children go to school: Connections and disconnections in literacy development in the year prior to school and the first year of school*. Canberra: DEETYA.

Hill, S., Comber, B., Louden, W., Rivalland, J., & Reid, J. (2002). *100 children turn 10: A longitudinal study of literacy development form the year prior to school to the first four years of school*. Canberra: DEETYA.

Hill, P. W., & Rowe, K. J. (1998). Modelling student progress in studies of educational effectiveness. *School Effectiveness and School Improvement, 9*(3), 310–333.

Hodgens, J. (1996). Editorial: Changing education. *Journal for Teachers and Administrators, 3*, 1–3.

Hogan, J. (2000). Numeracy – across the curriculum? *Australian Mathematics Teacher, 56*(3), 17–20.

Hollindale, P. (1995). Children's literature in an age of multiple literacies. *Australian Journal of Language and Literacy, 18*(4), 148–158.

House of Commons Education and Skills Committee. (2005). *Teaching children to read: Eighth Report of Session 2004–5*. London: The Stationery Office Limited.

Iano, R. P. (1986). The study and development of teaching: With implications for the advancement of special education. *Remedial and Special Education, 7*(5), 50–61.

Jacobs, J. E., & Paris, S. G. (1987). Children's metacognition about reading: Issues in definition, measurement, and instruction. *Educational Psychologist, 22*, 255–278.

Kameenui, E. J., & Carnine, E. W. (1998). *Effective teaching strategies that accommodate diverse learners*. Upper Saddle River, NJ: Merrill.

Kamler, B., & Comber, B. (1996). Critical literacy: Not generic – not developmental – not another orthodoxy. *Changing Education, 3*(1), 1–5.

Kilpatrick, J., Swafford, J., & Findell, B. (Eds.). (2001). *Adding it up: Helping children learn mathematics*. Washington, DC: National Academy Press.

Kimber, K., & Wyatt-Smith, C. M. (2009). Valued knowledges and core capacities for digital learners: Claiming spaces for quality assessment. In A. Burke & R. F. Hammett (Eds.), *Assessing new literacies: Perspectives from the classroom* (pp. 133–156). New York: Peter Lang Publishing, Inc.

Lankshear, C., Bigum, C., Durrant, C., Green, B., Honen, E., Morgan, Q., et al. (1997). *Digital rhetorics: Literacies and technologies in education – Current practices and future directions*. Canberra: DEETYA.

Leler, H. (1983). Parent education and involvement in relation to the schools and to parents of school-aged children. In R. Haskins & D. Adams (Eds.), *Parent education and public policy* (pp. 141–180). Norwood, NJ: Ablex.

Lerner, J. W. (2003). *Learning disabilities: Theories, diagnosis, and teaching strategies* (9th ed.). Boston: Houghton Mifflin.

Leu, D. J., Jr. (2000). Literacy and technology: Deictic consequences for literacy education in an information age. In M. L. Kamil, P. B. Mosenthal, P. D. Pearson, & R. Barr (Eds.), *Handbook of reading research, Vol III* (pp. 229–249). Mahwah, NJ: Lawrence Erlbaum Associates.

Leu, D. J., Jr. (2002). The new literacies: Research on reading instruction with the internet. In A. E. Farstrup, & S. J. Samuels (Eds.), *What research has to say about reading instruction* (pp. 310–336). Newark, DE: International Reading Association.

Leu, D., Kinizer, C., Coiro, J., & Cammack, D. (2004). Toward a theory of new literacies emerginf grom the Internet and other information and communication technologies. In R. Ruddell & N. Unrau (Eds.), *Theoretical models and processes of reading* (Vol. 5, pp. 1570–1613). Newark, DE: International Reading Association.

Leu, D. J., Jr., Mallette, M. H., Karchmer, R. A., & Kara-Soteriou, J. (2005). Contextualizing the new literacies of information and communication technologies in theory, research and practice. In R. A. Karchmer, M. H. Mallette, J. Kara-Soteriou, & D. J. Leu (Eds.), *Innovative approaches to literacy education: Using the Internet to support new literacies* (pp. 1–10). Newark, DE: International Reading Association.

Levine, A. (2006). *Educating school teachers*. Washington, DC: The Education Schools Project.

Louden, W., Chan, L., Elkins, J., Greaves, D., House, H., Milton, M., et al. (2000). *Mapping the territory*. Canberra: DETYA.

Louden, W., Rohl, M., Barratt-Pugh, C., Brown, C., Cairney, T., Elderfield, J., et al. (2005). *In teachers' hands: Effective literacy teaching practices in the early years of schooling*. Canberra: DETYA.

Louden, W., & Wildy, H. (2001). *Developing schools' capacity to make performance judgements*. Canberra: Department of Education, Training and Youth Affairs.

Luke, A., Woods, A., Land, R., Bahr, N., & McFarland, M. (2002). *Accountability: Inclusive assessment, monitoring and reporting*. Brisbane: Queensland Indigenous Education Consultative Body.

Marcon, R. (1999). *Impact of parent involvement on children's development and academic performance: A three-cohort study*. Paper presented at the meeting of the Southeastern Psychological Association, Savannah, GA. March, 1999.

McMeniman, M. (1997). One hundred years since Dewey: Back to the future for educational research? *Queensland Journal of Educational Research, 13*(2), 4–13.

Meichenbaum, D. (1980). *Cognitive behavior modification*. New York: Plenum Press.

Mellor, B., & Patterson, A. (2005). Critical literacy: Theory, pedagogy and the historical imperative. In R. Beach, J. Green, M. Kamil, & T. Shanahan (Eds.), *Multidisciplinary perspectives on literacy research* (pp. 455–480). Cresskill, NJ: Hampton Press.

Mills, K. (2005). Deconstructing binary oppositions in literacy discourse and pedagogy. *Australian Journal of Language and Literacy, 28*(1), 67–82.

Milton, M. (2000). Numeracy. In W. Louden, L. Chan, J. Elkins, D. Greaves, H. House, M. Milton, S. Nichols, J. Rivalland, M. Rohl, & C. van Kraayenoord (Eds.), *Mapping the territory*. Canberra, ACT: Department of Education, Training and Youth Affairs, Commonwealth of Australia.

Ministry of Education. (2005). *The numeracy story continued: What is the evidence telling us? Findings from the Numeracy development project*. Wellington: Author.

National Reading Panel. (2000). *Teaching children to read: An evidence-based assessment of the scientific research literature on reading and its implications for reading instruction*. Washington, DC: National Institute of Child Health and Human Development (NICHD).

New London Group. (1996). A pedagogy of multiliteracies: Designing social futures. *Harvard Educational Review, 66*(1), 60–92.

New London Group. (2000). *Multiliteracies: Literacy learning and the design of social futures*. London: Routledge.

OECD/Organisation for Economic Co-operation and Development. (1999). *Measuring student knowledge and skills: A new framework for assessment*. Paris: Author.

OECD/Organisation for Economic Co-operation and Development. (2005). *Teachers matter: Attracting, developing and retaining effective teachers*. Paris: Author.

O'Shea, L. J., O'Shea, D. J., & Algozzine, R. (1998). *Learning disabilities: From theory toward practice*. Upper Saddle River, NJ: Prentice-Hall.

Poplin, M. S. (1988a). The reductionistic fallacy in learning disabilities: Replicating the past by reducing the present. *Journal of Learning Disabilities, 21*(7), 389–399.

Poplin, M. S. (1988b). Holistic/constructivist principles of the teaching/learning process: Implications for the field of learning disabilities. *Journal of Learning Disabilities, 21*(7), 401–416.

Poplin, M. (1995). Looking through other lenses and listening to other voices: Stretching the boundaries of learning disabilities. *Journal of Learning Disabilities, 28*(7), 392–398.

Pressley, M. (2002). *Reading instruction that works: The case for balanced teaching* (2nd ed.). New York: Gilford.

Pressley, M. (2005). Balanced elementary literacy instruction in the United States. In N. Bascia, A. Cumming, A. Datnow, K. Leithwood, & D. Livingstone (Eds.), *International handbook of educational policy Part 2* (pp. 645–660). Dordrecht: Springer.

RAND Mathematics Study Panel. (2003). *Mathematical proficiency for all students: Towards a strategic research and development program in mathematics education*. Santa Monica, CA: RAND Education.

Rassool, N. (1999). Literacy: In search of a paradigm. In N. Rassool (Ed.), *Literacy for sustainable development in the age of information* (pp. 25–53). Clevedon, Philadelphia, PA: Multilingual Matters.

Rassool, N. (2002). Literacy: In search of a paradigm. In J. Soler, J. Wearmouth, & G. Reid (Eds.), *Contextualising difficulties in literacy development: Exploring politics, ethnicity and ethics* (pp. 17–46). London: Routledge Falmer.

Reid, D. K., Robinson, S. J., & Bunsen, T. D. (1995). Empiricism and beyond: Expanding the boundaries of special education. *Remedial and Special Education, 16*, 131–141.

Reid, D. K., & Valle, J. W. (2004). The discursive practice of learning disability: Implications for instruction and parent-school relations. *Journal of Learning Disabilities, 37*(6), 466–482.

Renshaw, P. D. (1996). A sociocultural view of the mathematics education of young children. In H. Mansfield, N. Pateman, & N. Descamps-Bednarz (Eds.), *Mathematics for tomorrow's young children: International perspectives on curriculum* (pp. 59–76). Dordrecht: Kluwer.

Rose, J. (2006). *Independent review of the teaching of early reading*. London: Department for Education and Skills.

Shaywitz, S. E., & Shaywitz, B. A. (2004). Neurobiologic basis for reading and reading disability. In P. McCardle & V. Chhabra (Eds.), *The voice of evidence in reading research* (pp. 417–442). Baltimore, MD: P. H. Brookes.

Snow, C. E., Burns, M. S., & Griffin, P. (Eds.). (1998). *Preventing reading difficulties in young children*. Washington, DC: National Academy Press.

Soler, J. (2002). Policy contexts and the debates over how to teach literacy. In J. Soler, J. Wearmouth, & G. Reid (Eds.), *Contextualising difficulties in literacy development: Exploring politics, ethnicity and ethics* (pp. 3–13). London: Routledge Falmer.

Sousa, D. A. (2001). *How the special needs brain learns*. Thousand Oaks, CA: Corwin Press.

Stanovich, K. E. (2000). Romance and reality. In N. Padak, T. V. Rasinski, J. K. Peck, B. W. Church, G. Fawcett, J. M. Hendershot, J. M. Henry, B. G. Moss, E. Pryor, K. A. Roskos, J. F. Baumann, D. R. Dillon, C. J. Hopkins, J. W. Humphrey, & D. G. O'Brien (Eds.), *Distinguished educators on reading: Contributions that have shaped effective literacy instruction* (pp. 28–46). Newark, DE: International Reading Association.

Steen, L. A. (1999a). *Seeking numeracy*. Keynote address delivered at the International Virtual Conference of the Australian Association of Mathematics Teachers.

Steen, L. A. (1999b). Numeracy: The new literacy for a data-drenched society. *Educational Leadership, 57*(2), 8–13.

The State of Queensland. (2000). *Literate futures report of the literacy review for Queensland state schools*. Brisbane: Education Queensland.

Torgesen, J. K. (2004). Lessons learned from the last 20 years of research on interventions for students who experience difficulty learning to read. In P. McCardle & V. Chhabra (Eds.), *The voice of evidence in reading research* (pp. 225–229). Baltimore, MD: P. H. Brookes.

Unsworth, L. (2002). Changing dimensions of school literacies. *The Australian Journal of Language and Literacy, 25*(1), 62–77.

Unsworth, L. (2008). Negotiating new literacies in English teaching. In L. Unsworth (Ed.), *New literacies and the English curriculum* (pp. 3–19). London: Continuum.

van Kraayenoord, C. E., & Goos, M. (2003). Metacognition. In P. Keeves & R. Watanabe (Eds.), *The international handbook of educational research in the Asia-Pacific region, Part 1* (pp. 477–494). Boston: Kluwer Academic.

Vincent, J., Stephens, M., & Steinle, V. (2005). *Numeracy research and development initiative 2001–2004: An overview of the numeracy projects*. Canberra: DEST.

Vygotsky, L. (1978). *Mind in society: The development of higher psychological processes*. Cambridge, MA: Harvard University Press.

Warner, M. M. (1993). Objectivity and emancipation in learning disabilities: Holism from the perspective of critical realism. *Journal of Learning Disabilities, 26*(5), 311–325.

West, A. (1992). Reading against the text: Developing critical literacy. In *Domains of literacy conference*. London: University of London.

Westwood, P. (2000). *Numeracy and learning difficulties: Approaches to learning and development*. Camberwell: Australian Council for Educational Research.

Wheldall, K., & Beaman, R. (2000). *An evaluation of MULTILIT, Making up lost time in literacy*. Canberra: DETYA.

Willis, S. (1998). Which numeracy? *Unicorn, 24*(2), 32–42.

Wong, B. Y. L. (1986). Metacognition and special education: A review of a view. *Journal of Special Education, 20*, 9–29.

Wyatt-Smith, C. M., & Bridges, S. (2008). *Meeting in the middle—Assessment, pedagogy, learning and students at educational disadvantage*. Evaluation for The Literacy and Numeracy in the Middle Years of Schooling Initiative Strand A. Retrieved July 30, 2008, from http://education.qld.gov.au/literacy/docs/deewr-myp-final-report.pdf

Wyatt-Smith, C. M., & Cumming, J. J. (2003). Curriculum literacies: Expanding domains of assessment. *Assessment in Education: Principles, policy and practice, 10*(1), 47–59.

Wyatt-Smith, C. M., & Elkins, J. (2008). Multimodal reading and comprehension in online environments. In J. Coiro, M. Knobel, C. Lankshear, & D. J. Leu (Eds.), *Handbook of research on new literacies*. New York: Lawrence Erlbaum Associates.

Wyatt-Smith, C. M., Elkins, J., Colbert, P., Gunn, S. J., & Muspratt, A. (2007). *Changing the nature of support provision—Students with learning difficulties: Interventions in literacy and numeracy project (InLaN)*. Canberra: Department of Education, Science and Training, Australian Government. Available from www.dest.gov.au/sectors/school_education/publications_resources/profiles/effective_teaching_learning_practices_stud_learn_difficult.htm#authors

Wyatt-Smith, C. M., & Gunn, S. (2009). Towards theorising assessment as critical inquiry. In C. Wyatt-Smith & J. Cumming (Eds.), *Educational assessment in the 21st century: Connecting theory and practice*. Dordrecht: Springer International.

Zevenbergen, R. (1995a). Constructivist approaches in mathematics education. *Unicorn, 21*(3), 76–81.

Zevenbergen, R. (1995b). Towards a socially critical numeracy. *Critical Forum, 4*(1), 82–102.

Zevenbergen, R. (2004). Reconceptualising numeracy for new times. *Curriculum Perspectives, 24*(3), 1–7.

Chapter 3
Researching the Opportunities for Learning for Students with Learning Difficulties in Classrooms: An Ethnographic Perspective

Judith Green, Maria Lucia Castanheira, and Beth Yeager

In the first chapter of this volume, Wyatt-Smith and Elkins argue that 'it is timely to review how different theoretical frameworks and methodologies provide different lenses through which to study students' learning needs'. By viewing different theoretical frameworks and methodologies as potentially complementary, Wyatt-Smith, Elkins and other authors in this volume move discussions beyond debates of which method is best, to a discussion of what different theoretical traditions contribute towards research on students' learning needs[1]. In this chapter, we seek to contribute towards this argument by demonstrating how multiple theoretical perspectives and methods can be included in a single research study as well as in programs of research that seek to explore common phenomena from different theoretical and methodological points of view (for example Green & Harker, 1988; Grimshaw, Burke, & Cicourel, 1994; Koschmann, 1999[2]; Cumming & Wyatt-Smith, 2001a).

In order to demonstrate how multiple theoretical and methodological traditions are central to studies of the learning needs of students, we constructed a new study from a continuing ethnographic research project. The project has been, and continues, exploring how opportunities for learning social and academic processes and practices, as well as content, are constructed in and through the actions of teachers with students, students with others and individual students for self (for example Floriani, 1993; Santa Barbara Classroom Discourse Group, 1992a, b; Tuyay, Jennings, & Dixon, 1995). In past studies, we have examined how, through these opportunities, students construct local and situated identities as learners (for example Castanheira, Green, Dixon, & Yeager, 2007; Putney, Green, Dixon, Durán, & Yeager, 2000; Rex, 2000), how language(s) are a resource for community development (for example Heras, 1993; Lin, 1993; Yeager, Green, & Castanheira, 2009) and how teachers construct with the class inclusive practices for linguistically

J. Green (✉)
University of California, Santa Barbara, CA, USA
e-mail: green@education.ucsb.edu

[1]In the United States, these students are referred to as having learning disabilities.

[2]The order of citations is listed in ascending order by date to show when different perspectives became available historically.

C. Wyatt-Smith et al. (eds.), *Multiple Perspectives on Difficulties in Learning Literacy and Numeracy*, DOI 10.1007/978-1-4020-8864-3_3,

diverse students who have special learning needs (for example Castanheira, 2004; Castanheira, Green, & Yeager, 2009).

Our goal in this chapter is to contribute to the discussions about what complementary methods enable us to see and understand about the social and academic processes in classrooms, and their consequences for learning for students, particularly those with special learning needs. By examining the developing theoretical and methodological decisions we made in constructing the case study of how Sergio, a student defined by the school as having a learning disability (see below), engaged in learning social science, we make transparent[3] how different theoretical traditions led to different methodological approaches at particular levels of an analytic scale, and to different ways of examining the interdependence of collective and individual learning and development. This approach is designed to make visible how each new theory adds descriptive or explanatory power to the framework of the study. Through this process, we demonstrate why Agar (2006) argues that ethnography is a non-linear system, and Anderson-Levitt (2006) argues that ethnography is a philosophy of inquiry, not a method.

Constructing a telling case

The case study constructed for this chapter is the fourth in a series of published ethnographic case studies (Mitchell, 1984) in which we have explored different dimensions of the learning experiences Sergio had in his linguistically and culturally diverse Grade 5 class. Our goal in tracing different dimensions of Sergio's opportunities for learning is two-fold: (1) to make visible the complex, multi-layered and multi-faceted nature of learning in classrooms, and (2) to create an understanding of how a multi-faceted, multi-theoretical research perspective is central in making visible the often invisible supports and constraints on learning, not only for students with special learning needs, but for all students as well as for their teachers. Through these two goals, we seek to highlight how situated studies of learning opportunities in classrooms provide new theoretical understandings of the developing and changing, almost fluid nature (c.f., Bauman, 2000) of educational reforms and their consequences for particular students and their teachers (for example McNeil & Coppola, 2006; Green, Heras, & Yeager, in press).

In order to meet these goals, we build on conceptual arguments about how ethnographic research provides a means for developing new theoretical inferences about particular dimensions of the social organisation, accomplishment and consequences of everyday life in classrooms[4]. Mitchell (1984) argues that the case study is one

---

[3]Both the American Educational Research Association (AERA) (2006) and the Economic and Social Research Council (ESRC) (2007) of the United Kingdom have created guidelines for empirical social science research that call for transparency for the logic of inquiry used. The purpose of such transparency is to make visible relationships between theory, method and interpretation.

[4]Hymes (1972) calls such studies *topic-centered* ethnographies.

form of ethnographic work that focuses on specific chains of events in order to make theoretical inferences

> [c]ase studies are the detailed presentation of ethnographic data relating to some sequence of events from which the analyst seeks to make some theoretical inference. The events themselves may relate to any level of social organization: a whole society, some section of a community, a family or an individual. What distinguishes case studies from more general ethnographic reportage is the detail and particularity of the account. Each case study is a description of a specific configuration of events in which some distinctive set of actors have been involved in some defined situation at some particular point of time. (p. 222)

From this perspective, ethnographic case studies constitute *telling cases*; that is cases that make possible theoretical inferences that focus on particular dimensions of the social and cultural life of members of particular social groups (see also Rex, 2006; Sheridan, Street, & Bloome, 2000).

The previous case studies focused on theorising Sergio's opportunities for constructing social and academic identities (Yeager, 2003), on how each developing event created identity potentials for both individuals and for the group (Castanheira et al., 2007), and on how Sergio constructed inclusive practices for self and others through his actions within and across events (Castanheira, 2000; Castanheira et al., 2007). Each of these studies provided a particular angle of analysis that permitted particular theoretical inferences about how participating in this Grade 5 class (a collective) and contributing to its construction was consequential for the individual students and the teacher as well as the class-as-a-collective. These studies also pointed to the need for further study of individual–collective relationships in order to make visible the interdependence of the two, and how each contributes to the learning potentials of the other in dynamic ways.

In this telling case study, we continue this pattern of theorising by once again tracing the opportunities for learning to be literate that the teacher constructed with Sergio and his peers in social science (see Mills, 1993 for a parallel in mathematics). The focus of the present study is grounded in questions left unresolved in the earlier studies: How can the interdependence of collective and individual development be explored and made visible? And, how can individual students' developing understandings be documented? These questions, unlike the earlier ones, focus on methodological concerns about the *nature of evidence*, as well as what can be learned through each analysis.

The present study, with its theoretical questions about the contributions of different theoretical perspectives, provides a means for making visible the complementary basis of the theoretical traditions guiding this ethnographically telling case. Additionally, by tracing the decisions we made in examining the roots and routes of a developing set of practices and understandings that Sergio and his partners developed in social science, we demonstrate how we explored the dynamics of the collective–individual relationships in social science across the school year. Through this analysis, we present an argument about the theoretical and methodological decisions needed to uncover different dimensions of these complex relationships. Finally, as part of this discussion of theory–method relationships, we provide a rationale for the warrants of our claim that the actions and practices of teacher and students are material resources (Bloome & Bailey, 1992; Gee & Green, 1998) that students read, interpreted and took up (or not) to guide their work individually and collectively across times and events of social science.

Selecting Sergio as a tracer unit

Before presenting our logic of inquiry for this case study, we present a brief description of Sergio to establish his status in the school as a learning disabled student and his position in the classroom as a contributing member of the class. Sergio entered Grade 5 with an identified learning disability, specifically in reading comprehension and writing, requiring regular support from the school's resource specialist. His heritage and first language was Spanish, and since kindergarten he had participated in a bilingual class, initially with Spanish reading instruction and English language development. Although orally bilingual by the time he reached Grade 5, Sergio had been placed in English reading instruction early in his school career, an action consistent with district and school 'resource'/Special Education policy and practice. The school's resource specialist argued that this early transition to English reading was based on an argument that bilingual students, who need resource instruction, would eventually be expected to function primarily in English in school, and that it was less confusing for them to focus solely on one language sooner rather than later.

Sergio entered Grade 5 with recorded test scores in English that placed him at grade equivalent levels of 1.8 (Grade 1, 8th month) in reading, 1.6 (Grade 1, 6th month) in language (writing) and 3.0 (Grade 3) in mathematics. These scores meant that he received focused support in reading and writing, as well as some in mathematics, from the resource specialist on a daily basis in a resource room, not in his classroom. These scores were noted by the teacher but were not used to place Sergio in groups in the class. Rather, as the telling case will make visible, the teacher created opportunities for inclusion for Sergio and all students, in order to support common access to academic work across subject matter. Additionally, Sergio's teacher had requested that the resource teacher work with him in the classroom; however, given the previous argument, this request was not honoured. Therefore, the resource specialist depended on standardised assessments of Sergio's reading and writing abilities, and had limited understanding of what Sergio was able to do in class. In contrast, the teacher (and ethnographers) was able to support how he actually wrote in social science, and how he read and interpreted the various oral, written and social texts, which demonstrated a much greater competence in reading and writing than was visible on standardised assessments.

This telling case, therefore, focuses not on how Sergio performed in traditional reading and writing events, but on how he was involved in developing processes and practices in social science that enabled him to read the requirements for being a social scientist, and to become a cultural guide for his social science partner, Jaime, who was new to the school. What is significant about the focus on this team is that Sergio guided the team's work, although Jaime was assessed as advanced in reading and writing of Spanish (test scores: 98th percentile in reading comprehension in

Spanish and in the 92nd percentile in language/writing). The telling case of Sergio, therefore, serves as a tracer unit (Castanheira et al., 2009; Green, 1983) to make visible how individual–collective relationships, jointly constructed by teachers with students, afforded Sergio (and Jaime) material resources for learning how to be a social scientist.

Theoretical assumptions governing the telling case

In this section, we present the developing set of complementary perspectives guiding our ethnographic system of analysis (c.f. Agar, 2006). Specifically, we present governing assumptions (Strike, 1974) that form an *orienting conceptual system* for the study of social construction of classroom life and its consequences for students and teachers. Our goal in presenting the theoretical perspectives is to make visible the conceptual system (c.f. Lakoff & Johnson, 1980) that we have developed over the past four decades. This conceptual system inscribes ontological stances about the consequential nature of classroom life, the roles and relationships developed among members of the class, constitutive nature of classroom discourse, and the situated and historical nature of communication and learning in classrooms (and other social settings).

In making transparent the conceptual system that guides our ethnographic decisions, we demonstrate Bateson's argument (cited in Birdwhistell, 1977) that theory is method and that method is theory. As part of this process, we illustrate how complementary conceptual theories guided particular methodological decisions, and how each provided a particular language about the interdependent, dynamic and consequential nature of individual–collective learning and development. Lima (1995) captures this conceptual argument succinctly, drawing on an international set of theoretical perspectives (for example Freire from Brazil, Vygosky from Russia and Wallon from France)

> It is precisely the experience of schooling that will transform the individual through the process of cultural development, enlarging the cultural capital of each one, and by this, transforming the cultural capital of the community... We have two dimensions of development: one that resides in the individual and the other in the collectivity. Both are interdependent and create each other. Historically created possibilities of cultural development are themselves transformed by the processes through which individuals acquire the cultural tools that are or become available in their context. (pp. 447–48)

This conceptual argument can be viewed as an overarching argument, and the governing assumptions that are presented in the following section constitute ways of conceptualising social and cultural processes that support the transformations and development captured in the above quote.

Roots and routes of the conceptual system

The conceptual system guiding the current case study is grounded in historical advances in work on discourse, ethnography and the social construction of everyday life. This system has expanded as new studies were undertaken within our research community and by others. As we will make visible, these studies led to new areas in need of theoretical exploration and new theoretical and methodological perspectives, which once uncovered, became part of the conceptual system (Heap, 1991). These governing assumptions can be viewed as constituting a language that frames the questions we ask as well as how we explore, read, interpret and represent the interconnected and, at times, interdependent, patterns of processes, practices, meanings and literate events that teachers, students and others construct in classrooms (or other social settings). Therefore, as you examine the governing assumptions that follow, consider the roots of the assumption and how each new assumption adds to the developing conceptual system, and how each expands the expressive potential of this developing system (c.f. Strike, 1974).

We present the conceptual system as phases of development over the past four decades. Although presented in phases, the development was not a linear progression but rather represents an overlapping series of developments. This conceptual system can be viewed as constituting a set of theoretically coherent and interconnected metaphors (Lakoff & Johnson, 1980) that we draw upon to inscribe a particular view of the nature of everyday life in classrooms (for example Green, Dixon, & Zaharlick, 2003; Santa Barbara Classroom Discourse Group, 1992a, 1992b).[5]

Two caveats need to be considered at this point. Although many of the governing assumptions that follow may be common to a number of other research communities (for example ethnomethodology, conversation analysis or sociology of knowledge), there is great variation among such traditions (for example Cameron, 1995; Cumming & Wyatt-Smith, 2001b; Green & Dixon, 1999; Mills, 1997; Rex & Green, 2007), given different disciplinary grounding. The second caveat focuses on the foregrounding of research in phases 1–2 that developed within the United States that is directly related to the National Institute of Education (NIE) report. In other countries, there were parallel bodies of work, some of which predate the NIE report. As indicated in the list of participants on the NIE panel that follows, Douglas Barnes and Ian Forsyth from the United Kingdom contributed to

[5]Although the governing assumptions presented are central to our conceptual system, the particular view of social construction of everyday life (for example Berger & Luckmann, 1967; Gergen, 1985) underlying our work is based on a series of complementary and, at times, parallel research traditions, including: *sociocultural and sociohistorical theories of learning* (for example Mercer & Littleton, 2007; Rieber & Carton, 1988; Rogoff, 2003; Wertsch, 1991; Ligorio & Pontecorvo, 2005), *theories of language and discourse(s)-in-use* (for example Bakhtin, 1986; Barnes & Todd, 1995; Barnes et al., 1969; Bernstein, 1973, 1990; Bloome & Clarke, 2006; Bloome et al., 2005; Cazden, 1988; Gee & Green, 1998; Green & Wallat, 1981; Gumperz, 1986; Wilkinson, 1982), *research on teaching* (for example Evertson & Green, 1986; Hudson & Schneuwley, 2007; Wittrock, 1986;) and *ethnography in education* (for example Gilmore & Glatthorn, 1982; Green & Wallat, 1981; Hammersley & Atkinson, 1995; Heath & Street, 2008; Heath, 1982; Spindler, 1982; Walford, 2008). In the United Kingdom, a similar report, the Bullock Report (1975) was entitled *A Language For Life.*

the development of this report between 1972 and 1974. Some of these influences have been reported in published work from our research community (for example Green & Dixon, 1993; Green & Dixon, 2007; Rex, Steadman, & Graciano, 2006; Rex & Green, 2007). Comprehensive cross-national examples of work on discourse, language and education are available in the *Encyclopedia of Language and Education* (Hornberger, 2008), in the *Handbook of Educational Linguistics* (Spolsky & Hult, 2008) and in recent volumes focusing on the influence of the work of Douglas Barnes (Mercer & Hodgkinson, 2008) and in cross-national explorations focusing on literacy (Cumming & Wyatt-Smith, 2001a, b). Additional arguments focusing on more traditional perspectives on classroom interaction have also begun to examine differences in traditions due to national location (for example Hudson & Schneuwly, 2007).

A review of these perspectives is beyond the scope of this chapter, but they need to be acknowledged as we describe the phases of influence on our work as a type of telling case of the scope and breadth of work available. Having framed these caveats and the work presented as *a telling case*, we now invite readers to consider how each angle we identified brings to the fore particular dimensions of schooling, while masking or backgrounding others, making it possible to trace our decisions and actions from one level of analytic scale to another (for example Green, Heras & Yeager, in press).

Phase 1: 1960–1980s in the United States context

The historical foundation for these governing assumptions is a panel report from the NIE,[6] which framed the call for research on teaching as a linguistic process in a cultural setting (Cazden, 1974) and analysis of ten studies (Green, 1983) funded in 1978 in response to a call from NIE for studies that address this area of research. The first set of assumptions was constructed by an international group of scholars on the panel convened by the NIE in 1974. On this panel were scholars representing a broad range of what might now be referred to as complementary perspectives from different disciplines: education (Courtney Cazden (Chair), B. O. Smith and Arno Bellack from the United States, and Douglas Barnes and Ian Forsyth from the United Kingdom); linguistics (Heidi Dulay, John Gumperz and Roger Shuy); psychology (Elsa Bartlett and William Hall) and cultural studies (Allan Tindall). These scholars developed research questions and conceptual assumptions that shaped a potential program of research entitled ‘Research on Teaching as a Linguistic Process in a Cultural Context’, which called for cross-disciplinary research that:

- identified rules governing classroom discourse and the relationship between classroom discourse and frame factors in the institutional setting of school
- examined the acquisition by students of rules for school discourse

[6]The National Institute of Education (NIE) is now the Institute of Education Science.

- determined ways in which differences in dialect, language style and interactional norms affect learning in classrooms
- compared children's interaction patterns in multiple settings
- determined how two languages or dialects are combined in a classroom, and how language and dialect differences are exploited for communicative ends through code and style switching
- explored science as a curriculum context for teaching children more context-independent speech
- analysed patterns of student–teacher communication in order to determine the effect of the social identity of participants on the ways in which teachers overtly and covertly presented information
- analysed the effect of such differential presentations on the acquisition of knowledge and skill
- specified critical components of characteristics of natural communication situations that are necessary for the acquisition of communicative skills in a second language and that encourage the maintenance of native language
- developed and field-tested materials and procedures to improve teaching and thereby learning, on the basis of knowledge about linguistic processes in classrooms.

In framing this call, the panel created a conceptual framework for studies of classroom discourse and communication in classrooms and its consequences for students, as well as what could be known in and through the language-in-use in classrooms with linguistically diverse students (Green, 1983). These questions guided early research across research communities (for example Gilmore & Glatthorn, 1982; Green & Wallat, 1979, 1981; Wilkinson, 1982) and are still relevant today.

Ten studies were funded to address particular questions and directions proposed by this panel. Those funded represented a diverse body of theories and methods from anthropology, education, linguistics, psychology and sociology research in classrooms. In her review of these ten studies, Green (1983) found that the diversity of perspectives was both a resource and a challenge, given that only 25% of the terms and constructs named in the studies overlapped. However, when she analysed the studies as ethnographic artifacts, she was able to identify theoretical assumptions for each study. Then, by engaging in a series of pair-wise contrasts of constructs central to each study, she identified a set of converging constructs that were common to 70–100% of the studies.

Common to all ten studies were the following conceptual arguments:

- Meaning is context specific.
- Inferencing is required for conversational comprehension.
- Contexts are constructed during interactions.
- Classrooms are communicative environments.

Six additional governing assumptions were common to 7–9 of the studies:

- Meaning is signalled verbally and nonverbally.
- All instances of a behaviour are not equal.

- Contexts constrain meaning.
- Meaning is determined by, and extracted from, observed sequences of behaviour.
- Communicative competence is reflected in appropriate behaviours.
- Frames of reference guide participation of individuals.

Once these governing assumptions were identified, Green sent them to each researcher or team to confirm that the assumptions attributed to their project were ones that the research team or researcher agreed represented their position(s). This latter step was important, given different theoretical and disciplinary traditions represented by the different research teams, as indicated previously.

Together, these two sets of governing assumptions created a conceptual framework in 1983 that framed subsequent studies in the United States and abroad in the 1980s. The governing assumptions from the period between the 1970s and 1980s therefore can be viewed as central to developing programs of research on classroom discourse.

Phase 2: 1980s–1990s: multiple-perspective research and expanding ethnographic studies in classrooms

In the 1980s and 1990s, ethnographic work in classrooms expanded, adding both theoretical understandings and new language to describe and study the social construction of everyday life and learning in classrooms. The following set of theoretical assumptions has continuity with the constructs framed in the previous sections, creating a conceptually expanding set of arguments. Like the previous work, the new studies build on conceptualisations of learning as a social construction in the contexts of teaching and schooling.

The following governing assumptions provide conceptual arguments about the relationships between historical, moment-by-moment and over-time communication in classrooms:

- Teachers and students construct an intertextual web of events and texts (Barr, 1987; Bloome & Bailey, 1992; Bloome & Egan-Robertson, 1993) that define what counts as (Heap, 1980, 1991) literate (and numeracy) practices within and across times and particular curriculum areas.
- In the moment-by-moment and over-time interactions among teacher and students, members of the class construct norms and expectations, roles and relationships, and rights and obligations, that constitute members' cultural knowledge of patterns of life in the classroom (for example Bloome & Theodorou, 1988; Cochran-Smith, 1984; Collins & Green, 1992; Cook-Gumperz, 1986; Corsaro, 1984; Edwards & Mercer, 1987; Erickson, 1986; Green, Weade, & Graham, 1988; Santa Barbara Classroom Discourse Group, 1992a).
- Classrooms can be viewed as cultures-in-the-making (for example Collins & Green, 1992; Green & Dixon, 1993; Santa Barbara Classroom Discourse Group, 1992b).

- Culture is not given but rather is a construct that represents what members of the sustaining group construct to shape what counts as ways of knowing, being and doing in a particular class or group within a class (for example Agar, 1994; Collins & Green, 1992; Gilmore & Glatthorn, 1982; Green, 1983; Heath, 1982; Santa Barbara Classroom Discourse Group, 1992a, b).

The governing assumptions that are the result of the 1980s and 1990s in the United States provide a (re)conceptualisation of how to understand core constructs, including: learning, classrooms as social systems and teacher–student relationships. From the perspective of the governing assumptions presented in this section, as well as in the previous sections, the focus has shifted from the observable moment of learning to learning as something that is visible over time in the performance of students. The focus of this research is the opportunities for learning that are constructed by teacher *with* students, rather than on individuals. This work has also made visible why the conceptual and social web of ideas, information and practices need to be traced to identify what students have access to, how the actions among members of the class support and/or constrain what is possible to know and do, and how and what students take up and are able to use in subsequent learning events.

This period, therefore, has shown that students contribute to both the construction of collective opportunities and construct possibilities for their own learning within the collective. Therefore, a shift in conceptualisation of individual as the individual-within-the-collective has been proposed to capture the relational dimension of this body of work (for example Cushman, 1991; Gergen, 1985). Finally, the historical nature of ideas, actions and information were shown to be important to examine, moving discussions from concepts such as *background variables* to historical processes as visible in particular moments and events through the discourse used by teacher and students. In 1986, Erickson provided a conceptual argument for this way of understanding teaching–learning relationships and their implications for research methods.

Phase 3: curriculum, discourse and the social construction of knowledge

The final set of governing assumptions was identified from studies of the intersection of curriculum, discourse and the social construction of knowledge. Although they represent different theoretical angles of vision on the issues, when taken together with the other governing assumptions, they add expressive potential to our orienting framework, provide insights into how curriculum is a construction, not a given, and raise questions about what counts as disciplinary knowledge afforded to students in classrooms. The following set of governing assumptions provides a way to view what is accomplished in and through the communication and actions in the classroom:

- Curriculum is constructed in and through the communication and meanings members propose, read, interpret and take up (or not). From this perspective content is not a given or held in particular textbooks, but is dynamic and developing in and through the communication among members (for example Barnes & Todd, 1995; Barnes, 1976; Chandler, 1992; Mercer & Hodgkinson, 2008; Weade, 1987).
- In and through the discourse of classrooms, local, situated understandings of what counts as learning and as disciplinary knowledge are socially constructed (for example Beach, Green, Kamil, & Shanahan, 1992, 2005; Lemke, 1990; Mills, 1993; Guzzetti & Hynd, 1998; Kelly & Chen, 1999; Kieran, Sfard, & Forman, 2003; Roth, 2005; Street, Baker, & Tomlin, 2005; Ford & Forman, 2006; César & Kumpulainen, 2009; Green & Luke, 2006; Greeno, 2006; Kelly, Luke, & Green, 2008; Kumpulainen, Hmelo-Silver, & César, 2008; Lemke, Kelly, & Roth, 2006). For a discussion of this tradition in mathematics, see Brown, this volume.
- Through the moment-by-moment and everyday actions that take place over time in classrooms, members of a class construct common knowledge (for example Edwards & Mercer, 1987; Mercer & Hodgkinson, 2008) or local knowledge (Agar, 1994, 2006; Geertz, 1973).
- People provide contexts for each other (Erickson & Schultz, 1981), and reading the world is critical in order to read the word (Freire & Macedo, 1987).
- Ideas and meanings are first formulated between people (the intersubjective space) and then are (re)formulated for self, and then when used to communicate with others are again (re)formulated for others (Rieber & Carton, 1988; see also Cole, John-Steiner, Scribner & Souberman, 1978; Moll, 1990; Wertsch, 1991; Lee & Smagorinsky, 2000; Putney & Wink, 1998).
- Identities are not fixed but are constructed within and across the events as participants interact with particular groups of people in particular ways for particular purposes (Castanheira et al., 2007; Holland & Cole, 1995; Holland, Lachicotte, Skinner, & Cain, 2001; Ivanic, 1998).
- Participants within a developing event, social group or social setting make decisions (consciously and unconsciously) about when to participate, in what ways, what to take up and with whom to work, for what purpose(s), and under what conditions; thus, participants are viewed as agentive and the world in which they interact is malleable (for example Giddens, 1989).
- Children are not socialised to adult norms, but rather contribute to the developing social world as they interact with, and are responded to, by adults and others (Fernie, Kantor, & Klein, 1988; Gaskins, Miller, & Corsaro, 1993; Kantor & Fernie, 2003).

Although this set of governing assumptions is still in progress, those included provide a sketch map of those that are central to discourse, sociocultural and social constructionist perspectives, guiding the construction of the telling case that follows.

Decision point 1: from whose point(s) of view will the telling case be constructed?

In order to explore individual–collective relationships, as discussed previously, the question that we faced was: *From whose perspective will this telling case be constructed, the teacher's or a student's?* Several decisions needed to be made. Given our governing assumptions about the relational and interdependent nature of teacher and students in classrooms, the choice of one over the other is a false decision. The cumulative argument across the three sets of governing assumptions is the interrelated nature of the social and discursive construction of everyday life. Having said that, it is not possible to view the developing social world from both actors' perspectives simultaneously. Therefore, we needed to select one angle of analysis on the joint construction to trace over time. As indicated previously, we focused on Sergio's journey.

This choice required a two-step process. The first involved transcribing and representing what the teacher with students, or students working together, constructed as the text, social actions and event (for example Castanheira, Crawford, Dixon, & Green, 2001). Once this textual (re)presentation was constructed from the video record and related materials in the archive, we then (re)read the text through the perspective, not perceptions, of Sergio. Questions we asked of the text included: *What could we see Sergio contributing? What was made available to Sergio to read, interpret and take up (or not)?* When Sergio was visible, we focused more closely on the contextualisation cues (Gumperz, 1992) that made visible what he was focusing on as well as how he contributed to the developing event. In this way, we sought to explore the part of the *world* that Sergio was reading (see Table 3.2, p. 70).

The logic of inquiry presented in this section made visible the interrelationship between the two forms of analysis and (re)presentation necessary to examine the opportunities for learning that Sergio participated in constructing. The first focused on how he participated in constructing the text of the event that constituted collective activity. The second (re)visited the event and examined what Sergio did as others were contributing. This section made visible how the collective and individual-within-the-collective communication and actions are central to the analyses in the sections below. It also makes visible why we argued that dichotomising these two different angles is to mask their interdependence.

The archive as text: bounding the telling case

Having selected Sergio as the tracer unit for the present telling case, the next questions we faced, focused on our search and retrieval of relevant records from the ethnographic archive, were: *What counted as opportunities for learning social science, and thus social science curriculum?* These questions formed the ground for exploring what counted as being literate in social science. The reason for asking the question, *what counted as...* a question guided by work in ethnomethodology (Heap, 1980, 1985, 1991), is captured in the following governing assumption about the constructed nature of curriculum identified in Phase 3: *Curriculum is constructed in and through the communication and meanings members propose, read, interpret and take up (or not).* From this perspective, social science is neither given, nor held in particular textbooks, but is dynamic and developing in and through the communication among members in interaction with the material resources (for example textual [oral, written, visual] and artifactual [objects, textbooks, products, multimedia resources] made available to and constructed by members).

Using an *if... then... logic* once again, this assumption led us to build on the map of social science presented in Fig. 3.1 to identify potential cycles of activity for the telling case. Our goal was to select an anchor artifact within an event marked as a key event (Gumperz, 1986) by Sergio. While Fig. 3.1 provided a map of social science cycles across the school year, in order to identify which was significant to Sergio and might serve as an anchor event for analysis of the processes and practices involved in being literate, we searched a previous article on inclusive practices (Castanheira et al., 2009). In that article, Castanheira and colleagues had analysed the end-of-the-year essays in the archive and had identified the Island History Project as an 'important project'. Based on this claim by Sergio in his 'Dear Reader' letter, we re-entered the archive and selected a range of records, beginning with the first day of school (based on information in Fig. 3.1) and concluding with the end-of-the-year essays. We then selected all records available that directly related to the Island History Project, informed by prior studies undertaken in earlier years and a cross-year analysis (for example Castanheira et al., 2009; Floriani, 1993; Yeager et al., 2009).

This process, therefore, was not linear but one that required decisions to be made throughout in order to trace the roots of the processes underlying the construction of the event and the routes or pathways leading to its construction and from it for future work in social science. Thus, the *if... then...* logic guided the decisions we made in constructing a purposeful data set that permitted the exploration of the literate practices inscribed in the artifact selected (for example Lakoff & Johnson, 1980). Central to this conceptualisation of curriculum as socially constructed is Bakhtin's (1986) argument that

> Sooner or later what is heard and actively understood will find its response in the subsequent speech or behavior of the listener. In most cases, genres of complex cultural communication are intended precisely for this kind of actively responsive understanding with delayed action. Everything that we have said here also pertains to written and read speech, with the appropriate adjustments and additions. (p. 60)

This provides a theoretical argument about why we needed to trace both the collective construction and Sergio's take-up over time. This argument also supports a view of curriculum processes, practices and substance as genres of complex cultural communication that are constructed within and across events and become material resources for future acts of communicating, whether in writing, speech or multimodal representations (for example Jewitt, 2006; Sefton-Green, 2006). For Bakhtin (1986), genres are not pre-existing structures but rather are speech (writing, reading) patterns constructed by members of social circles, or small-world constructs.

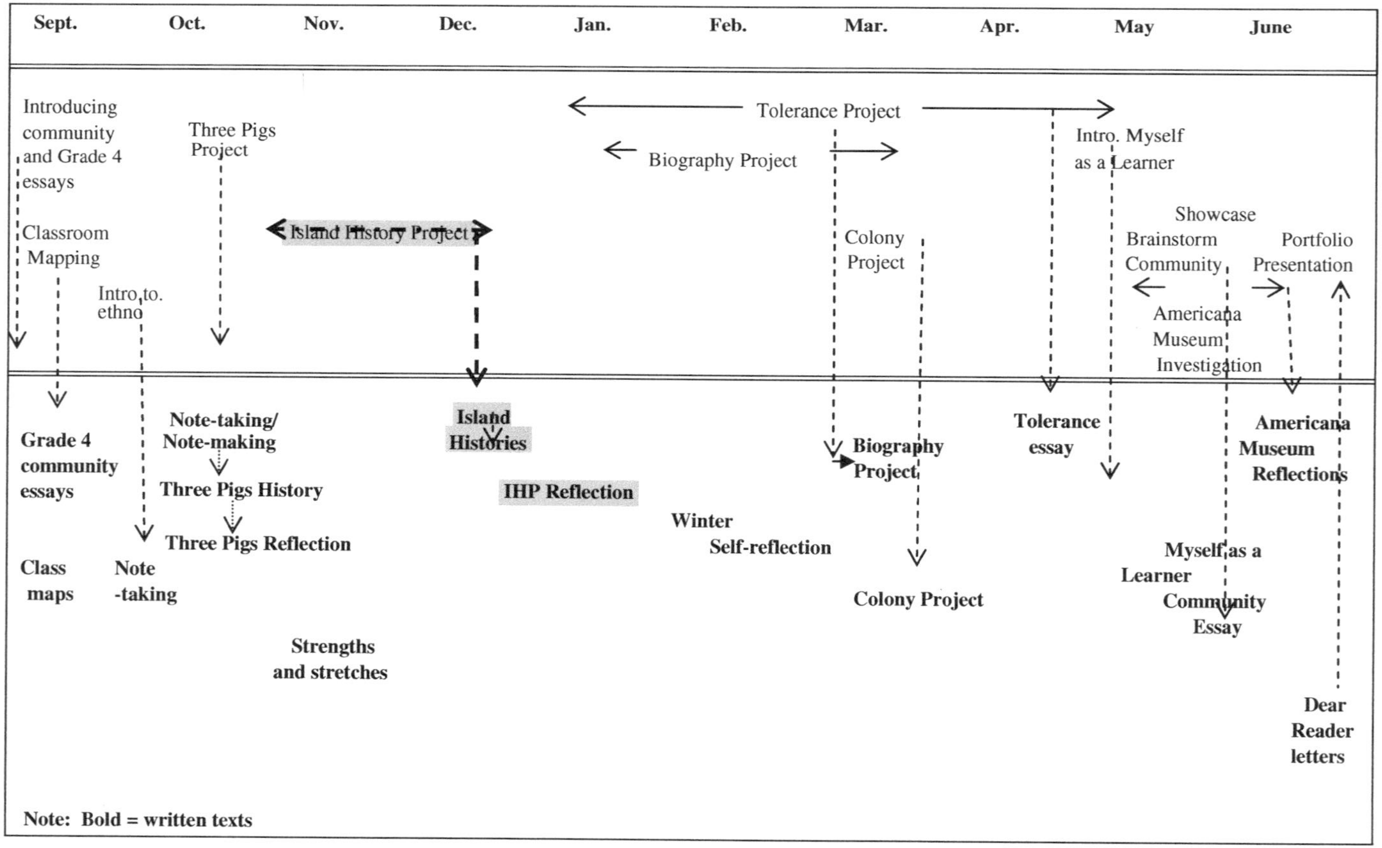

Fig. 3.1 Intertextually tied cycles of social science across the school year

> In each epoch, in each social circle, in each small world of family, friends, acquaintances, and comrades in which a human being grows and lives, there are always authoritative utterances that set the tone—artistic, scientific, and journalistic works on which one relies, to which one refers, which are cited, imitated, and followed. (p. 88)

Although not specific to classrooms, we posit that this argument can apply to ways of being literate in social science within the small world of the classroom. Support for this argument comes from an analysis that Skukauskaite and Green (2004) undertook of a recent unpublished article by Bakhtin (2004). In that article, Bakhtin documented and described how, as a secondary school teacher, he analysed then-existing forms of grammar teaching in professional arguments as well as in student work (homework and writings) in a secondary school class he was teaching. He developed a dialogic approach to the teaching of a form of Russian grammar (parataxis) that engaged students through use of novels and dialogue about the texts and the work of grammar. Through his arguments and descriptions, Bakhtin made visible how in his class, with particular groups of students, he created new forms of grammar work that in turn led students to take up and use these forms for their own work (Skukauskaite & Green, 2004).

Building on these readings of Bakhtin, we argue that what counts as curriculum processes and practices constructed by students and their teacher can be viewed as *authoritative utterances and genres* that constitute the opportunities for learning to be literate in social science (and other subject areas). These opportunities, in turn, set the tone for artistic, scientific and curricular works upon which individual students and the collective group can rely, or refer to, cite, take up and follow in subsequent work.

Additionally, by viewing the class as *a small world in which a human being (a student) grows and lives*, we add to our understanding of what it means to claim that a class (not a classroom) is a *culture-in-the-making*. Bakhtin's arguments about authoritative works as created within such small worlds also provides theoretical confirmation about the approach we took in identifying what was significant to Sergio. Thus, the Island History Project became an anchor (an authoritative work) that served as a *rich point* (Agar, 1994, 2006) for the construction of the telling case.

Constructing an anchor for the telling case: the Island History Project essay

Given our interest in how Sergio contributed towards and took up the opportunities for learning to be literate in social science, we selected Sergio's Island History Project essay that he wrote with a Spanish-dominant partner, using English and Spanish (Castanheira et al., 2009). Drawing on Lakoff & Johnson's (1980) argument that the choice of words and configuration of words inscribe the ways in which the author views the world, we saw this artifact and the events surrounding its production as a source for analysis of the literate processes and practices that Sergio drew upon to write this essay, an authoritative genre in this class. We also viewed the processes and practices he drew upon to construct the essay as representing common knowledge (for example Edwards & Mercer, 1987) of how to be literate in this class, not just a personal view of what it meant to write this type of essay (for example Floriani, 1993; Putney et al., 2000).

For this telling case, therefore, we decided to explore how this anchor essay (a rich point), provided a grounding for examination of collective–individual relationships at multiple levels of analytic scale. The particular question that guided this new analysis was: *How does the essay provide evidence of how Sergio used the previous opportunities afforded him from prior work to create opportunities for himself, his partner and others?* Therefore, through the (re)analyses in this telling case, we demonstrate how the individual–collective relationships across times and events afforded students cycles of opportunities for learning to be literate and for introducing information and practices needed for subsequent events and areas of the developing curriculum (Barnes, 1976; Barr, 1987; Bloome & Egan-Robertson, 1993; Dixon, Green, & Brandts, 2005; Fernie et al., 1988).

Reading the world(s) of the classroom: multiple actors, multiple readers and multiple points of viewing

Once we had selected the anchor artifact and identified the boundaries of the telling case, our next task was to frame each level of analysis within the telling case. As in the case of the retrieval of data from the archive, we decided to revisit the *governing assumptions* identified previously, to construct an *if... then...* logic for the first level of analysis of this telling case. Before turning to the guiding assumptions, we need to discuss how we view the difference between records and data. From our theoretical stance, records are not data until the researcher acts on them and uses particular theoretical perspectives to turn the 'bit of life' recorded on the record (written, graphic or audio/video) into data for analysis of the questions under study[7] (for complementary arguments about transcribing, see Green, Franquíz, & Dixon, 1997 and Psathas, 1995).

Although the essay served as an anchor for the construction of this telling case, we elected to start not with the essay but with the events of the first day that initiated the process of developing patterns of classroom life that members used to construct local authoritative genres. We also elected to start with the first morning to explore

[7] To examine how this works within our research community, see Yeager, 2003. In her dissertation, Yeager drew upon an analysis of the first morning by Castanheira (2000) and (re)analysed the data through her questions, which differed from those of Castanheira. The two sets of analyses of a common period of time make visible how the questions guiding the research lead to overlapping (re)presentations of the work of the teacher and students. The unique dimensions of each analysis show why (re)analysis is productive when each is guided by additional theoretical arguments and new questions.

what world(s) were constructed that Sergio had available to read from in his first moments of school in this school year.

Using the *if... then...* logic once again, we reasoned that *if* Freire and Macedo (1987) have captured a basic relationship between reading words and the social and historical context in which they are embedded, *and* people must learn to read the world in order to read the word, *then* several questions arise: *What is the world of the classroom that was available to be read? Who constructed what world with whom, when, where, for what purposes? What were the literate practices of that world that were part of Sergio's repertoire for action in this class?* These questions are guided by the governing assumption that people are contexts for each other (for example Erickson & Schultz, 1981; McDermott, 1976).

In order to address the question of what is the world that is available to be read, we started our analysis of the archived materials, not with the first moments of communication between teacher and students, but with decisions the teacher made prior to student entry. The analysis that follows, therefore, focuses first on the construction of the physical world that students entered, and then shifts the angle of analysis to Sergio as a tracer unit to make visible the individual–collective construction of the events of the first day of school. Through these analyses, we describe different methodological decisions that were needed to (re)present and analyse particular moments in time; moments that varied in time scale.

Constructing the physical world

We focus first on decisions that the teacher made as she constructed a particular physical and material world (Gee & Green, 1998) prior to student entry. This analysis raises the question: *When does class begin?* Most research on teaching begins with moments of interaction or entry into an already constructed physical world of a classroom. In this study, we ask this question so that we can uncover what resources were afforded the teacher and what resources she brought to the students. One guiding assumption for this analysis is that decisions are made by those beyond the classroom door, including who the students are in that class, and constitute the *hand the teacher is dealt* (Barr & Dreeben, 1983). However, as the analysis of the decisions Sergio's teacher made will show, the decisions prior to the student entry are complex and involve multiple actors (for example Dixon, Green, Yeager, Baker, & Franquiz, 2000; Green, Heras, Yeager, Castanheira, & Dixon, in press).

Our ethnographic work over the previous 7 years with the teacher and her students, and the participation of the teacher in development of this chapter, provided a historical record that made visible the teacher's and her colleagues' agency in making decisions about resources and student placements. The district and the administration had a policy of site-based management and participatory leadership. The teacher was not only a member of a collaborative team, but also the school liaison to the university's teacher education program, a teacher fellow of the South Coast Writing Project (SCWriP) and a Fellow of the Carnegie Academy for the

Scholarship of Teaching and Learning (CASTL K–12). She was also a published author, as indicated by citations of Yeager in the previous sections.

Given this history, we cannot view the physical and material world as what she was dealt; rather, we view it as a construction that represents her goals, district and state curriculum goals, and her reading and interpretation of students assigned to her class. In addition, this year, the district committed along with the school to bilingual instruction, and as a credentialed bilingual teacher, she made decisions with her colleagues about the placement of students and the nature of the programs for language learning afforded all students (English as a second language, Spanish as a second language and Spanish for native Spanish speakers).

In this way, the teacher and her colleagues, like the students, were active agents in the construction of their work, as they took up and constructed a world for their students and themselves, using the decisions and resources available beyond the classroom door (for example Barr & Dreeben, 1983; Kelly & Green, 1998; McNeil & Coppola, 2006; Green et al., in press). The material and social resources provided to the teacher in this year included an assigned physical space, a classroom below the school's bell tower, in this case, with a physical layout that included a main room with extended work area, an entry room and an attached workroom. Within this space, she then designed a series of spaces for students (workspaces, personal desks at a table, visual materials, texts and other forms of educational resources). Thus, what was available to read on this first day of school was a physically structured space.

As students (and, in some cases, parents and siblings) entered, the teacher greeted them in English or Spanish. The teacher then invited the students to select their name card and select a seat that was unoccupied at one of the six table groups. The students were then asked to decorate their name cards in a way that represented themselves to others in the class. As the students took up and acted upon what was proposed, the textual world of the classroom expanded. Each action was available for others to see and, at times, to hear. From this perspective, students took up a role of active constructor of the class, the table group and their own space within the class. They also, at times, took up the role of *overhearing (seeing) audience* (Larson, 1995) as they observed what others were doing, and through this process of reading the world that was developing, they were able to explore what languages were valued in the classroom, who could talk with whom and how members were taking up and interpreting the common task, among other actions visible.

From this perspective, as members were *structuring the world* through the *flow of conduct* (Giddens, 1989) between, and among, actors, they created, and simultaneously made visible through their actions, communication and visual/multimedia texts, what counted as ways of knowing, being and doing that constituted the developing cultural practices and processes of classroom life. Thus, as students entered individually, in small groups, or accompanied by a parent (and siblings), this world became (re)formulated as a living space through the actions of those who were entering as members as well as through the readings and interpretations of those already part of the class. The class, therefore, was a dynamic and developing world, one that ended officially when the school year ended. However, our work across years has shown that students often maintained contact with each other and with the teacher, thus suggesting that for some it ends, and for others who take up the opportunity to continue contact, the relationships continue, as do opportunities for learning from each other.

The class as a developing text: What Sergio contributed towards and had available to read

In this section, we provide two (re)presentations of the developing text of Sergio's class. The first is a (re)constructed fieldnote, written by (re)viewing the video record of the first moments of Grade 5 in 1996. This fieldnote was a (re)construction or, rather, a 're'presentation in written form of the chains of action of Sergio and those with whom he communicated and interacted. The second is a (re)presentation that takes the form of an *event map* (for example Castanheira et al., 2001; Green & Meyer, 1991) of the ebb and flow of activity and the events produced on the first morning of school in 1996.

These two forms of mapping the developing classroom life focus upon different levels of analytic scale: (1) individual-within-the-collective, and (2) the collective accomplishments. Through these (re)presentations, we make visible the logic of inquiry that moves from theoretical arguments to methodological representations to analyses guided by a series of conceptually driven decisions. At each level, the focus is on a particular dimension of the social life of the group that, when juxtaposed or connected to others, makes visible how a small world is being socially constructed in and through the intentional communication of actors within a developing social system (for example Castanheira, 2004; Heap, 1991).

Sergio as a tracer unit: uncovering the first chain of events of the school day

The reconstructed fieldnote, shown in Table 3.1, provided a way to capture what we call a *running record* of chains of developing action and activity (for example Castanheira et al., 2001; Kelly, Crawford, & Green, 2001). The following *reconstructed fieldnote* focuses on what Sergio could be seen doing on the video record (a visual form of fieldnote)[8] of the first morning of school, given that he was not the original point of focus for the video in the classroom. In the reconstructed fieldnote, we have three different types of notes, each presented in a different font style: *methodological notes* (MN), *fieldnotes* (FN) and *interpretive*

[8] As argued by Baker, Green, & Skukauskaite, 2008, a video record is a form of fieldnote, recorded by an ethnographer from a particular angle of vision. It is not a record of the event, the whole of classroom life, or even the event itself. It constitutes a recording of a 'bit of life' (Hymes, 1982) from a particular angle of vision that can then be (re)read for particular purposes (see also, Barnes, Britton, Rosen, 1969).

Table 3.1 Reconstructed fieldnote

MN: During the period from 8:15 to 8:45, we were able to observe Sergio for only a portion of the time, given where the camera was pointed. Therefore, the description of what Sergio did, how he created opportunities to position himself and to take up positions in relationship to others, and how others responded to him is a partial picture. However, by tracing what was visible on the video record, we are able to gather a range of possible actions that individuals could and did take as they engaged in the tasks set by the teacher
FN: During this period, we were able to see that Sergio engaged in the activity of producing his name card, using materials provided by teacher to each table group. He shared his drawing and writing on the name card with table group members and talked with them about their cards. He also showed his work to the teacher, teacher aide and student teacher when they came to his table group. He engaged in conversations about computer games and summer vacation with two of his table group members and adults in the class. He also talked briefly with a girl who was sitting close to him in another table group. In the transitioning moment that marked the end of the sub-event 'Welcoming to the Tower' (WC), we also saw that Sergio attended to teacher's signal (chime) and re-oriented to WC interactional space, responding to her greeting as other students did (8:45). In this way, he contributed to establishing the end of the first sub-event and the beginning of the next sub-event
IN: By following Sergio's actions, we were able to make visible that all members entering the class engaged in the same chain of activity that Sergio did, and those in his table group responded to his initiations or engaged him in dialogue. In this way, we were able to identify how individuals took up the opportunities they were afforded by the teacher and created local and situated opportunities for exploring self and others within these activities. Therefore, the use of a tracer unit provides a systematic way to identify not only the work of an individual, but also all of those with whom the individual interacts or who are present in the same or contiguous interactional spaces

notes (IN). The first two types of notes, along with personal notes (PN) and theoretical notes (TN), not represented in this text, were proposed by Corsaro (1981) as a means of distinguishing the different forms of work that ethnographers do during a study.

As indicated in this fieldnote, we elected to record the *developing social world* that was visible to students, teacher and others, by tracing the chain of actions of Sergio and those with whom he had contact. As described in the *methodological note* (MN) and the *fieldnote* (FN) (Corsaro, 1981), what is available to be recorded was limited by what we could see and interpret, the angle of vision recorded on the video.

These different forms of notetaking make visible the dynamic and interrelated processes of describing, recording, interpreting, responding and making meaning(s) of bits and pieces of the developing lifeworld(s) of teachers, students and others in classrooms. For this case study, we added IN to our descriptive notes to represent the *interpretive* nature of reading the world of action. Given that one of the goals of this telling case study was to make visible theory–method relationships, and the complementary nature of different theoretical perspectives and associated methods of analysis, we elected to use IN at this level of analysis. From this perspective, we view interpreting as a form of theorising. By adding IN to other forms of notes, we make visible the continuing *processes of interpretations, decision making, hypothesising* and *meaning making* that we undertook across times and events of this case study.

As we will demonstrate, through other forms of (re)presentation, these interpretive actions are recursive and iterative practices (Agar, 2006) of the abductive reasoning processes that ethnographers engage in at multiple levels of scale and multiple points in time within a telling case. As this fieldnote shows, the decision of *angle of vision or (re)presentation* foregrounds particular dimensions of the developing world and backgrounds others. Therefore, we did not rely on the written fieldnotes but rather used these as a sketch map of the chains of activity visible on the video records.

In this way, the fieldnote level of representation and analysis provided a focus for identifying the chains of action, patterns of organisation and individual–collective activity that is (re)presented in the next level of mapping, *the event map.* An event map (re)presents the chain of actions that were the basis for the teacher to guide students in constructing a series of differentiated events (Castanheira et al., 2001; Green & Meyer, 1991; Green & Wallat, 1979, 1981). This level of mapping is a description of the chains of actions signalled by the teacher (or other designated actors) and makes visible the ebb and flow of collective activity and through this the construction of events. Central to this level is the assumption that events are produced in and through the interactions among members and are not pre-existing entities, even when planned (for example Chandler, 1992; Weade, 1987). Table 3.2 is a (re)presentation of the events identified through the construction of running records of the flow of conduct.

As indicated in Table 3.2, shifts in the flow of conduct are visible in changes in the types of action being taken and the topics being constructed. These actions are presented as *present continuous verbs (participles).* Information included in this table also includes the order of language used for each action (that is English/Spanish or Spanish/English), and the pattern of physical organisation of participants. The sub-event and event columns represent the types of activity and activity shifts that were accomplished by members. Once again, the decisions that we made about what to include were guided by the governing assumptions supporting particular types of analyses and interpretations of the work of members but not others.

Our goal, in representing various types of information on the table, was to provide a text that represented different dimensions of the unfolding work of teacher and students, as well as students with others, from the first moments of entry to the end of the first academic event, the Name Game, an insider term. Representing the actions as verbs, rather than as behaviours, was purposeful, as was using the emic or insider terms attributed to the actions and events.

The construction of this table addressed the questions posed at the beginning of this section; questions that focused our thinking on what was being proposed, and thus socially constructed (the opportunities column), in what ways (the actions column), with whom (interaction space column) and under what conditions (language column). These columns formed the basis for examining the flow of conduct and

Table 3.2 Exploring people, actions and spaces as texts: opportunities constructed in the first two events of the first day

Time	Speaker	Actions	Language	Interaction Space	Sub-event	Event	Opportunities to explore self, others and physical environment as texts
8:10	St/P	***arriving*** in the classroom		T-I	ENTERING THE TOWER	ONSET OF COMMUNITY	- observing and 'reading' what others are doing
	St/P	***meeting*** teacher		I-TG			- re-establishing contact with friends
	T	***greeting*** St/P	S/E				- meeting other class members
	St/P	***responding*** to T	S/E				- listening to English and Spanish being spoken
	T	***orienting*** students to	S/E				- speaking English or Spanish
		finding name card,					
		choosing place to sit					
	St	***choosing*** where to sit		I-I			- meeting other members of the class
	St	***decorating*** name square	S/E				- getting acquainted with others
(45')	St	***talking*** to classmates sitting at table group	S/E				- choosing language to interact with others
	T/T	***talking*** to Sts at table groups	S/E				
	St	***talking*** to T/T.A. and St. Teacher.	S/E				
8:55	T	***introducing*** chime as a sign	S/E	WC	WELCOMING TO THE TOWER COMMUNITY		
	T	***welcoming*** participants	S/E	St/WC			
	T	***celebrating*** the languages of the Tower	S/E				- re-situate self within whole group
(40')		community: Spanish and English					- getting support from adults and classmates
	T	***explaining*** way of using Spanish and					- helping student teacher learn her job
		English in the classroom	S/E				- becoming an ethnographer
	T	***introducing*** adult members to students	S/E				- knowing local community ways of leaving
	T	***introducing*** ethnography as community	S/E				- making decisions about routine aspects of norms being
		practice					established for the class
	T	***talking*** about basic routines: drinking water,	S/E				- becoming a Tower community member
		signing up for lunch, bathroom, recess, etc.					- exploring own knowledge and experience in
							constructing Tower community in 96/97
	T	***exploring*** students' knowledge about	S/E				- defining uses and exploring multiple spaces
		Tower community					- hearing S/E and speaking language of choice
	T	***introducing*** Tower as community with traditions					
	T	***presenting*** multiple physical spaces of Tower					- exploring physical spaces of Tower as classroom
		classroom					

Table 3.2 (continued)

Time	Speak	Actions	language	Interaction Space	Sub-event	Event	Opportunities for Exploring Self, Others, and Physical Environment as Texts
9:35	T	***explaining*** that students would meet each other	E/S	**WG**	CHOOSING ADJECTIVE	NAME GAME	
	T	***explaining*** students would help each other	S/E	**TG**			- Establishing contact among classroom participants
(28')	T	***describing*** appropriate/inappropriate actions	S/E	**I-TG**			- Establishing relations between space and actions
	T	***exploring*** students' knowledge of adjectives	E/S				- Understanding what counts as material resource within the classroom spaces
	T	***presenting*** examples of procedures for choosing adjectives and support others	S/E				- Modeling ways of describing self
	T	***opening*** the possibility for classroom ethnographer to use Portuguese	S/E E/S				- Using language(s) of choice in order to participate
	T	***emphasizing*** expectation that members of tables groups help each other	E/S				- Establishing others as resources
	St	***choosing*** adjectives in table groups	S/E				- Exploring with others possible ways of Naming/describing self and others
	T, TA/ St T	**helping** students on request	S/E				
	T	***extending*** time on request of student	E/S				- Establishing time for learning as flexible
9:58	T	***discussing*** next activity	S/E	**WC**	TEA PARTY		
	T	***explaining*** what 'introducing themselves' in 'Tea Party' would look and sound like	E/S	**I-I**			- Broadening the basis for establishing contact, from individuals in TG, to individuals within the whole Group.
12'	T'	***providing*** examples (student in skit)	S/E	**WC**			
	T	***ringing*** chime to signal beginning of Tea Party	E/S				- Positioning individuals as members of the larger collective *as* a collective
	All	T/S Teacher/T Assistant/R/Sts ***performing*** introductions	E/S				
	T	***asking*** students to reach others they did not know	E/S				- Opening possibilities of including new people
	T	***ringing*** chime to end performance of Tea Party	S/E				- Engaging in a collective work *for* the collective (meeting others, reaching out to new people)
	T	***discussing*** 'community' in context of Tea Party	E/S				
	T	***exploring*** the diversity of the Tower	S/E				- Framing Tower as diverse group
10:10	St	***attempting*** to name as many names with adjectives as possible (volunteers)	S/E	**WC**	NAMING MEMBERS		- Taking risk within classroom activity
	T	***clarifying*** expectations for playing Name Game	E/S	**St-WC**			- Acknowledging others and being acknowledged
		attempting to name as many names with adjectives as possible (volunteers)	S/E				- Picturing classroom as constituted by a large number of members
(8')	T	***re-stating*** names and adjectives of all students	S/E				- Using others as texts for learning

Key: **I S**: Interactional Space; **T-I**: Teacher-Individual Student; **I -TG**: Teacher – Table Group; **I-I**: Individual – Individual; **St-WC:** Student – Whole Class the social accomplishment of coordinated actions. Through this process, we represented the developing social world and how the way it was developing provided signals to students about what counted as possible and appropriate actions. Thus, this table provides a basis for examining what the collective was structuring with the guidance of the teacher and others in the class.

We constructed this table from the angle of analysis of the developing collective, but read the actions from the point of view of students, including Sergio. The column labelled *opportunities for exploring self, others and physical spaces* makes visible the angle of analysis we took, that of Sergio and his peers. Each action and event was read interactively to identify the range of opportunities constructed by, and thus afforded to, students, teacher and others, including the ethnographer.

Guiding our reading and interpretation of the textual representations of social actions and activity presented in this table was a governing assumption located in the Phase 2 section of the history of governing assumptions as follows:

- In the moment-by-moment, and over-time interactions among teacher and students, members of the class construct norms and expectations, roles and relationships, and rights and obligations that constitute members' cultural knowledge of patterns of life in the classroom.

Our goal in this analysis was to begin to hypothesise ways of being and doing that were possible in the classroom. At this point, we elected *to background* ways of knowing particular academic materials in order to focus on how social knowledge was being constructed. This process made visible a developing body of common knowledge that includes the norms and expectations, and roles and relationships, and referential system of the classroom being constructed among members of a class (for example Barnes, Britton, & Rosen, 1969; Edwards & Furlong, 1978; Edwards & Mercer, 1987; Green & Wallat, 1981; Lin, 1993).

The question guiding this analysis was: *What ways of being and doing were signalled to and by students as represented in the opportunities for exploring self, others and physical environment as texts?* In reading the chain of possible actions in the opportunities column, we were able to identify a range of possible actions: students were able to re-establish contact with friends, to engage with people that they did not know, get support from adults and classmates, support others, make decisions about where to sit and other routines, as well as explore self and others through collective activity. This led us to construct a prediction that if this pattern was to become a practice, not just a first-day activity, then one or more of these actions would repeat in subsequent events constructed in the class. The prediction (or hypothesis) that we constructed from these patterns is as follows:

Prediction: Students will engage in chains of tasks that are proposed by the teacher to the whole group, making public the goal and required actions. The teacher will then engage students in opportunities that enable them to explore information personally, collectively (small and/or large group) as well as publicly. This chain of activity serves as a common basis for organising cycles of activity in particular subject areas. These processes and practices, *if an organising principle of* *practice*, will *then* be used in iterative and recursive ways, creating anticipated forms of organisation of work across subject matter.

The first test of the prediction can be seen in the chains of action visible in the first two sub-events. It was also visible in the third event, the Name Game, suggesting that this was a potential pattern of practice that would, across times and events, become an expected way of engaging in subject-matter learning.

Testing the prediction: mapping the flow of conduct of the first day of the Watermelon Project

In order to test our prediction, we focused on the first academic cycle of activity, the Watermelon Project, which introduced mathematics from a problem-based and inquiry perspective. In a previous analysis, Mills (1993) examined the processes and practices involved in doing mathematics in a Grade 6 class for this teacher. She argued that students were engaged in a process of becoming mathematicians. Her analysis provided a *point of triangulation* (Corsaro, 1984), one that permitted us to test our prediction for Sergio and his colleagues.

In order to triangulate the processes in practices constructed by, and engaged in, by Sergio and others in the class, we engaged in a new mapping process. Rather than repeating the level of analysis in Table 3.1, we read through the transcript of this developing chain of actions, noting how the teacher configured groupings within the class as well as what actions were taken by whom, in what ways and for what purposes. This process enabled us to identify iterative and recursive practices and processes as well as the pathways that were constructed through these actions. Figure 3.2 represents the pathways of this practice and what Sergio and his peers were engaged in at each point across days and times on the first day of school.

As indicated in Fig. 3.2, the flow of conduct moved among whole-group (collective), individual and small-group dynamics and then shifted from individual-within-the-collective (table group) to a public sharing of the table group's decisions about the weight and cost of their watermelon. The public sharing made visible the small group's decisions and reasoning processes, thus foregrounding the contrast in processes and practices. However, during this event, in the morning, the contrast was primarily verbal, while during the afternoon, and on subsequent days, the differences in group estimates would be contrasted publicly with the actual weight and cost. However, on this day, in the morning, the pattern ended with two members of the group reporting their estimates and processes to the class, in English and in Spanish. Following their presentation, students then returned to their table group to record individually the process in which they had engaged in order to construct a personal record of their thinking and actions leading to their individual and table group estimate.

For each physical space, we examined the processes used and the connections or pathways proposed and then taken up. Once again, we used the convention of

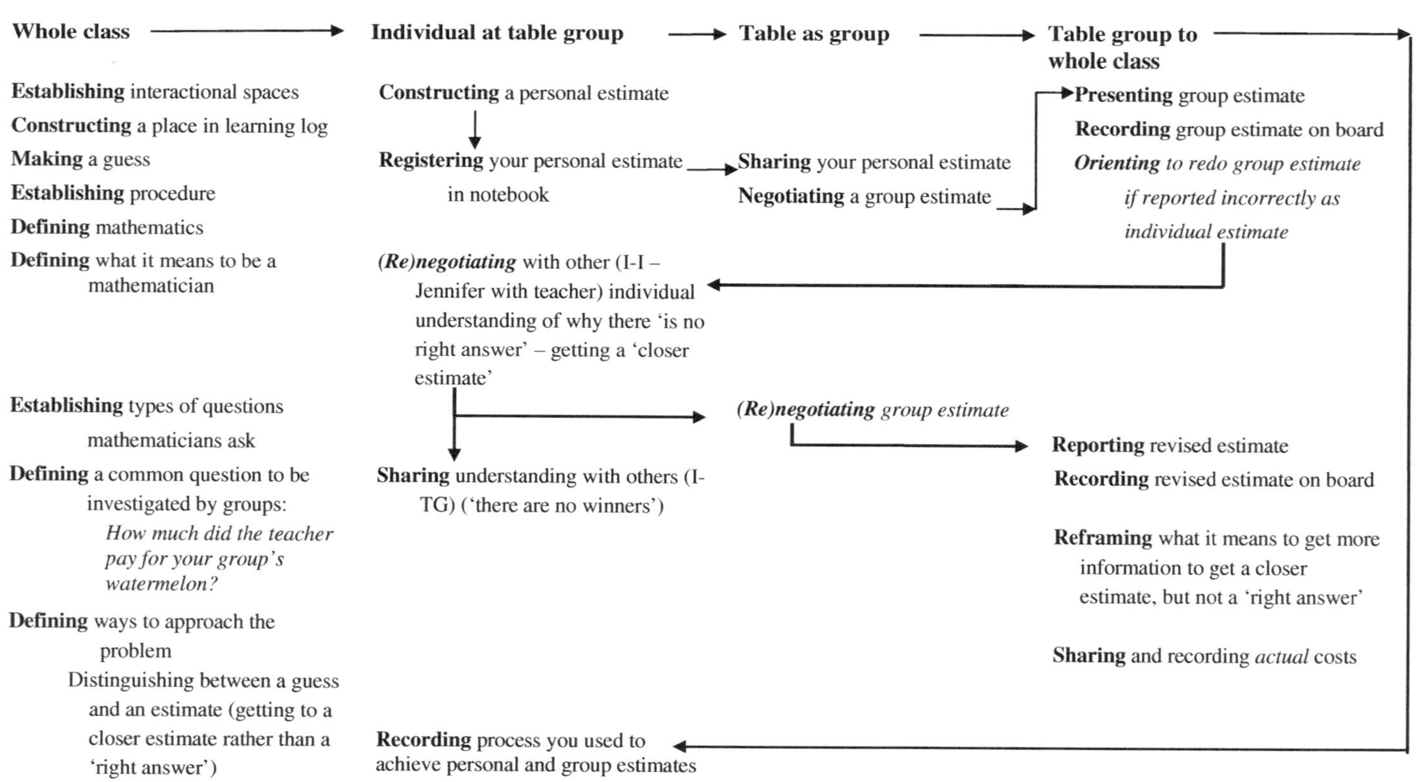

Fig. 3.2 Patterns of structuring participation: creating multiple contrastive opportunities for negotiating understandings and potential understandings across time on day 1

present continuous verbs to map the flow of activity and the intertextual (Bloome & Egan-Robertson, 1993) ties constructed in and through a series of iterative and recursive processes. These processes and practices, when examined as texts to be read by students, teacher and ethnographers alike, made visible how the patterns of organisation created in the previous events of the morning were used by the teacher to create opportunities for learning to be a mathematician in this class on this day.

Viewed in this way, the teacher initiated particular patterns in particular events or sub-events that were then used as a material resource for subsequent work, in which content and substance of the work varied, while the pattern of action was recursive and at times iterative. In order to explore this hypothesis more closely, we examined the practices used across the 6 days of the Watermelon Project cycle of activity on the first 6 days of school. This level of analysis built upon event maps of each day, through which we identified the flow of conduct (Giddens, 1989), the organisational patterns, the events and sub-events, and through this, the patterns of activity.

Thus, in Table 3.3, we draw upon the earlier analyses to construct a map of when practices in four areas of interest were introduced and used by the group and/or individuals-within-the-group. This level of analysis made visible the range of opportunities afforded students to explore inquiry processes, to construct literate practices and to begin to develop identities as mathematicians and as ethnographers (Mills, 1993; Yeager, Floriani, & Green, 1998). This table, therefore, makes visible the distribution of opportunities that recurred across time and where new ones were introduced to the group for particular purposes. The table makes visible when and where the practices constructed on a particular day, in a particular event, were taken up and used on subsequent days, indicating that they were material resources that the group used to explore new topics or to expand the current cycle of activity.

Forward mapping: the Island History Project

To further test our prediction that the patterns of practice constructed in one event of the Watermelon Project, a cycle of activity, became material resources for students to take up and use to guide their work in subsequent areas of the curriculum, we moved forward in time to the anchor artifact that we identified in the Island History Project. As indicated previously, this project was the one that Sergio stated was important in his Dear Reader Letter at the end of the school year (Castanheira et al., 2009). As argued earlier in this chapter, this artifact inscribed a series of actions, practices and processes that Sergio took up and supported his partner, a new student, in using to construct their Island History Essay. This analysis drew upon analyses that Castanheira (2000) and Yeager (2003) had previously conducted, as well as an analysis of the practices introduced and used for this cycle of activity.

Table 3.3 Range of practices initiated across six days of the Watermelon Project essay

Practice	Day 1	Day 2	Day 3	Day 4	Day 5	Day 6
Situating events, activities, or practices in history (e.g. history as tradition, participants' histories) for purposes of drawing on histories as resource in new ways	X	X				
Orienting to work in and through different interactional spaces (e.g. whole group, table groups, student–student, teacher/adult–individuals, teacher/adult–table group, teacher–whole group)	X	X	X	X	X	X
Literate practices						
Choice/use of languages (e.g. Spanish/English)	X	X	X	X	X	X
Labelling, dating log and data entries	X	X	X	X	X	
Taking/recording notes	X	X		X		
Talking with others (e.g. I-I, I-T, table groups) to communicate information, share ideas, reach consensus	X	X	X	X	X	X
Drawing on others as resource (e.g. multiple adults, peers, etc.)	X	X		X	X	X
Writing to learn (e.g. explaining a process, interpreting information)			X			X
Reading data (e.g. reading a graph)	X			X	X	X
Reporting data/presenting in public space	X		X	X		X
Inquiry practices						
Observing for different purposes, from different perspectives/angles of vision	X	X				
Gathering information/data from multiple sources	X	X				
Recording data	X	X	X			
Supporting with evidence	X	X		X		X
Determining a problem/question	X					
Investigating a problem/solving a problem	X		X	X	X	X
Estimating/predicting	X					
Interpreting data	X	X	X	X	X	X
Representing data in different ways, for different purposes (e.g. graphing, charting)		X	X	X	X	
Understanding/taking different points of view/angles of vision	X	X	X	X	X	X
Potential academic identities						
Doing the work of mathematicians	X		X	X	X	X
Doing the work of ethnographers		X				

Table 3.4 presents the Island History essay and the analysis of the patterns inscribed by Sergio in his essay. The interpretation of patterns column is a (re)formulation of the written text through Bakhtin's (1986) argument that genres are constructed through speech events and actions among members and that communication is a complex process in which response often occurs across time, not in the moment. Additionally, we drew upon those governing assumptions that curriculum practices constructed at a particular point in time are part of a web of potential actions that can be taken up and used at other points in time, creating what Bakhtin called authoritative genres.

Table 3.4 Evidence of take-up of practices of Social Science: Sergio & Jaime's Island History essay

Essay text by paragraph	Interpretation of patterns
Somos historiadores. We are studying a mysterious island. *La isla es un misterio porque cuando llegamos, hallamos las ruinas de un pueblo, pero ninguna gente. Parece que se desaparecieron y estamos estudiando para saber por que (a)*. We want to know what happened to these people after they made such a big voyage	Talking with others (e.g. I-I, I-T, table groups) to communicate information, share ideas, reach consensus Choosing a language in which to write Code-switching between tied segments of text Writing to learn (explaining a process; interpreting information) Determining a problem Working in different interactional spaces: writing as a group
Creemos que llegaron 10 personas en la isla (b). We are studying when this happened, but we know it is in the past. *Nosotros estamos estudiando cuando pasó, pero sabemos que fué en el pasado. Creemos que se hundió un barco y nada más sobrevivieron 10 personas. Hallamos 10 diferentes huellas en toda la isla y zapatos diferentes como chiquitos, grandes y medianos que parecían que vinieron de los años tempranos y no más tarde en la isla.* (c)	Investigating a problem Gathering information/data from multiple sources Code-switching between tied segments of text
When we went to get evidence, we found shoes, footprints and beds and we knew that this happened in the past, because the things were from the past	Gathering information Warranting how they knew what happened
We believe the people spent their first weeks trying to survive. *Sobrevivieon por modo de comer frutas y se durmieron en la playa (d)*	Proposing hypotheses Interpreting data Supporting interpretations with evidence Code-switching between tied segments of text
During their first 2 years, they moved from the beach to the waterfall. During these years they had children. Our evidence is that we found bones and small clothes	Proposing hypotheses Supporting interpretations with evidence

Table 3.4 (continued)

Essay text by paragraph	Interpretation of patterns
By 2 years they had also discovered fire. They fished in the beach water. Our evidence is that we found sticks with a little string and with a pointy rock tied on the end of the string. We found this in a house. We know that it was a fishing stick because it had a little piece of fish on it	Proposing hypotheses Supporting interpretations with evidence
Durante este tiempo, creemos que ellos dividieron los trabajos entre ellos. Tal vez pasó que dos personas no podían hacer todo el trabajo y por eso se lo dividieron en todos los que estaban en la isla sin hacer nada (e). We believe that they divided the jobs between them	Proposing hypotheses Code-switching between tied segments of text
During the 10 years on the island, the people changed. They put mud on themselves to protect them from the animals. They sharpened rocks to make weapons and tools. By 10 years the people were planting vegetables and fruits to eat and they made holes. They covered them with old sticks and they put dried grass and leaves to cover the hole. This was a trap for the animal. We know this because we found evidence. We found a hole with old sticks and we found old grass and old leaves. We found the hole in the middle of the island	Proposing hypotheses Supporting with evidence
Durante los 10 años, las personas cambiaron y pusieron el lodo para protegerse (g)	Reiteration of the ideas of the first two sentences of the preceding paragraph
By 10 years, the people had a big village by the waterfall, but we know that something caused everyone to die some time after that. We have figured out how the people died. There were little insects that went in the fruit and when the people ate the fruit, they died, because the little insect was poisonous. People may have had stomachaches and headaches, but did not know how poisonous the little insect could be. Our evidence is that we found a fruit tree and picked a mango. We cut it in half and the little insect was in the mango. We tested the insect and we noticed that it is very, very poisonous. Everyone seems to have died. The name we put on the insect was *pilinche* (made up name). We tested the people's skeletons and so we believe	Raising hypotheses Gathering information Using data from multiple sources to construct evidence Assessing evidence Supporting interpretations with evidence
Based on our study of the island, we're now ready to tell everyone our theory of how these people could have such a good village and then disappear	Reporting data Presenting in public space

Finally, we draw upon Bloome & Egan-Robertson's (1993) argument that people propose, recognise, acknowledge and mark as socially—and we add academically—significant, texts that were interactionally accomplished. To these arguments we add Floriani's (1993) adaptation of Bloome and Egan-Robertson's (1993) argument, in framing the concept intercontextuality—ways of constructing and being with texts. Floriani argued that the actions of creating the text are part of what is learned by students and that, when taken up, these actions are themselves material resources that members use in future events.

Table 3.4 provides a representation of the text and the actions inscribed in the text, and the ways in which the boys structured the information in the text. This analysis involved asking the same type of question about the text that we did about the developing class: *What was inscribed, in what ways, using what language(s), for what purposes? And, what evidence do the two boys provide about this text and its place in the cycle of activity known as the Island History Project?* In this way, we applied our ethnographic perspective to analysis of this text (for example Dixon et al., 2005; Putney et al., 2000; Skukauskaite & Green, 2004).

Analysis of the column entitled 'interpretations of patterns' makes visible a number of iterative processes within the text, foremost of which was the alternating use of two languages. This pattern is one that was made visible (Tables 3.1 and 3.2) in the ways in which students were greeted and how they were able to choose the language in which they greeted and spoke with others in the first sub-event of the first day of school. It was also visible across all sub-events (re)presented in Table 3.2. The use of the two languages in this essay also mirrors the form of code-switching that the teacher used, a form in which students were expected to listen across languages: the teacher *did not* reiterate what was said in one language in a literal form in the other language. Thus, the pattern of language use in this essay represents this complex genre of discourse and communication.

Additionally, the form that this essay takes is one that reconstructs the processes that they used to study the island's history. This genre focuses on their actions in constructing the essay, not on the people on the island. In this way, they positioned themselves as investigators who were studying the island, searching for evidence of why the people disappeared from the island: *Based on our study of the island, we're now ready to tell everyone our theory of how these people could have such a good village and then disappear.* In the essay, they inscribed an iterative process of investigating, gathering evidence, hypothesising and constructing a theory based on evidence. If we juxtapose the practices they identify with those presented in the analysis of the chain of activity in the Watermelon Project (Fig. 3.2) and the literate and inquiry practices identified across days in the Watermelon Project (Table 3.3), we see the roots of the practices and genres used in this essay.

Although this analysis provides only a sketch map of the intertextual and intercontextual resources members constructed, it demonstrates how, as ethnographers, we move between different forms of analysis and different types of text to identify common processes and practices. It also demonstrates how a student, defined as having learning difficulties, took up the opportunity to guide a new student, one whose reading levels exceeded his own. The analysis also shows how Sergio, as an individual learner, drew upon knowledge he constructed with collective events, to support his work with his partner in accomplishing this complex task. Furthermore, in the ways in which he and his partner structured their text, they made visible the literate practices that Sergio identified as necessary for this new task. Thus, this analysis makes visible the interdependence of collective and individual learning and development for Sergio. It also foreshadows how the text that these two ethnographers wrote will contribute theoretically to the work of the collective.

Complementary perspectives as material resources: some final comments

We began this chapter by arguing that complementary perspectives, not just methods, were resources for studies over time of learning and development in classrooms as the outcome of the interdependence between individual and collective. To make visible this complex relationship, we selected Sergio, a student defined as having special learning needs, as a tracer unit. Sergio served as an anchor for constructing a *telling case*. By focusing on how Sergio and his colleagues jointly constructed the developing social world of the classroom and then took up (or not) the texts, actions and social accomplishments of social science, we created a telling case that enabled us to construct theoretical inferences about the dynamic and developing nature of individual–collective relationships.

Central to this process and approach to analysis was an ethnographic approach that had at its core a coherent set of theoretical perspectives that supported the analysis undertaken. Using an *if... then...* approach, we made visible the logic of inquiry guiding each level of analysis, guided by particular theoretical perspectives. The process made visible how different levels of analytic scale required different conceptual arguments to guide the ethnographic work that uncovered how Sergio and his colleagues drew on the chains of historical actions and text construction to participate in and accomplish subsequent tasks. For each set of analyses, we also demonstrated how we (re)presented the work of members of the class and how these (re)presentations became texts that we read, analysed and interpreted to construct a grounded argument about what was available to be learned.

Finally, by using a non-linear approach (Agar, 2006), we traced the roots and routes of particular levels of events. Through these different forms and levels of analysis, we made visible what each contributed to the grounded argument about the interdependence of theory and method, and collective and individual learning and development. Thus, through this telling case, we constructed an intertextual web (Bloome & Egan-Robertson, 1993) of theoretical inferences that grounded subsequent analyses and how we engaged in a form of hypothesis testing, what we called predictions, within and across times and events. Through this logic of inquiry, we sought to add insights into how, why and under what conditions, for what purposes, multiple theoretical lenses were necessary in exploring the complexity of everyday life in this classroom.

In describing how we identified the anchor artifact for the telling case and then identified the boundaries of this case, we demonstrated how, in a program of research that includes multiple studies from a common set of records, analyses produced for one study are part of the archived materials that can be (re)visited, (re)analysed, (re)read and (re)interpreted across studies. In the construction of this case, we also showed how these previous analyses constitute a form of *indexing* that can be used to locate potential candidates for further analysis. From this perspective, a program of research provides a basis for constructing a synthesis of both theories and outcomes across studies. Therefore, through this telling case, we seek to show how complementary perspectives within a study enhance the expressive potential of the conceptual system guiding the ethnographer's work and support testing of hypotheses (predictions or questions) within a case through purposeful (re)analysis of a common data set.

Essential next questions

How can researchers build programs of research that use complementary methods to examine the impact of decisions and actions within and across times, actors and events that support and constrain opportunities for learning and inclusive practices for teachers and students?

In this chapter, we focused on how complementary perspectives were needed to trace how students took up and used the events, texts and social actions and discourse constructed by members to guide subsequent work in classrooms. The potential that this form of complementary research holds for research on opportunities for learning of students with special learning needs will require further exploration across levels of schooling as well as across linguistically and culturally diverse students. This approach provides ways of uncovering the range of processes and practices jointly constructed by the teacher with students in innovative curriculum projects as well as across times and events in classrooms. Without research on multiple levels of analytic time scale, the researcher and/or the teacher may not be able to make visible how what an individual is afforded at one point in time becomes a material resource (Gee & Green, 1998) across times and events, or how the intertextual web of texts provides resources that students view as socially and academically significant for successful participation and learning in particular classroom events (for example Bloome & Egan-Robertson, 1993; Bloome, Carter, Christian, Otto, & Shuart-Faris, 2005).

How and where can the everyday work and accomplishments of students in classrooms that make visible differing levels of competence enter into the assessment process for students and how can these accomplishments be related to the opportunities for learning afforded them in classrooms?

In tracing how the actions by teacher, students (the class-as-a-collective, pairs of students, table groups) in the classroom provided oral, written and visual texts that supported Sergio as well as how Sergio supported others, we identified how Sergio was afforded opportunities for learning to read the world of the classroom, and how these processes and practices enabled him to participate in ways not represented on standardised assessments of his reading ability. As the analysis in the classroom showed over time, Sergio was able to read the world, to take up and use the text of the classroom, along with the norms and expectations for their production and performance, to support a new student. In his Island History essay, he drew upon and used these texts, practices and processes to create a text that involved two speakers, in order to successfully accomplish the task of communicating theory to others about the island's history (the object of theorising in social science). Thus, by tracing Sergio's work and contributions across times, actors, events and types of artifacts he created (written and oral texts), we provided evidence of the level of success and understanding he developed about the work of social science. The levels of understanding identified and how he used these to support a student with reading abilities beyond his own could not be assessed by standardised, discrete point tests that do not consider the intertextual relationships that shaped and were shaped by Sergio's performance. The ability to trace performances across times and events is central to documentation of complex processes that lead to transformations of understandings in classrooms (for related arguments see Mercer & Hodgkinson, 2008; Ligorrio & Pontecorvo, 2005; Bloome et al., 2005 and Walford, 2008).

How might new theoretical and technological resources be used by both teacher and students to help students, teachers, administrators and policy makers see the developing competencies?

The issues raised for policy makers are ones that authors in many of the chapters have raised, particularly ones raised by Ray Brown about how learning difficulty is a socially constructed category. Like Brown, we argue for more complex assessments that are based on a (re)formulation of what counts as support for students and how such supports require changes not only in beliefs about ability, but also about the capacity building nature of schooling. The multi-faceted and multi-theoretical approach presented in our chapter demonstrates the need to examine the impact of policy actions on what is possible in classrooms and to document the actions of teachers and students over time (McNeil & Coppola, 2006).

The different levels of analysis presented in this chapter were possible given the archived records from the classroom across the year. Today, e-portfolios are being developed that make possible such analyses by teachers and students alike that will provide evidence of development of understandings within and across subject areas over time. If policy is to build on practice (McNeil & Coppola, 2006), not merely mandated changes, in ways that enhance the capacity of the teacher, students and system itself, then new ways of documenting and making visible the relationships between the opportunities for learning and for student performance will need to be developed and a means of articulating them to different audiences will also need to be developed. In this chapter, we proposed potential ways that such documentation can be developed through ethnographic research across times, actors and participants.

The next steps need to explore how those directly involved can use such theories and approaches to document their own work in classrooms. Although not reviewed here, research by Sergio's teacher, a co-author on this chapter (Yeager, 2003) demonstrates how teacher as researcher and researcher as teacher are positions that can inform each other, creating a potential for reflexive actions (see also Yeager, 2006; Yeager & Green, 2008). The step that follows from these arguments for teacher development is one that is represented in work in a book by Carolyn Frank (1999), a member of our research community, entitled *Through Ethnographic Eyes: A Teacher's Guide to Classroom Observation.* Frank captures teachers' use of this approach, and the book has been used in education classes as well as anthropology classes. It builds new ways of exploring classrooms as cultures for learning and what students learn in their communities.

The questions that we propose for this chapter are but a beginning of a dialogue with readers of this volume. The arguments in this chapter complement those raised by other authors and by the editors. We look forward to continuing this dialogue and to examining how complementary perspectives and the methods generated will enhance the expressive potential of the field, not just of a particular research approach. The arguments by Wyatt-Smith and Elkins move us forward in innovative ways, ways that have helped us (re)consider what counts as learning disabilities (in the United States) and students with special learning needs (in Australia). We look forward to the next steps that this volume generates.

References

Agar, M. (1994). *Language shock: Understanding the culture of conversation.* New York: William Morrow and Company, Inc.

Agar, M. (2006). An ethnography by any other name. *Forum: Qualitative Social Research, 7*(4). Retrieved from http://www.qualitative-research.net/index.php/fqs/article/view/177

American Educational Research Association (AERA). (2006). Standards for reporting on empirical social science research in AERA publications. *Educational Researcher, 35*(6), 33–40.

Anderson-Levitt, K. (2006). Ethnography. In J. Green, G. Camilli, & P. Elmore (Eds.), *Handbook of complementary methods in education research* (pp. 279–296). Washington, DC/Mahwah, NJ: AERA and Lawrence Erlbaum Assoc.

Baker, W. D., Green, J., & Skukauskaite, A. (2008). Video-enabled ethnographic research: A micro ethnographic perspective. In G. Walford, (Ed.), *How to do educational ethnography* (pp. 76–114). London: Tufnell Press.
Bakhtin, M. M. (1986). The problem of speech genres. In C. Emerson & M. Holmquist (Eds.), *Speech genres and other late essays* (pp. 60–102). Austin, TX: University of Texas Press.
Bakhtin, M. M. (2004). Dialogic origin and dialogic pedagogy of grammar: Stylistics as part of Russian language instruction in secondary school. *Journal of Russian and East European Psychology, 42*, 12–24.
Barnes, D. (1976). *From communication to curriculum*. London: Penguin Books.
Barnes, D., Britton, J., & Rosen, H. (1969). *Language, the learner, and the school* (1st ed). London: Penguin Books Ltd.
Barnes, D., & Todd, F. (1995). *Communication and learning revisited: Making meaning through talk*. London: Boynton/Cook.
Barr, R. (1987). Classroom interaction and curricular content. In D. Bloome, , *Literacy and schooling*. Norwood, HJ: Ablex.
Barr, R., & Dreeben, R. (1983). *How schools work*. Chicago: University of Chicago Press.
Bauman, Z. (2000). *Liquid modernity*. London: Polity Press.
Beach, R., Green, J., Kamil, M. & Shanahan, T. (Eds.) (1992). *Multidisciplinary perspectives on literacy research*. Urbana, IL: National Conference on Research in English/National Council of Teachers of English.
Beach, R., Green, J., Kamil, M. & Shanahan, T. (Eds.) (2005). *Multidisciplinary perspectives on literacy research* (2nd ed). Cresskill, NJ: Hampton Press.
Berger, P., & Luckmann, L. (1967). *The social construction of reality: A treatise in the sociology of knowledge*. Landover Hills, MD: Anchor.
Bernstein, B. (1973). *Class, codes and control, Vol. 1*. London: Routledge/Kegan Paul.
Bernstein, B. (1990). *Class, codes and control*, Vol. 4: The structuring of pedagogic discourse. London: Routledge.
Birdwhistell, R. (1977). Some discussion of ethnography, theory and method. In J. Brockman (Ed.), *About Bateson: Essays on Gregorian Bateson* (pp. 103–144). New York: E.P. Dutton.
Bloome, D., & Bailey, F. (1992). Studying language and literacy through events, particularity and intertextuality. In J. L. Green R. Beach, M. Kamil, & T. Shanahan (Eds.), *Multidisciplinary perspectives on literacy research* (pp. 181–209). Urbana, IL: NCTE.
Bloome, D., Carter, S. P., Christian, B. M., Otto, S., & Shuart-Faris, N. (2005). *Discourse analysis & the study of classroom language & literacy events: A microethnographic perspective*. Mahwah, NJ: Lawrence Erlbaum Associates, Publishers.
Bloome, D., & Clarke, C. (2006). Discourse-in-use. In J. Green, G. Camilli, & P. Elmore (Eds.), *Handbook of complementary methods in education research* (pp. 227–242). Washington, DC/Mahwah, NJ: AERA & Lawrence Erlbaum.
Bloome, D., & Egan-Robertson, A. (1993). The social construction of intertextuality in classroom reading and writing lessons. *Reading Research Quarterly, 28*(4), 305–333.
Bloome, D., & Theodorou, E. (1988). Analyzing teacher-student and student-student discourse. In J. Green & J. Harker (Eds.), *Multiple perspectives analyses of classroom discourse* (pp. 217–248). Norwood, NJ: Ablex.
Bullock, A. (1975). *Bullock report: A language for life*. London: Her Majesty's Stationery Office. Retrived from http://www.dg.dial.pipex.com/documents/docs1/bullock.shtml
Cameron, D. (1995). *Verbal hygiene*. London: Routledge.
Castanheira, M. L. (2000). *Situating learning within collective possibilities: Examining the discursive construction of opportunities for learning in the classroom*. Unpublished dissertation, University of California at Santa Barbara.
Castanheira, M. L. (2004). *Aprendizagem contextualizada: discurso e inclusão na sala de aula*. Belo Horizonte: Autêntica.

Castanheira, M., Crawford, T., Dixon, C., & Green, J. (2001). Interactional ethnography: An approach to studying the social construction of literate practices. In J. Cumming & C.Wyatt-Smith(Eds.), *Linguistics and Education: Analyzing the Discourse Demands of the Curriculum, 11*(4), 353–400 (Special issue).
Castanheira, M. L., Green, J., Dixon, C., & Yeager, B. (2007). (Re)Formulating identities in the face of fluid modernity: An interactional ethnographic approach. *International Journal of Education Research, 46*(3–4), 172–189.
Castanheira, M. L., Green, J. L., & Yeager, E. (2009). Investigating inclusive practices: An interactional ethnographic approach. In K. Kumpulainen & M. César (Eds.), *Investigating classrooms – Methodologies in action* (pp. 145–178). Rotterdam: Sense Publishers.
Cazden, C.(1974). Panel 5, *Report on research on teaching in a cultural context*. Washington, DC: National Institute of Education.
Cazden, C. (1988). *Classroom discourse: The language of teaching and learning*. Portsmouth, NH: Heinemann.
Cazden, C., John, V., & Hymes, D. (Eds.) (1972/1985). *Functions of language in the classroom* (1st & 2nd eds.). Long Grove, IL: Waveland Press.
César, M. & Kumpulainen, K. (Eds.) (2009). *Social interactions in multicultural settings*. Rotterdam: Sense Publishing.
Chandler, S. (1992). Learning for what purpose? Questions when viewing classroom learning from a socio-cultural curriculum perspective. In H. H. Marshall (Ed.), *Redefining student learning: Roots of educational change* (pp. 33–58). Norwood, NJ: Ablex.
Cochran-Smith, M. (1984). *The making of a reader*. Norwood, NJ: Ablex.
Cole, M., John-Steiner, V., Scribner, S. & Souberman, E. (Eds.) (1978). *L.S. Vygotsky: Mind in Society: The Development of Higher Psychological Processes*. Cambridge, MA & London, England: Harvard University Press.
Collins, E., & Green, J. (1992). Learning in classroom settings: Making or breaking a culture. In H. Marshall (Ed.), *Redefining learning: Roots of educational restructuring* (pp. 59–86). Norwood, NJ: Ablex.
Cook-Gumperz, J. (Ed.) (1986). *The social construction of literacy*. New York: Cambridge University Press.
Corsaro, W. (1981). Entering the child's world: Research strategies for field entry and data collection in a preschool setting. In J. L. Green & C. Wallat (Eds.), *Ethnography and llanguage in eseducational settings* (pp. 117–146). Norwood, NJ: Ablex Publishing Co.
Corsaro, W. (1984). *Friendship and peer culture of the young child*. Norwood, NJ: Ablex.
Cumming, J. & Wyatt-Smith, C. (Eds.). (2001a). *Literacy and the curriculum: Success in senior secondary schooling*. Melbourne: ACER.
Cumming, J., & Wyatt-Smith, C. (Eds.). (2001b). *Analyzing the Discourse Demands of the Curriculum, Linguistics & Education, 11*(4) (Special Issue).
Cushman, P. (1991). Ideology obscured: Political uses of the self in Daniel *Stern's infant. American Psychologist, 46*, 206–219.
Dixon, C., Green, J., & Brandts, L. (2005). Studying the discursive construction of texts in classrooms through Interactional Ethnography. In R. Beach, J. Green, M. Kamil, & T. Shanahan (Eds.), *Multidisciplinary perspectives on literacy research* (pp. 349–390). Cresskill, NJ: Hampton Press/National Conference for Research in Language and Literacy.
Dixon, C., Green, J., Yeager, B., Baker, D., & Franquiz, M. (2000). 'I used to know that': What happens when reform gets through the classroom door. *Bilingual Education Research Journal (ERJ), 24*(1&2), 113–126.
Economic and Social Research Council (ESRC). (2005). *Research ethics framework*. Retrieved May 12, 2007, from www.esrcsocietytoday.ac.uk/ESRCInfoCentre/opportunities/research_ethics_framework/?data=%2fFrXH
Edwards, A., & Furlong, V. (1978). *The language of teaching*. London: Heinemann.
Edwards, D., & Mercer, N. (1987). *Common knowledge*. New York: Methuen.
Erickson, F. (1986). Qualitative research. In M. Wittrock (Ed.), *Third handbook for research on teaching* (pp. 119–161). New York: MacMillan.

Erickson, F., & Schultz, J. (1981). When is a context? Some issues and methods in the analysis of social competence. In J. L. Green & C. Wallat (Eds.), *Ethnography and language in educational settings* (pp. 147–150). Norwood, NJ: Ablex.

Evertson, C., & Green, J. L. (1986). Observation as inquiry and method. In M. Wittrock (Ed.), *Handbook of research on reading* (3rd ed., pp. 160–213). New York: MacMillan.

Fernie, D., Kantor, R., & Klein, E. (Eds.) (1988). Becoming students. *Theory Into Practice*, *27*(1).

Floriani, A. (1993). Negotiating what counts: Roles and relationships, texts and contexts, content and meaning. *Linguistics and Education*, *5*(3/4), 241–274.

Ford, M.J. & Forman, E.A. (2006). Redefining learning in classroom contexts. In J. Green & A. Luke (Eds.), *Review of Research in Education* 30 (pp. 1–32). Washington DC: American Educational Research Association.

Frank, C. (1999). *Ethnographic eyes: A teacher's guide to classroom observation*. Portsmouth, New Hampshire: Heinemann.

Freire, P., & Macedo, D. (1987). *Literacy: Reading the world and the word*. Westport, CT: Bergin & Garvey.

Gaskins, S., Miller, P. J., & Corsaro, W. A. (1993). Theoretical and methodological perspectives in the theoretical study of children. In W. A. Corsaro & P. J. Miller (Eds.), *Interpretive approaches to children's socialization* (pp. 5–23). San Francisco: Jossey Bass.

Gee, J. P., & Green, J. (1998). Discourse analysis, learning and social practice: A methodological study. *Review of Research in Education.*, *23*, 119–169.

Geertz, C. (1973). *The interpretation of cultures*. New York: Basic Books.

Gergen, K. J. (1985). The social constructionist movement in modern psychology. *American Psychologist*, *40*, pp. 266–275.

Giddens, A. (1989). *Sociology*. London: Polity Press.

Gilmore, P. & Glatthorn, A. A. (Eds.) (1982). *Children in and out of school: Ethnography and education*. Washington DC: Center for Applied Linguistics.

Green, J. L. (1983). Research on teaching as a linguistic process: A state of the art. In E. W. Gordon (Ed.), *Review of Research in Education* (Vol. 10, pp. 151–254). Washington, DC: American Educational Research Association.

Green, J., & Dixon, C. (1993). Talking knowledge into being: Discursive and social practices in classrooms. *Linguistics and Education*, *5*(3 & 4), 231–239.

Green, J., & Dixon, C. (1999). Discourse analysis. In B. Guzetti (Ed.), *Literacy in America: An encyclopedia*. Denver, CO: ABC-CLIO.

Green, J., & Dixon, C. (2007). Classroom interaction and situated learning. In M. Martin-Jones & A. M. de Mejía (Eds.), *Encyclopedia of language and education* (Vol. 3), Discourse and Education. New York: Springer.

Green, J., Dixon, C., & Zaharlick, A. (2003). Ethnography as a logic of inquiry. In J. Flood, S. B. Heath, & D. Lapp (Eds.), *The handbook for research in the English language arts*. Hillsdale, NJ: Erlbaum & Associates.

Green, J., Franquíz, M., & Dixon, C. (1997). The myth of the objective transcript. *TESOL Quarterly*, *31*, 172–176.

Green, J. L., & Harker, J. O. (1988). *Multiple perspectives analyses of classroom discourse*. Norwood, NJ: Ablex.

Green, J., Heras, A. I., Yeager, B., Castanheira, M.L., & Dixon, C. (in press). Identities in shifting educational policy contexts: The consequences of moving from two languages, one community to English only. In G. Lopez Bonilla & K. Englander (Eds.), *Discourses and identities in contexts of educational change*. New York: Peter Lang.

Green, J. & Luke, A. (Eds.). (2006). Rethinking learning: What counts as learning and what learning counts. *Review of Research in Education*, *30*(1), xi–xiv. Washington, DC: AERA.

Green, J., & Meyer, L. (1991). The embeddedness of reading in classroom life: Reading as a situated process. In C. Baker & A. Luke(Eds.), *Towards a critical sociology of reading pedagogy*. Philadelphia, PA: John Benjamins Publishing.

Green, J. L., & Wallat, C. (1979). What is an instructional context? An exploratory analysis of conversational shifts across time. In O. Garnica & M. L. King, *Language, children & society* (pp. 159–188). Elmsford, NY: Pergamon Press.

Green, J. L., & Wallat, C. (1981). Mapping instructional conversations: A sociolinguistic ethnography. In J. L. Green & C. Wallat (Eds.), *Ethnography and language in educational settings* (pp. 161–205). Norwood, NJ: Ablex.

Green, J. L., Weade, R., & Graham, K. (1988). Lesson construction and student participation: A sociolinguistic analysis. In J. L. Green & J. Harker (Eds.), *Multiple perspective analyses of classroom discourse* (pp. 11–48). Norwood, NJ: Ablex.

Greeno, J. (2006). Theoretical and practical advances through research on learning. In J. Green, G. Camilli, & P. Elmore (Eds.), *Handbook of complementary methods in education research* (pp. 795–822). Washington, DC/Mahwah, NJ: AERA/LEA.

Grimshaw, A., Burke, P., & Cicourel, A. (Eds.). (1994). *What's going on here? Complementary studies of professional talk.* Norwood, NJ: Ablex.

Gumperz, J. (1986). Interactive sociolinguistics on the study of schooling. In J. Cook-Gumperz (Ed.), *The social construction of literacy* (pp. 45–68). New York: Cambridge University Press.

Gumperz, J. (1992). Contextualization and understanding. In A. Duranti & C. Goodwin (Eds.), *Rethinking context: Language as an, interactive phenomenon* (pp. 229–252). Cambridge: Cambridge University Press.

Guzzetti, B., & Hynd, C. (1998). *Theoretical perspectives on conceptual change.* Hillsdale, NJ: Erlbaum.

Hammersley, M., & Atkinson, P. (1995). *Ethnography: Principles in practice* (2nd ed.). London: Routledge.

Heap, J. (1980). What counts as reading? Limits to certainty in assessment. *Curriculum Inquiry, 10*(3), 265–292.

Heap, J. (1985). Discourse in the production of classroom knowledge: Reading lessons. *Curriculum Inquiry, 1*(5), 245–279.

Heap, J. (1991). A situated perspective on what counts as reading. In C. Baker & A.Luke, *Towards a critical sociology of reading pedagogy* (pp. 103–139). Philadelphia, PA: John Benjamins Publishing.

Heath, S. B. (1982). Ethnography in education: Defining the essentials. In P. Gillmore & A. A. Glatthorn (Eds.), *Children in and out of school: Ethnography and education* (pp. 33–35). Washington, DC: Center for Applied Linguistics.

Heath, S. B., & Street, B. (2008). *Ethnography: Approaches to language and literacy research.* New York: Teachers College & National Council on Research in Language and Literacy.

Heras, A. (1993). The construction of understanding in a sixth-grade bilingual classroom. *Linguistics and Education, 5*(3 & 4), 275–300.

Holland, D., & Cole, M. (1995). Between discourse and schema: Reformulating a cultural-historical approach to culture and mind. *Anthropology and Education Quarterly, 26*(4), 478–489.

Holland, D., Lachicotte, W., Skinner, D., & Cain, C. (2001). *Identity and agency in cultural worlds.* Cambridge, MA: Harvard.

Hornberger, N. (Ed.) (2008). *Encyclopedia of language & education* (2nd ed). New York: Springer.

Hudson, B. & Schneuwly, B.(Eds.). (2007). Special Issue of the *European Educational Research Journal* on Didactics–Learning and Teaching in Europe, No. 2. Retrieved 8/30/09 http://www.wwwords.co.uk/EERJ/content/pdfs/6/issue6_2.asp

Hymes, D. (1972). Introduction: Functions of language in the classroom. In A. Cazden, V. John, & D. Hymes (Eds.), *Functions of language in the classroom.* Prospect Heights, IL: Waveland Press, Inc.

Hymes, D. (1982). What is ethnography? In P. Gilmore & A. A. Glathorn (Eds.), *Children in and out of school: Ethnography and education.* Washington, DC: Center for Applied Linguistics.

Ivanic, R. (1998). *Writing and identity: The discoursal construction of identity in academic writing.* Amsterdam/Philadelphia: John Benjamins Publishing.

Jewitt, C. (2006). Multimodeality and literacy in school classrooms. *Review of Research in Education, 32*, 241–267.

Kantor, R., & Fernie, D. (2003). *Early childhood classroom processes.* Cresskill, NJ: Hampton Press.

Kelly, G. J., & Chen, C. (1999). The sound of music: Constructing science as sociocultural practices through oral and written discourse. *Journal of Research in Science Teaching, 36*, 883–915.

Kelly, G., Crawford, T., & Green, J. (2001). Common task and uncommon knowledge: Dissenting voices in the discursive construction of physics across small laboratory groups. *Linguistics and Education*. Special Issue on Language and Cognition, *12*(2), pp. 135–252.

Kelly, G., & Green, J. (1998). The social nature of knowing: Toward a socioculturalp perspective on conceptual change and knowledge construction. In B. Guzzetti & C. Hynd (Eds.), *Theoretical perspectives on conceptual change* (pp. 145–181). Mahwah, NJ: Lawrence Erlbaum Associates.

Kelly, G. J., Luke, A., & Green, J. (Eds.). (2008). What counts as knowledge in educational settings: Disciplinary knowledge, assessment, and curriculum. *Review of Research in Education, 32*(Special themed volume).

Kieran, C., Sfard, A., & Forman, E. (2003). *Learning discourse: Discursive approaches to research in mathematics education*. New York: Springer.

Koschmann, T. (Ed.). (1999). Meaning making: *Discourse Processes, 27*(2), A special issue.

Kumpulainen, K., Hmelo-Silver, C. E., & César, M. (Eds.). (2008). *Investigating classroom interaction: Methodologies in action*. Rotterdam: Sense Publishers.

Lakoff, G., & Johnson, M. (1980). *Metaphors we live by*. Chicago: University of Chicago Press.

Larson, J. (1995). Talk matters: The role of pivot in the distribution of literacy knowledge among novice writers. *Linguistics and Education, 7*(4), 277–302.

Lee, C. D., & Smagorinsky, P. (Eds.). (2000). *Vygotskian perspectives on literacy research: Constructing meaning through collaborative inquiry*. New York: Cambridge University Press.

Lemke, J. (1990). *Talking science: Language, learning and values*. Norwood, NJ: Ablex.

Lemke, J., Kelly, G. J., & Roth, W.-M. (2006). Forum: Toward a phenomenology of interviews. *Cultural Studies of Science Education, 1*, 83–106.

Ligorio, M.B., & Pontecorvo, C. (2005). Why and how we learn. *European Journal of School Psychology*, 3(1), 257–264.

Lima, E. (1995). Culture revisited: Vygotsky's ideas in Brazil. *Anthropology & Education Quarterly, 26*(4), 443–457.

Lin, L. (1993). Language of and in the classroom: Constructing the patterns of social life. *Linguistics and Education, 5*(3 & 4), 367–409.

McDermott, R. (1976). *Kids make sense: An ethnographic account of the interactional management of success and failure in one first grade classroom*. Unpublished dissertation. Stanford University.

McNeil, L. M., & Coppola, E. M. (2006). Official and unofficial stories: Getting at the impact of policy on educational practice. In J. Green, G. Camilli, & P. Elmore (Eds.), *Handbook of cmer, complementary methods in education research* (pp. 681–700). Washington, DC/Mahwah, NJ: AERA/Lawrence Erlbaum.

Mercer, N. & Hodgkinson, S. (Eds.). (2008). *Exploring talk in school*. London: Sage.

Mercer, N., & Littleton, K. (2007). *Dialogue and the development of children's thinking: A sociocultural approach*. Abingdon: Routledge.

Mills, H. (1993). Becoming a mathematician: Building a situated definition of mathematics. *Linguistics and Education, 5*(3 & 4), 301–334.

Mills, S. (1997). *Discourse: A new critical idiom*. London: Routledge.

Mitchell, C. J. (1984). Typicality and the case study. In R. F. Ellen (Ed.), *Ethnographic research: A guide to general conduct* (pp. 238–241). New York: Academic Press.

Moll, L. C. (Ed.). (1990). *Vygotsky and education*. Cambridge: Cambridge University Press.

Morine-Dershimer, G. (1988b). Comparing systems: How do we know? In J. Green & J. Harker (Eds.), *Multiple perspective analyses of classroom discourse* (pp. 195–214). Norwood, NJ: Ablex.

Psathas, G. (1995). *Conversation aanalysis: The study of talk in interaction* (Vol. 35). London: Sage.

Putney, L., Green, J., Dixon, C., Durán, R., & Yeager, B. (2000). Consequential progressions: Exploring collective-individual development in a bilingual classroom. In C. D. Lee & P. Smagorinsky (Eds.), *Vygotskian perspectives on literacy research: Constructing meaning through collaborative inquiry* (pp. 86–126). New York: Cambridge University Press.
Putney, L. G., & Wink, J. (1998). Breaking rules: Constructing avenues of access in multilingual classrooms. *TESOL Journal*, *7*(3), 29–34.
Rex, L., Steadman, S. & Graciano, M. (2006). Researching the complexity of classroom interaction. In J. Green, G. Camilli, & P. Elmore (Eds.) *Handbook of Complementary Methods for Research in Education* (pp. 727–772). Washington DC: American Educational Research Association.
Rex, L., & Green, J. (2007). Classroom discourse and interaction: Reading across the traditions. In B. Spolsky & F. Hult (Eds.), *Handbook of eleducational linguistics* (pp. 571–584). Oxford: Blackwell.
Rex, L., & McEachen, D. (1999). 'If anything is odd, inappropriate, confusing, or boring, 'it's probably important': The emergence of inclusive academic literacy through English classroom discussion practices. *Research in the Teaching of English*, *34*(1), 65–129.
Rex, L. A. (2001). The remaking of a high school reader. *Reading Research Quarterly*, *36*(3), 288–314.
Rieber, R., & Carton, A. (1988). *The collected works of L. S. Vygotsky: Volume 1: Problems of general psychology, including the volume thinking and speech (Cognition and language: A series in psycholinguistics)*. New York: Springer.
Rogoff, B. (2003). *The cultural nature of human development*. Oxford: Oxford University Press.
Roth, W.-M. (2005). *Talking science: Language and learning in science classrooms*. Lanham, MD: Rowman and Littlefield.
Santa Barbara Classroom Discourse Group. (1992a). Constructing literacy in classrooms: Literate action as social accomplishment. In H. Marshall (Ed.), *Redefining student learning: Roots of educational restructuring* (pp. 119–150). Norwood, NJ: Ablex.
Santa Barbara Classroom Discourse Group. (1992b). Do you see what we see? The referential and intertextual nature of classroom life. *Journal of Classroom Interaction*, *27*(2), 29–36.
Sefton-Green, J. (2006). Youth, technology, and media cultures. *Review of Research in Education*, *30*, 279–306.
Sheridan, D., Street, B., & Bloome, D. (2000). *Writing ourselves: Mass-observation and literacy practices*. Language & Social Processes. Cresskill, NJ: Hampton Press.
Skukauskaite, A., & Green, J. L. (2004). A conversation with Bakhtin: On inquiry and dialogic thinking. *Journal of Russian and East European Psychology*, *42*(6), 59–75.
Spindler, G. (1982). *Doing the ethnography of schooling*. New York: Holt, Rinehart & Winston.
Spolsky, B., & Hult, F. (Eds.). (2007). *Handbook of educational linguistics*. Oxford: Blackwell.
Street, B., Baker, D., & Tomlin, A. (2005). *Navigating numeracies: Home/school numeracy practices* (Vol. 4), Liverhulme Numeracy Research Programme. Dordrecht: Kleuwen.
Strike, K. (1974). On the expressive potential of behaviorist language. *American Educational Research Journal*, *11*(2), 103–120.
Tuyay, S., Jennings, L., & Dixon, C. (1995). Classroom discourse and opportunities to learn: An ethnographic study of knowledge construction in a bilingual third grade classroom. *Discourse Processes*, *19*(1), 75–110.
Weade, R. (1987). Curriculum 'n'instruction: The construction of meaning. *Theory Into Practice*, *26*(1), 15–25.
Wertsch, J. (1991). *Voices of the mind: A sociocultural approach to mediated action*. Cambridge, MA: Harvard University Press.
Wilkinson, L. (1982). *Communicating in the classroom*. New York: Academic Press.
Wittrock, M. C. (Ed.) (1986). *Handbook of research on teaching* (3rd ed). New York: Macmillan.
Yeager, E. (2003). *' I am a historian': Examining the discursive construction of locally situated academic identities in linguistically diverse settings*. Dissertation. University of California, Santa Barbara.

Yeager, B. (2006, Summer). Teacher as researcher/researcher as teacher: Multiple angles of vision for studying learning in the context of teaching. *Language Arts Journal of Michigan*, *22*(1), 26–33.

Yeager, B. & Green, J. (2008). 'We have our own language as well as the languages we bring': Constructing opportunities for learning through a language *of* the classroom. In J. Scott, D. Straker, & L. Katz (Eds), *Affirming Students' Right To Their Own Language: Bridging Language Policies and Pedagogical Practices*, pp. 153–175. NY: Routledge with National Council of Teachers of English.

Yeager, B., Floriani, A., & Green, J. (1998). Learning to see learning in the classroom: Developing an ethnographic perspective. In D. Bloome & A. Egan-Robertson (Eds.), *Students as inquirers of language and culture in their classrooms* (pp. 115–139). Cresskill, NJ: Hampton Press.

Yeager, E., Green, J. L., & Castanheira, M. L. (2009). Two languages, one community: On the discursive construction of community in bilingual classrooms. In M. César & K. Kumpulainen (Eds.), *Social interactions in multicultural settings* (pp. 235–268). Rotterdam: Sense Publishers.

Chapter 4
The New Literacies of Online Reading Comprehension: New Opportunities and Challenges for Students with Learning Difficulties

Jill Castek, Lisa Zawilinski, J. Greg McVerry, W. Ian O'Byrne, and Donald J. Leu

Improving the ability to read and comprehend, especially for students who struggle with reading, is one of today's most pressing educational priorities (Fuchs & Fuchs, 1994; Snow, Burns, & Griffin, 1998). Despite extensive efforts, many students continue to struggle with reading, increasing the possibility that they will drop out of school (Hammond, Linton, Smink, & Drew, 2007). Unless we develop more effective instructional approaches to address this problem, we will fail to develop the potential these students possess to improve their individual lives, as well as our collective lives. It is essential that all students become fully prepared for a wide range of reading experiences so that each and every individual can reach their potential, fulfil their goals, and make our world a better place through their accomplishments.

Increasing reading achievement is now doubly challenging. As the Internet becomes a central aspect of daily life (Dede, 2007), any approach seeking to improve students' reading ability must recognise that the nature of reading comprehension has expanded in the 21st century (International Reading Association, 2009). In order to participate in the global information age, students must now be able to read and comprehend information on the Internet at high levels (International Reading Association, 2009; Organization for Economic Co-operation and Development, n.d.).

The largest and most recent review of research on reading comprehension concluded that reading on the Internet requires additional comprehension skills beyond those required to read traditional print texts (RAND Reading Study Group, 2002). Online reading is not isomorphic with offline reading (International Reading Association, 2002, 2009; Leu et al., 2007a). Information and communication technologies (ICTs), such as the Internet, require new literacies to fully exploit their potential (Coiro, Knobel, Lankshear, & Leu, 2008b). An emerging body of

J. Castek (✉)
University of California, Berkeley, CA, USA
e-mail: jcastek@berkeley.edu

C. Wyatt-Smith et al. (eds.), *Multiple Perspectives on Difficulties in Learning Literacy and Numeracy*, DOI 10.1007/978-1-4020-8864-3_4,

work (Bråten, Strømsø, & Britt, 2008; Coiro, Knobel, Lankshear, & Leu, 2008a; Goldman, Wiley, & Graesser, 2005; Graesser et al., 2007; Kuiper & Volman, 2008; Kulikowich, 2008; Lawless & Schrader, 2008; Leu et al., 2005; Mayer, 2008; Rouet, 2006) is beginning to define these new literacies, increasing our understanding of the skills required to read, comprehend and learn online.

Since online reading comprehension appears to differ in important ways from offline reading comprehension, a central issue has become whether adolescent readers can benefit from instruction with online reading comprehension skills before they are fully proficient with offline reading comprehension. The general bias, from research with struggling readers in offline contexts, suggests that instruction should be carefully sequenced (National Reading Panel, 2000; Snow et al., 1998). Such approaches would logically presume that successful offline reading skills be developed among students who struggle in reading before embarking with instruction in online reading comprehension. Though approaching literacy instruction in this way may be common in many school classrooms, we argue that students of all ability levels benefit from experiences with online reading comprehension, especially struggling readers. Providing online reading opportunities for struggling readers has important implications on two levels. First, it extends valuable experiences through which online reading comprehension can develop. Second, developing these skills builds students' capacity to learn how to learn. The Internet is now a central source of information, and learning is dependent on the ability to read and comprehend complex information at high levels (Alexander & Jetton, 2002; Bransford, Brown, & Cocking, 2000).

This chapter explores the opportunities and challenges posed by the new literacies of online reading comprehension among students with learning difficulties. It will begin with a review of a model of online reading comprehension advanced by Castek et al. (2008), Leu, Kinzer, Coiro, and Cammack (2004), and Leu et al. (2008).

Then, in the sections that follow, we present four case studies of struggling readers from a recent study of one-to-one laptop use in a Grade 7 science classroom (Leu et al., 2005). The students we profile in this chapter were selected from a population of 89 students in Grade 7, including 42 males and 47 females. They attended a Grade 6–8 middle school, located in a rural New England town in the northeastern United States, which has a total enrolment of 416 students.

These brief case studies demonstrate both the nature of online reading comprehension and the opportunities and challenges for struggling readers that are inherent in this new context for reading. Each case study describes unique online reading comprehension strategies these students used that may have led to higher-than-expected online reading comprehension performance. In the final section we speculate as to why a pattern of low offline reading and high online reading performance may have occurred among this population. Implications for online reading instruction for students who struggle with reading are also addressed.

Theoretical perspective

Many literacy scholars are beginning to look past the technological aspects of the Internet to analyse the underlying social practices it serves (Lankshear & Knobel, 2006; Leander, 2007; Street, 1998, 2003). This work has helped the research community view the Internet not as a technology, but rather as a powerful context for literacy. The Internet is no more a technology than is a book; its functional affordances define it more than its technological affordances.

Recognition of the use of the Internet as a literacy issue has prompted individuals from many disciplines to begin a collaborative approach to theory building (cf. Coiro et al., 2008b). This approach is coming to be referred to as New literacies theory (International Reading Association, 2002, 2009; Leu et al., 2004; Leu, O'Byrne, Zawilinski, McVerry, & Everett-Cocapardo, 2009). It takes an open-source approach to theory development, at the highest level, inviting everyone who studies the Internet's impact on our literacy lives to contribute towards theory development and to benefit from others' contributions.

New literacies: a dual-level theory

The term 'new literacies' means many different things to many different people. To some, new literacies are seen as new social practices (Street, 1998, 2003). Others see new literacies as new strategies and dispositions essential for online reading comprehension, learning and communication (Castek, 2008; Coiro, 2003; Henry, 2006; Leu et al., 2004). Still others see new literacies as new discourses (Gee, 2003) or new semiotic contexts (Kress, 2003; Lemke, 2002). Still others see literacy as differentiating into multiliteracies (The New London Group, 1996) or multimodal contexts (Hull & Schultz, 2002). Some see a construct that juxtaposes several of these orientations (Lankshear & Knobel, 2006). When one includes these different definitions of new literacies with terms such as ICT literacy (International ICT Literacy Panel, 2002) or informational literacy (Hirsh, 1999; Kuiper & Volman, 2008; Webber & Johnson, 2000), the construct of new literacies becomes even broader. In this breadth, however, there is an opportunity to benefit from the richness of these different perspectives, as the research community develops theory to direct our collective understanding of Internet usage in school settings.

New literacies theory (Coiro et al., 2008b; Leu et al., 2004; Leu et al., 2009) operates on two levels: upper case (New Literacies) and lower case (new literacies). *New literacies*, as the broader, more inclusive concept, benefits from work taking place in the multiple, lower-case dimensions of *new literacies*. This is seen as an advantage, not a limitation. It enables the larger theory of New literacies to keep up with the richness and continuous change that will always define the Internet. Lower-case theories explore either a specific area of new literacies, such as the social communicative transactions occurring with text messaging (for example Lam, 2006), or a focused disciplinary base, such as the semiotics of multimodality in online media (for example Kress, 2003). Each body of work contributes to the larger, continually changing theory of New literacies.

What defines this larger theory of New literacies? A recent review (Coiro et al., 2008b) concludes that most lower-case new literacies perspectives share four common elements that define the larger theory of New literacies:

1. New literacies include the new skills, strategies, dispositions and social practices that are required by new technologies for information and communication.
2. New literacies are central to full participation in a global community.
3. New literacies regularly change as their defining technologies change.
4. New literacies are multi-faceted and our understanding of them benefits from multiple points of view.

Research efforts focused towards understanding the new literacies of online reading comprehension (Leu et al., 2004; Leu et al., 2007b) demonstrate how lower-case new literacies theory have already begun to enrich upper-case New literacies theory. The new literacies of online reading comprehension is a theoretical frame that views online reading as a process of problem-based inquiry involving the new skills, strategies, dispositions and social practices of the Internet.

The new literacies of online reading comprehension

Online reading comprehension differs from traditional print comprehension in that new skills and strategies are required by a process of self-directed text construction and problem solving (Coiro & Dobler, 2007; Leu et al., 2007b). At least five processing practices are required when reading on the Internet: (a) reading to construct useful questions; (b) reading to locate information; (c) reading to evaluate information critically; (d) reading to synthesise information and (e) reading and writing to communicate information. Within these five areas reside the skills, strategies and dispositions that are distinctive to online reading comprehension, as well as others that are important for offline reading comprehension.

Reading to construct useful questions. Consider first the initial phase of online reading comprehension—we read on the Internet to solve problems and answer questions. How a problem is framed or how a question is understood is a central aspect of online reading comprehension. Recent work by Taboada and Guthrie (2006) within traditional texts suggests that reading initiated by a question differs in important ways from reading that does not. A central component of online reading involves using the Internet to prompt and refine useful questions.

Reading to locate online information. Another critical component of successful online reading is the ability to locate information that meets one's needs (Broch, 2000; Guinee, Eagleton, & Hall, 2003; Eagleton, Guinee, & Langlais, 2003; Educational Testing Service, 2002; Sutherland-Smith, 2002). New online reading skills and strategies appear to be required; for example to generate effective keyword search strategies (Bilal, 2000; Eagleton & Guinee, 2002; Kuiper & Volman, 2008); to read and infer which websites listed may be most useful within a set of search engine results (Henry, 2006) and to efficiently scan for relevant information within websites (McDonald & Stevenson, 1998; Rouet, 2006). The reading ability required to search for and locate information on the Internet may very well serve as a gatekeeper skill, since readers who are unable to locate information online are unable to solve their problem. In short, if one cannot locate information online, one cannot read.

Reading to critically evaluate online information. A third component of online reading comprehension is the ability to critically evaluate information on the Internet (Burbules & Callister, 2000; Leu et al., 2004). During online reading comprehension, one must evaluate the level of accuracy, reliability and bias of information (Burbules & Callister, 2000; Fitzgerald, 1997; Illinois Mathematics and Science Academy, 2006; Kirk, 1996; Kurland, 1996). Critical evaluation on the Internet presents challenges quite different from traditional print and media sources, since the content of online information may be even more diverse (Tillman, 2003) and commercially and ideologically biased (Fabos, 2008) than that of most print sources we typically encounter.

Reading to synthesise online information. Successful online reading comprehension also requires the ability to read and synthesise information from multiple online sources (Leu et al., 2004; Jenkins, 2006). Synthesis, or the integration of separate and unique ideas (Irwin, 1990), is thought to be the most challenging of offline comprehension strategies (Keene & Zimmerman, 1997). Employing this strategy effectively requires the reader to bring an awareness of the reading process together with an understanding of the text (Dole, Duffy, Roehler, & Pearson, 1991). The Internet introduces additional challenges to be coordinated because vast amounts of information must be synthesised. These texts are often presented in multiple media formats, from a nearly unlimited and disparate set of sources (Gilster, 2000; Jenkins, 2006; Rouet, 2006). Bulger (2006) shows how the ability to manage, process and filter multiple electronic documents is a highly challenging component of online literacy.

Reading to communicate information using new technologies. A final component of successful online reading comprehension is the ability to communicate on the Internet while reading (Britt & Gabrys, 2001). Online reading and writing are so closely connected that it is not possible to separate them; we read online as authors and we write online as readers (Huffaker, 2004, 2005; McVerry, 2007; Zawilinski, 2009). Online communication involves the use of texting, blogs, wikis, video, shared writing spaces such as Google docs and social networks such as Nings. Emerging research suggests that the interactive processes of reading and communicating have become so intertwined on the Internet that they often happen simultaneously (Leu et al., 2004). Thus, the communication processes involved in using a range of online tools to ask and answer questions on the Internet appear to be inextricably linked to aspects of online reading comprehension (Boyd & Ellison, 2008; Forte & Bruckman, 2006; Lewis & Fabos, 2005).

Why do struggling readers sometimes perform well during online reading?

Students who are challenged with offline reading typically are neither fluent when reading offline nor when reading online. Some of these students, however, read surprisingly well online and can teach us important lessons. For example, they make good decisions at crucial points in the online reading comprehension process and access useful digital features. We have found several common patterns to their online reading comprehension that intersect with the supportive nature of online information.

These students are often good at locating information

Though struggling readers may find it difficult to locate information in offline texts, struggling readers who are good online readers typically know how to use search engines and do so effectively on the Internet. They actually read search engine results rather than simply clicking and looking their way down a list of search results, a pattern common with less-effective online readers (Leu et al., 2007b). Though their reading of search engine results may be slow and deliberate, these students ultimately make informed choices about what to skim and what to read more carefully.

The Internet requires the reading of shorter text units, a benefit to struggling readers

Weaker readers often shut down when confronted with long units of text offline, such as a chapter in a science or social studies text. During online reading comprehension, units of text are much shorter as readers follow informational links from one location to another to seek information that will help them solve their informational problem. Online readers may read a search engine entry to decide where to go, read a screen to decide which link to follow, or skim a page for information. Reading shorter units of text leads to more sustained reading by struggling readers.

The Internet permits readers to construct their own texts, a benefit to struggling readers

Reading online requires a complex set of choices, as one follows a series of different links and moves from one webpage to another, seeking information related to the problem that initiated the process. Online readers, in essence, construct the texts that they read by the links they choose to follow. Struggling readers are often more engaged during online reading, largely because they control the informational paths they follow online.

The Internet provides supportive multimedia features for struggling readers

Online texts contain multimedia supports that benefit struggling readers in important ways. Especially useful are multimedia features such as images, animations, video and audio. Struggling readers may be especially effective at using these supports strategically to scaffold their own meaning construction during online reading.

Struggling offline readers often develop their online reading skills at home

Students profiled in the case studies that follow came to school with a host of online reading skills and strategies acquired outside of school. Their experience with online texts provided an important foundation for online reading comprehension skills and strategies introduced in school. Interviews with students who struggled with offline texts but did well with online text indicated that they used the Internet at home, typically for important individual purposes. One example was of a student who liked to 'pretend shop' for items online that she could not afford now, but hoped to be able to later in life. This suggests that using students' outside-of-school Internet knowledge to extend academic learning in school can be particularly useful.

Electronic organisational tools often provide important supports for struggling readers

Reading across texts often requires students to develop new organisational and management skills. Struggling readers, in particular, appear to benefit from online organisational tools that the Internet provides. Organising and bookmarking favourite websites, for example, make it possible for readers to quickly find the resources they need and encourages the ability to review them as needed by toggling between multiple windows.

Case studies of struggling readers

The sections that follow introduce four challenged offline readers who were successful at completing an online reading comprehension assessment activity that focused on science content. The description of each reader's skills and strategies was analysed by playing back a 30-minute, screen capture video that documented their reading processes as they completed an online reading comprehension assessment (ORCA). The ORCA was made up of three online comprehension tasks and utilised a blog for posting information requests and collecting student responses. The online reading comprehension tasks evaluated students' abilities to locate, critically evaluate, synthesise and communicate information on the Internet. Videos of students' reading comprehension performances were scored using a specially designed rubric with good psychometric properties (see Leu et al., 2005). ORCA scores for the 89 students in our sample ranged from 0 to 30 out of 32 possible points, with a sample mean of 12.84 points.

We also describe these students' offline reading achievement using scores from state reading tests. The Connecticut Mastery Test (CMT-Total Reading) includes four strands of reading comprehension: (1) forming an initial understanding, (2) developing an interpretation, (3) demonstrating a critical stance and (4) the Degrees of Reading Power (DRP). These comprehension subtests are reported as one composite score. The raw scores range from a low of 100 to a high of 400. Scores ranging from 100 to 207 indicate 'below basic' reading proficiency; scores ranging from 208 to 221 indicate 'basic' reading proficiency; scores ranging from 222 to 238 indicate 'proficient' reading; scores ranging from 239 to 294 indicate 'meeting grade level reading goals' and scores ranging from 295 to 400 indicate 'advanced' reading proficiency. The sample mean for the CMT-Total Reading was 263.

We selected to profile four students who had learning difficulties. Some had identified learning disabilities and print-processing challenges such as difficulty in writing, reasoning, recalling or organising information. Others were struggling readers whose academic achievement was limited to below grade-level reading ability.

Most school systems in the United States use what is called a discrepancy formula to determine whether an individual is learning disabled and eligible for Special Educational services. In order to qualify, there must be a discrepancy between the students' potential (IQ) and academic performance. Many struggling readers may have learning difficulties, but are not considered learning disabled.

Students with learning difficulties are often described in terms of what they cannot do (O'Brien, 2006). In contrast, we have structured the descriptions that follow around these students' strengths. Carefully analysing their performance through this lens draws attention to their skills, capabilities and competencies as online readers, transforming perceptions of their overall academic abilities.

Michael

Michael had a documented learning disability in reading and writing. Though his learning disability made typical Grade 7 reading materials challenging, Michael was fully mainstreamed into regular classes. Michael's teachers accommodated his learning disability by providing structured study guides. They also modified assignments and provided additional time to complete tests and projects. Although Michael read below grade level, he fully engaged in instructional activities and participated actively in class discussions.

Offline reading. Michael's below-level reading skills were a documented component of his learning disability. His score on the CMT-Total Reading assessment sustained this evaluation. Michael's total raw score of 216 fell within the 'basic' range on this assessment. His lack of confidence and skill in offline reading caused him to rush through typical reading assignments with little self-monitoring for comprehension.

Online reading. You may view the screen-capture video of Michael's online reading episode and the rubric used to evaluate his reading performance at <www.newliteracies.uconn.edu/casestudies.html>. The video shows how Michael was highly engaged in his online reading, perhaps due to his active construction of the texts that he read through the links he followed, some of which led to multimedia enhancements. The online reading comprehension evaluation showed that Michael scored 17.5 points out of 32 on the ORCA, placing him in the top half of the sample. Michael used a number of efficient reading strategies during the assessment. His familiarity with the Internet outside of school is likely to have led to his successful performance on this activity.

Managing multiple windows. Michael used multiple windows to manage the many sources of information that he encountered: one for the task document, one for the browser and later an additional window for blog comments. His moving back and forth between the task document and other windows, a self-monitoring strategy, enabled him to focus on and remember the task directions. These strategies were apparent across each of the three ORCA tasks.

Searching for information. The video also shows Michael's advanced skills with locating information online. When searching, Michael began by locating a familiar search engine and easily determined which keywords to use. Michael effectively used copy-and-paste strategies to select the appropriate search terms from the task and move them into the search box. He also demonstrated effective use of Boolean operators such as 'and' and '+' to create phrases while searching with a search engine. At other times, he chose to type terms into the appropriate search box. He also understood and used a specific feature of a search engine, the 'did you mean' link, which recognised misspelled words in his search terms and offered a reasonable alternative.

Adjusting reading rate. Michael also adjusted his reading rate during online reading. As indicated by the cursor he used to track the text he was reading, Michael read search-result descriptions slowly and carefully to decide which links to follow and which search results to explore. At a new webpage, he quickly scanned for keywords or an image required by the task. When he determined that a page had merit, he read more closely, comparing the text on the page to the task requirements. The choices he made after carefully reading descriptions typically brought him to the correct site.

Self-monitoring while communicating. After locating the necessary sites, Michael composed a response that showed an awareness of audience and purpose. Michael was skilled in knowing what information to include in the Name, E-mail and Comment fields on the blog where responses were collected. He continued to traverse windows, re-reading the task and copy-and-pasting webpage titles as he composed. He also reread his comments during the composing process, correcting mistakes as he typed. His revisions not only corrected misspellings, but also added detail as he reworded text for clarity. He began one entry with, 'I think that this site is the best becase [sic] this gives so much more detail than the,' but stopped, and reworded the text to say, 'I think that site A is the best becase [sic] this gives so much more detail than site B.' This revision and others he consciously made suggest Michael's awareness of his audience's informational needs.

Checking the accuracy of information on a website. Another of Michael's strengths was his understanding of how to check the accuracy of information on a website. When asked about the reliability of the site he found, Michael wrote, 'To check that the information on the site is good information you can compare this information to a nuther [sic] site or book.'

Leslie

Leslie had a documented learning disability. Due to her difficulties with reading and writing, she was given modified assignments and received support from a paraprofessional in her science classroom.

Offline reading. Leslie's below-level reading skills were evident in her classroom performance and documented on results of state reading tests. Leslie's total reading comprehension score of 221 on the CMT-Total Reading was in the 'basic' range. Though reading was challenging for her, Leslie worked diligently to complete assignments given to her, even when they were seemingly beyond her ability level.

Online reading. The online reading comprehension evaluation showed that Leslie scored 18.5 out of 32 points on the ORCA, placing her in the top half of the sample. You may view the screen-capture video of Leslie's online reading episode and the rubric used to evaluate her reading performance at <www.newliteracies.uconn.edu/casestudies.html>.

Although considered a struggling reader with offline materials, Leslie demonstrated several strengths when completing the ORCA assessment. These strengths were noted in the area of searching for information and included: (1) flexibility with the use of three different search engines, (2) the implementation of key-word search strategies and (3) determining when to continue sifting through a set of search results and when to initiate a new search using different search terms. Leslie reported using the Internet at home to explore personal interests as well as to complete homework assignments.

Flexibility with the use of three different search engines. At different points during the assessment task, Leslie used three different search engines. To complete part A of the first task, she opted to use the *Ask.com* search engine and appropriately framed her query as a question: 'What are some good human anatomy online websites?' The structure of this question indicated that she was cognisant of target language from the task (the phrase 'human anatomy' online) that would yield fruitful search results. Despite a focused search and a question-framed query, her initial attempt to locate the correct information did not yield a productive list of results. Leslie then opted to use a key-word strategy on her second try. To maximise efficiency, Leslie returned to the search box and deleted all but the phrase 'human anatomy online' from her first query. This brought the desired website to the top of the search results, which Leslie immediately recognised as she exclaimed, 'I found something!'

During the second task, Leslie persevered for an extended period as she tried different combinations of key words to locate the target site. In contrast to her first set of searches, she used the meta-search engine Dogpile on subsequent searches. After using different search phrases, such as 'man having an x-ray on his arm', 'x-ray' and 'educators and staff' without locating the target site, she opted to try a third search engine, Google. Leslie strategically combined search terms from her previous attempts, entering 'educators, staff, human body systems' into the search box. When the search results were displayed, she quickly recognised that the bolded words 'educators, staff and human body systems' matched those in her search string with those in the task request. Leslie selected this website, glanced over an image of a man having an x-ray of his arm, and left the site without further examination. Although it matched the description she was given, Leslie didn't recognise that this was the site she was seeking. This oversight does not point to a lack of skill in searching, but rather indicates an inability to determine whether the site requested and the site found matched the description given.

Implementation of key-word search strategies. Although Leslie was unable to locate any of the sites requested on her first try, she made multiple attempts and generated alternative search terms. She verbalised why she avoided sites highlighted at the top of her Google search results, saying, 'because they're commercial, have ads, and are meant to sell you things'. These responses indicated that Leslie could distinguish which sites were more reliable and less commercially biased than others.

Determining when to continue sifting through a set of search results and when to initiate a new search using different search terms. Leslie's search processes appeared to be iterative. Over the course of subsequent searches, she carried over what she learned from one search attempt to inform the next. This strategy may indicate that she was learning more about searching and how to make adjustments to subsequent searches from the act of searching itself. Although Leslie was unable to locate the man having an x-ray of his arm on the same page where the words 'educators and staff' appeared, she made several attempts that may have taught her useful search strategies, such as using key words in combination and linked with '+', monitoring what sites had been previously visited and switching between searching the web and searching images to locate specific kinds of graphics.

During Leslie's completion of the final task, she critically evaluated the website she located. Before posting her response to the task, she examined the site and thought about what made it reliable, saying 'Yes, knowing who created this site will help you know if the information is accurate because if it is from say a 4th grader, then you know this information probably isn't the most accurate because 4th graders haven't learned much.' This response indicated an attempt to evaluate the background of the author as a way to determine if the information on the site was reliable.

Since Leslie did not locate the information required in both parts of the first request, she was unable to determine which of these sites would best meet the needs of the Grade 7 class requesting the information. As a result, researchers had limited information upon which to evaluate her ability to synthesise across multiple online resources.

Jessica

Jessica had a documented learning disability and was assigned an instructional aid to support her academic achievement. This paraprofessional worked with Jessica individually to complete the modified assignments she was given by her regular education teachers.

Offline reading. Because Jessica's reading level was far below grade level, her progress was tracked using individualised achievement measures that were administered yearly in conjunction with her individualised education plan (IEP), and she was exempted from completing standardised achievement tests. There were no CMT-Total Reading scores available to document Jessica's performance as an offline reader. One can assume, however, that her reading was substantially below average.

Online reading. Accommodations listed on Jessica's individual educational plan (IEP) were also employed during the online assessment. The researcher read the directions, supported Jessica by reading aloud sections of text from specific websites and also redirected her by referring her back to directions provided in the task. The online reading comprehension evaluation showed that Jessica scored 9.5 points out of 32 on the ORCA, placing her near the mean of the sample. You may view the video of Jessica's online reading episode and the rubric used to evaluate her reading performance at <www.newliteracies.uconn.edu/casestudies.html>.

Jessica was able to independently accomplish several sections of the assessment. For example, she used a search engine to locate websites and posted responses containing useful information to the blog interface. However, when prompted to critically evaluate the information she found, these sections were left unanswered. Due to the need to standardise the amount of time given to complete the assessment, she may have been limited in what she could accomplish.

Key-word searching. Jessica conducted key-word searches and located information on many different web pages. She routinely sifted through search engine results and selected sites to examine more closely without researcher support. She determined when the information on the site she found was not useful for her purposes. Employing this strategy improved her efficiency in sifting through information and afforded her more time to complete a more careful preview of the sites that she found useful.

When synthesis skills were needed to evaluate which was a better source of information for a Grade 7 class, Jessica was not as skilled. She also lacked skill in critical evaluation and did not provide an analysis beyond the initial judgment of whether the information was useful or not.

Sifting through information. Jessica showed a great deal of persistence in locating and sifting through online information. Present in her workspace were several toolbars, links, pop-ups and updates needed for various programs. She patiently navigated through these impediments to complete the series of tasks.

Use of browser tools. Jessica took advantage of the tools offered by Internet browsers. Frequently, the key words she used as search terms were misspelled. She used the 'did you mean' hints given at the top of the search results page to correct the spelling of the intended terms.

Persistence with searching. Jessica exhibited persistence when looking for an annotated site with information on the respiratory system. When initiating a search, she attempted to employ the 'did you mean' strategy when she misspelled 'respiratory'. She used this strategy three times, each with different variants of spellings for respiratory, all with no success. Recognising that using different combinations of key words can be used to locate specific sites, she revised search terms and began searching for 'good three body systems'. Jessica quickly skimmed search results, determined they were not useful for the audience and purpose, and added 'kids' to her search terms. Within these results she chose a site that contained general information on human body systems meant for kids. From there, she found the correct spelling of the word 'respiratory' and went back to successfully finish the task.

Communicating using the blog interface. Jessica exhibited the skills needed to log into the blog and post responses. Each response was posted underneath the relevant information requested. Although this interface was a somewhat novel context, she exhibited a strong desire to share her ideas with others and persevered through the particulars of the interface to do so.

An impediment to Jessica's success in completing the assessment was her lack of experience in navigating between windows of the browser. Numerous times during the assessment she would use the back button repeatedly to return to the task. It is not known if Jessica used this strategy to keep track of the multitude of information she encountered, or whether this strategy is further evidence of the admirable persistence shown while reading online.

Larry

Larry was identified as an at-risk student prior to enrolment in school for his Grade 7 year. Larry was seldom engaged in instructional activities. However, he thrived when he was involved in Internet-based group activities in his science class. His science teacher remarked at his increased class participation during Internet-based activities and commented that during the period of time when the internet was used, his grades had improved.

Offline reading. Larry's score of 232 on the CMT-Total Reading fell within the proficient range, but was well below the class average. His reading comprehension scores on the Degrees of Reading Power (DRP) sub-test of the CMT - Total Reading assesment, prior to participation in Internet intensive science instruction (36), was significantly lower than the post DRP score he earned after the intervention (48).

Online reading. Although Larry was known to be a struggling reader when it came to offline reading comprehension, his score on the ORCA assessment was 25.5 out of a total of 32 points, placing Larry in the top third of his class in online reading comprehension proficiency. You may view the video on Larry's online reading episode and the rubric used to evaluate his reading performance at <www.newliteracies.uconn.edu/casestudies.html>. Larry completed the ORCA and achieved a high overall score because he was efficient, highly skilled and managed time spent on each of the three tasks well. The video shows how he skilfully located what was requested, evaluated his choices and communicated his ideas on the blog interface. He also carefully monitored his work on the activity. Across all three tasks, he referred back to the instructions and made sure he not only found all the required websites, but also that he communicated all the requested information succinctly.

Use of efficient search strategies. Larry employed efficient search strategies and used a variety of searching techniques that made it possible for him to locate all the relevant information needed to complete each of the tasks. In addition, he demonstrated several strategies for critically evaluating information. Finally, he was able to negotiate the particulars of posting to a blog so that his responses could be communicated online to others.

Careful reading of the task supported the selection of search terms. Larry exhibited the skills a strong online reader would need to locate information. He began each of the three tasks by carefully reading the directions, as demonstrated by his use of the cursor to follow the words as he read. When Larry encountered words that could be used as possible search terms, he paused. This may indicate he was reflecting on the individual parts of the task in order to identify appropriate key words to use.

Use of advanced search features. Larry was knowledgeable about using Boolean searching techniques such as linking key words. In each of the tasks he used a '+' between relevant search terms. For example, on the first search task his key words were 'anatomy+cardiovascular and educators', on the second he used 'animated graphic+respiratory' and on the third, he used 'digestion+7th grade'.

Scanning for information. Once he arrived at a search-results page, Larry skilfully scanned the descriptions under each website title. He did not rely on a simple 'click and look' strategy. Larry scanned through the results, looking for website descriptions that contained the bold words he used in his search. When he did not find the first page of search results helpful, Larry moved onto to the second or third page before ultimately selecting a website to read. This is a rare occurrence with most students who tend to examine only the first page of search results.

Skimming the text. Once Larry chose a website he was skilled at locating information within that page. The first thing Larry did, as evident by cursor movements, was skim the entire website, quickly scrolling up and down. If the website did not have any relevant information Larry would quickly hit the back button and go to the search results page. If the website looked like it might contain the information he was looking for, Larry would spend time reading the page. Finally, if Larry thought the webpage might contain useful links to other webpages containing the information he was looking for, Larry would explore them.

Using organisational tools. Larry skilfully took advantage of organisational affordances provided within the online environment. Often, struggling readers cannot manage multiple offline texts. They may lose directions, notes or even books. In contrast, Larry was able to quickly organise and move between multiple windows in the online environment. While completing the ORCA, Larry saved relevant links he found in the favourites folder. On the first task, when he had to compare two websites, Larry paged back between websites and even re-entered search terms in Google to find websites he had already visited. Then, halfway through, he clicked on the favourites star in Internet Explorer. This opened up a bookmark panel on the right, where Larry added the two websites he found. He then minimised all his windows and noted where the download directions and the open browser pages were in his task bar. Organising the online workspace in this way allowed Larry to quickly navigate between his sources and refer back to the task to make sure he completed each task in its entirety.

Reconsidering online reading instruction for struggling readers

Common wisdom may suggest that struggling readers must become proficient with offline reading before initiating online reading comprehension instruction. These case studies, however, suggest just the opposite—many struggling readers appear to benefit in important ways from online reading experiences and instruction in the new literacies of online reading comprehension. Surprisingly, some of our weaker offline readers may be among our better online readers. Schools, teachers and parents need to recognise this possibility. Far too often, students skilled with the use of online information go undetected in schools that continue to focus on the more traditional text experiences of offline reading (Chandler-Olcott & Mahar, 2003).

These cases demonstrate that shorter units of text, characteristic of online reading, may encourage more active attempts at reading comprehending, and thus avoiding tendencies to shut down when confronted with lengthy, continuous text segments such as those typically found in a chapter or text book. The opportunity to make choices and construct their own texts through the links they follow, may also engage struggling readers more actively in their own reading and meaning construction experiences. These cases also show how challenged offline readers, who read effectively online, have important search skills in place that were often developed during online reading experiences outside of school. Moreover, these students are effective at communicating information with online tools. These students appear to be supported by the graphic and other multimedia supports available online as well as the organisational tools found there. Finally, in many of these cases we find surprisingly high levels of comprehension monitoring taking place, including the awareness of audience needs during communication. It may be that the more supportive context of online information frees up attention to focus on comprehension monitoring. Or, it may be that continuous feedback, provided during interactions with the Internet, prompts monitoring of the many decisions online readers must make. In any case, thinking more deeply about information as a result of these interactions may be an especially positive aspect inherent when struggling readers engage in online reading comprehension.

Essential next questions

It is increasingly important that we find ways to move struggling readers away from the periphery of our classrooms and engage them in all facets of literacy learning, especially online reading comprehension instruction. However, little is known about how to structure classroom activities that empower these students specifically. Which instructional approaches might best help students with learning difficulties to develop the new literacies of online reading comprehension? How can we organise our classrooms so that these students become active participants and contributors to online reading comprehension instruction? These are important questions to address as we seek to integrate the Internet more fully into literacy instruction.

If today's learners are to become successful at literacy and life in a global information economy, a concerted effort must be put forth to transform classroom practice and fully integrate online reading comprehension within the curriculum. Most of the reading students are expected to do in classrooms today is designed to help them acquire offline reading competency. How can we support teachers in integrating online reading comprehension skills and strategies required in the 21st century?

In today's world, all students must become efficient information managers and reflective thinkers who can collaborate and communicate effectively in new and complex online contexts that are constantly changing. Will providing more Internet activities help students with learning difficulties improve their academic achievement? How can we ensure this occurs? These and other questions are essential to address so that all students can become fully literate and achieve their lifelong goals.

New literacies provide new opportunities for struggling readers. While students with learning difficulties have cognitive needs that require attention, they benefit from being engaged in complex and rich literacy instruction (Alvermann, Hinchman, Moore, Phelps, & Waff, 2006). As educators, we need to take advantage of these learning opportunities by integrating the Internet more fully into our literacy and content learning curricula, and including all students in online reading comprehension instruction.

References

Alexander, P. A., & Jetton, T. L. (2002). Learning from text: A multidimensional and developmental perspective. In M. L. Kamil, P. Mosenthal, P. D. Pearson, & R. Barr (Eds.), *Handbook of reading research* (Vol. 3, pp. 285–310). Mahwah, NJ: Erlbaum.

Alvermann, D. E., Hinchman, K. A., Moore, D. W., Phelps, S. F., & Waff, D. R. (Eds.) (2006). Reconceptualizing the literacies in adolescents' lives. Mahwah, NJ: Erlbaum.

Bilal, D. (2000). Children's use of the Yahooligans! web search engine: I. Cognitive, physical, and affective behaviors on fact-based search tasks. *Journal of the American Society for Information Science, 51*, 646–665.

Boyd, D., & Ellison, N. B. (2008). Social network sites: Definition, history, and scholarship. *Journal of Computer-Mediated Communication, 13*, 210–230.

Bransford, J., Brown, A., & Cocking, R. (2000). How people learn: Brain, mind, experience, and school: Expanded edition. Washington, D.C.: National Academy Press.

Bråten, I., Strømsø, H. I., & Britt, M. A. (2008). Trust matters: Examining the role of source evaluation in students' construction of meaning within and across multiple texts. *Reading Research Quarterly, 44*, 6–28.

Britt, M. A., & Gabrys, G. L. (2001). Teaching advanced literacy skills for the world wide web. In C. R. Wolfe (Ed.), *Learning and teaching on the world wide web* (pp. 74–91). San Diego, CA: Academic Press.

Broch, E. (2000). Children's search engines from an information search process perspective. *School Library Media Research*. Retrieved January 20, 2007, from www.ala.org/ala/aasl/aaslpubsandjournals/slmrb/slmrcontents/volume32000/childrens.htm#cognition

Bulger, M. (2006). *Beyond search: A preliminary skill set for online literacy*. Retrieved March 19, 2007, from http://transliteracies.english.ucsb.edu/post/research-project/research-clearinghouse-individual/research-papers/beyond-search-a-preliminary-skill-set-for-online-literacy

Burbules, N. C., & Callister, T. A., Jr. (2000). *Watch IT: The risks and promises of information technologies for education*. Boulder, CO: Westview Press.

Castek, J. (2008). *How do 4th and 5th grade students acquire the new literacies of online reading comprehension? Exploring the contexts that facilitate learning*. Unpublished doctoral dissertation: University of Connecticut.

Castek, J., Leu, D. J., Jr., Coiro, J., Gort, M., Henry, L. A., & Lima, C. (2008). Developing new literacies among multilingual learners in the elementary grades. In L. Parker (Ed.), *Technology-Based Learning Environments for Young English Learners: In and out of school connections* (pp. 111–153). Mahwah, NJ: Lawrence Erlbaum Associates.

Chandler-Olcott, K., & Mahar, D. (2003). 'Tech-savviness' meets multiliteracies: Exploring adolescent girls' technology-mediated literacy practices. *Reading Research Quarterly, 38*, 356–385.

Coiro, J. (2003). Reading Comprehension on the Internet: Expanding our understanding of reading comprehension to encompass new literacies. Reading Online. [Article reprinted from *The Reading Teacher, 56*, 458–464]. [OnlineSerial]. Retrieved on March 15, 2009, from: www.readingonline.org/electronic/elec_index.asp?HREF=/electronic/rt/2-03_Column/index.html

Coiro, J., & Dobler, E. (2007). Exploring the online reading comprehension strategies used by sixth-grade skilled readers to search for and locate information on the internet. *Reading Research Quarterly, 42*, 214–257.

Coiro, J., Knobel, M., Lankshear, C., & Leu, D. J. (Eds.). (2008a). *Handbook of research in new literacies*. Mahwah, NJ: Erlbaum.

Coiro, J., Knobel, M., Lankshear, C., & Leu, D. J. (2008b). Central issues in new literacies and new literacies research. In J. Coiro, M. Knobel, C. Lankshear, & D. J. Leu (Eds.), *The handbook of research in new literacies* (pp. 1–22). Mahwah, NJ: Erlbaum.

Dede, C. (2007). Reinventing the role of information and communication technologies in education. In L. Smolin, K. Lawless, & N. C. Burbules (Eds.), *Information and communication technologies: Considerations of current practices for teachers and teacher educators: 106th Yearbook of the National Society for the Study of Education, Part 2* (pp. 11–38). Malden, MA: Blackwell.

Dole, J. A., Duffy, G. G., Roehler, L. E., & Pearson, P. D. (1991). Moving from the old to the new: Research on reading comprehension instruction. *Review of Educational Research, 61*, 239–264.

Eagleton, M. B., & Guinee, K. (2002). Strategies for supporting student internet inquiry. *New England Reading Association Journal, 38*, 39–47.

Eagleton, M. B., Guinee, K., & Langlais, K. (2003). Teaching internet literacy strategies: The hero inquiry project. *Voices from the Middle, 10*, 28–35.

Educational Testing Service. (2002). *Digital transformation. A framework for ICT literacy: A report of the International ICT Literacy Panel*. Ewing, NJ: Author.

Fabos, B. (2008). The price of information: Critical literacy, education, and today's internet. In J. Coiro, M. Knobel, D. Leu, & C. Lankshear (Eds.), *Handbook of research on new literacies* (pp. 839–870). Mahwah, NJ: Erlbaum.

Fitzgerald, M. A. (1997). Evaluating information: An information literacy challenge. *School Library Media Research*. Retrieved February 15, 2007, from www.ala.org/ala/aasl/aaslpubsandjournals/slmrb/slmrcontents/volume21999/vol2fitzgerald.cfm

Forte, A., & Bruckman, A. (2006). *From Wikipedia to the classroom: Exploring online publication and learning*. Proceedings of the 7th international conference on Learning sciences, pp. 182–188.

Fuchs, D., & Fuchs, L. S. (1994). Inclusive school movement and the radicalization of special education reform. *Exceptional Children, 60*(4), 294–305.

Gee, J. (2003). *What video games have to teach us about learning and literacy*. New York: Palgrave.

Gilster, P. (2000). Digital literacy. In R. Pea (Ed.), *The Jossey-Bass reader on technology and learning*. San Francisco: Jossey-Bass, Inc.

Goldman, S. R., Wiley, J., & Graesser, A. C. (2005, April). *Learning in a knowledge society: Constructing meaning from multiple information sources*. Paper presented at the annual meeting of the American Educational Research Association, Montreal, Canada.

Graesser, A. C., Wiley, J., Goldman, S. R., O'Reilly, T., Jeon, M., & McDaniel, B. (2007). SEEK Web tutor: Fostering a critical stance while exploring the causes of volcanic eruption. *Metacognition and Learning, 2*, 89–105.

Guinee, K., Eagleton, M. B., & Hall, T. E. (2003). Adolescents' internet search strategies: Drawing upon familiar cognitive paradigms when accessing electronic information sources. *Journal of Educational Computing Research, 29*, 363–374.

Hammond, C., Linton, D., Smink, J., & Drew, S. (2007). *Dropout risk factors and exemplary programs*. Clemson, SC: National Dropout Prevention Center, Communities In Schools, Inc.

Henry, L. A. (2006). SEARCHing for an answer: The critical role of new literacies while reading on the internet. *The Reading Teacher, 59*, 614–627.

Hirsh, S. G. (1999). Children's relevance criteria and information seeking on electronic resources. *Journal of the American Society for Information Science, 50*, 1265–1283.

Huffaker, D. (2004). Spinning Yarns around a Digital Fire: Storytelling and dialogue among youth on the Internet. *First Monday*. Retrieved from http://firstmonday.org/htbin/cgiwrap/bin/ojs/index.php/fm/article/view/1110/1030

Huffaker, D. (2005). The educated blogger: Using weblogs to promote literacy in the classroom. *AACE Journal, 13*(2), 91–98.

Hull, G., & Schultz, K. (2002). *School's Out: Bridging out-of-school literacies with classroom practice*. New York: Teachers College Press.

Illinois Mathematics and Science Academy. (2006). 21st Century information fluency project. Retrieved October 5, 2006, from http://21cif.imsa.edu/

International ICT Literacy Panel. (2002). *Digital transformation: A framework for ICT literacy*. Educational Testing Service. Retrieved June 9, 2009, from www.ets.org/Media/Tests/Information_and_Communication_Technology_Literacy/ictreport.pdf

International Reading Association. (2002). *Integrating literacy and technology in the curriculum: A position statement*. Newark, DE: International Reading Association.

International Reading Association. (2009). *IRA Position Statement on New Literacies: A position Statement*. Newark, DE: International Reading Association.

Irwin, J. (1990). *Teaching reading comprehension processes* (2nd ed). Boston, MA: Pearson.

Jenkins, H. (2006). *Confronting the challenges of participatory culture: Media education for the 21st century*. Chicago, IL: The John D. & Catherine T. Macarthur Foundation.

Keene, E., & Zimmerman, S. (1997). *Mosaic of thought*. Portsmouth, NH: Heinemann.

Kirk, E. (1996). *Evaluating information found on the internet*. Retrieved July 14, 2001, from Purdue University, Online Writing Lab. Retrieved April 10, 2006, from http://milton.mse.jhu.edu:8001/research/education/net.html

Kress, G. (2003). *Literacy in the new media age*. London: Routledge.

Kuiper, E., & Volman, M. (2008). The web as a source of information for students in K-12 education. In J. Coiro, M. Knobel, C. Lankshear, & D. Leu (Eds.), *Handbook of research on new literacies*. (pp. 241–266). Mahwah, NJ: Erlbaum.

Kulikowich, J. (2008). Experimental and quasi-experimental approaches to the study of new literacies. In J. Coiro, M. Knobel, C. Lankshear, & D. Leu (Eds.), *Handbook of research on new literacies* (pp. 179–208). Mahwah, NJ: Erlbaum.

Kurland, D. J. (1996). *The net, the web, and you*. Belmont, CA: Wadsworth Publishing.

Lam, W. (2006). Culture and learning in the context of globalization: Research directions. *Review of Research in Education, 30*, 213–237.

Lankshear, C., & Knobel, M. (2006). *New literacies* (2nd ed.). Maidenhead: Open University Press.

Lawless, K. A., & Schrader, P. G. (2008). Where do we go now? Understanding research on navigation in complex environments. In J. Coiro, M. Knobel, C. Lankshear, & D. Leu (Eds.), *Handbook of research on new literacies* (pp. 267–296). Mahwah, NJ: Erlbaum.

Leander, K. (2007). 'You won't be needing your laptops today': Wired bodies in the wireless classroom. In M. Knobel & C. Lanskhear (Eds.), *A new literacies sampler*. New York: Peter Lang Publishing.

Lemke, J. (2002). Travels in hypermodality. *Visual Communication, 1*, 299–325.

Leu, D. J., Castek, J., Hartman, D., Coiro, J., Henry, L., Kulikowich, J., et al.(2005). *Evaluating the development of scientific knowledge and new forms of reading comprehension during online learning*. Final report presented to the North Central Regional Educational Laboratory/Learning Point Associates. Retrieved May 15, 2006 from www.newliteracies.uconn.edu/ncrel.html

Leu, D.J., Coiro, J., Castek, J., Hartman, D., Henry, L. A., Reinking, D. (2008). Research on instruction and assessment in the new literacies of online reading comprehension. In K. Collins Block and S. Parris (Eds). *Comprehension instruction: Research-based best practices*. (321-346). New York, NY: Guilford.

Leu, D. J., Jr., Kinzer, C. K., Coiro, J., & Cammack, D. (2004). Toward a theory of new literacies emerging from the internet and other information and communication technologies. In R. B. Ruddell & N. Unrau (Eds.), *Theoretical Models and Processes of Reading* (5th ed., pp. 1568–1611). Newark, DE: International Reading Association. Retrieved October 15, 2005 from www.readingonline.org/newliteracies/lit_index.asp?HREF=/newliteracies/leu

Leu, D. J., O'Byrne, W. I., Zawilinski, L., McVerry, J. G., & Everett-Cocapardo, H. (2009). Expanding the new literacies conversation. *Educational Researcher, 38*, 264–269.

Leu, D., Reinking, D., Carter, A., Castek, J., Coiro, J., Henry, L.A., et al. (2007a) Defining online reading comprehension: Using think aloud verbal protocols to refine a preliminary model of Internet reading comprehension processes. In D. Alvermann (Chair) *21st Century Literacy: What is it, How do students get it, and how do we know if they have it? A Symposium presented at American Education Research Association*. Chicago, IL. Retrieved from: http://docs.google.com/Doc?id=dcbjhrtq_10djqrhz

Leu, D. J., Zawilinski, L., Castek, J., Banerjee, M., Housand, B., Liu, Y., & O'Neil, M. (2007b). What is new about the new literacies of online reading comprehension? In J. Eakle, L. Rush, & A. Berger (Eds.) *Secondary school literacy: What research reveals for classroom practice*. (37-68). Urbana, IL: National Council of Teachers of English (NCTE). Website: http://www.newliteracies.uconn.edu/reading.html

Lewis, C., & Fabos, B. (2005). Instant messaging, literacies, and social identities. *Reading Research Quarterly, 40*, 470–501.

Mayer, R. E. (2008). Multimedia literacy. In J. Coiro, M. Knobel, C. Lankshear, & D. Leu (Eds.), *Handbook of research on new literacies* (pp. 359–376). Mahwah, NJ: Erlbaum.

McDonald, S., & Stevenson, R. J. (1998). The effects of text structure and prior knowledge of the learner on navigation in hypertext. *Human Factors, 40*, 18–27.

McVerry, J. G. (2007). Forums and functions of threaded discussions. *New England Reading Association Journal, 43*(1), 79–85.

National Reading Panel(2000). Teaching children to read: An evidence-based assessment of the scientific research literature on reading and its implications for reading instruction [online]. Retrieved June 9, 2009, from www.nichd.nih.gov/publications/nrp/report.cfm

O'Brien, D. (2006). Struggling adolescents' engagement in multimediating: Countering the institutional construction of competence. In D. Alverman, K. Hinchman, D. Moore, S. Phelphs, & D. Waff (Eds.), *Reconceptualizing the literacies in adolescents' lives* (2nd ed., pp. 29–46). New York: Erlbaum.

RAND Reading Study Group. (2002). *Reading for understanding: Toward an R&D program in reading comprehension*. Santa Monica, CA: Rand.

Rouet, J. F. (2006). *The skills of document use: From text comprehension to Web-based learning*. Mahwah, NJ: Erlbaum.

Snow, C. E., Burns, M. S., & Griffin, P. (Eds.). (1998). Preventing reading difficulties in young children. Washington, DC: National Academy Press.

Street, B. (1998). New literacies in theory and practice: What are the implications for language in education? *Linguistics and Education*, *10*(1), 1–24.

Street, B. (2003). What's new in new literacy studies? *Current Issues in Comparative Education*, *5*(2), 1–14.

Sutherland-Smith, W. (2002). Weaving the literacy Web: Changes in reading from page to screen. *The Reading Teacher*, *55*, 662–669.

Taboada, A., & Guthrie, J. T. (2006). Contributions of student questioning and prior knowledge to construction of knowledge from reading information text. *Journal of Literacy Research*, *38*, 1–35.

The New London Group. (1996). A pedagogy of multiliteracies: Designing social futures. *Harvard Educational Review*, *66*(1), 60–92.

Tillman, H. N. (2003, March 28). *Evaluating quality on the internet*. Retrieved March 28, 2007, from www.hopetillman.com/findqual.html

Webber, S., & Johnson, B. (2000). Conceptions of information literacy: New perspectives and implications. *Journal of Informational Science*, *26*, 381–397.

Zawilinski, L. (2009). HOT Blogging: A framework for blogging to promote higher order thinking. *Reading Teacher*, *62*(8), 650–661.

Organization for Economic Co-operation and Development [OECD]. (n.d.) *The OECD programme for international student assessment*. Retrieved August 15, 2007, from www.oecd.org/dataoecd/51/27/37474503.pdf

Chapter 5
Literacy, Technology and the Internet: What Are the Challenges and Opportunities for Learners with Reading Difficulties, and How Do We Support Them in Meeting Those Challenges and Grasping Those Opportunities?

Colin Harrison

In this chapter I ask and attempt to answer three questions:

1. What are the skills that readers need to acquire in the 21st century, and how should they be taught?
2. What types of software can support the teaching of literacy and the development of literacy, bearing in mind the particular needs of less able readers?
3. How do new technologies change and extend the teacher's role?

In answering these questions, I provide examples of the use of programs or web-based resources that make gradually increasing use of the interactive properties of the Internet. These all have the potential to be used in ways that support underachieving readers as well as successful ones.

Introduction—'there's just no rabbits on the Internet'

When Katie said to her teacher, 'Miss, there's just no rabbits on the Internet', the teacher knew immediately that Katie had a big problem. Her teacher had asked a group of students to pretend that they had a friend who had a sick rabbit, and to try to use the Internet to help their friend get some information that might be useful (Dwyer & Harrison, 2008). Katie was nine when she tackled this task, and she and two other students had spent 10 minutes on the computer, trying to find information they needed. Despite the fact that on the day they tackled this task a simple search on the Google search engine would have returned 74 000 000 results for the entry

C. Harrison (✉)
University of Nottingham, Nottingham, UK
e-mail: colin.harrison@nottingham.ac.uk

C. Wyatt-Smith et al. (eds.), *Multiple Perspectives on Difficulties in Learning Literacy and Numeracy*, DOI 10.1007/978-1-4020-8864-3_5,

'rabbit' (many of which would have offered a link to the 'House Rabbit Society Rabbit Care Guide'), Katie and her friends had not been able to find the information they needed.

Katie was at a school in Dublin, Ireland, but her problems are common ones across the English-speaking world. Katie's needs, and the implications for her teacher, raise a number of broader issues related to reading and the acquisition of literacy that I try to address in this chapter.

- What are the skills that readers need to acquire in the 21st century, and how should they be taught?
- What types of software can support the teaching of literacy and the development of literacy particularly for those experiencing difficulties in learning literacy?
- How do new technologies change and extend the teacher's role?

These questions, and the answers to them, are interrelated, but it is nevertheless worth giving some attention to each, since readers, software and teachers are all crucial elements in the pedagogic equation.

The theoretical positions that underpin this chapter have been expressed in some detail in my book on reading development (Harrison, 2004), and some detail on the practicalities for teaching that follow from them are presented in the next section. My starting point is Keith Stanovich's Interactive Compensatory Model of the reading process (Stanovich, 1980). Stanovich argues for approaches to the teaching of reading that encourage rapid, context-free word recognition, in order to free up processing capacity for dealing with text integration and comprehension. This is not an argument for 'death by phonics', but in my view it does imply the need to develop phonological and then phonemic awareness (and supporting those who do not develop these skills rapidly and easily). I began my teaching career as a secondary school teacher of English. In teaching, the traditional goals of encouraging a love of books and reading, and of developing a sense of oneself as a reader, are just as important as developing an understanding of letter–sound relationships. As I state later in this chapter, all of these skills and understandings need to be allied with the power to critically evaluate what is read, whether it is in a book, a text message on a mobile phone, or on the Internet, and this final challenge, of developing critical literacy, is one that we need to take very seriously, and that should be in every teacher's consciousness.

What are the skills that readers need to acquire for the 21st century, and how should they be taught?

Most of the skills that readers need to acquire are not 'new', in that we have known about their importance for some years. Michael Pressley (2000) shared what many still regard as the definitive list of reading skills in his review paper for the *Handbook of Research in Reading*. Pressley's list was influential because it was based on a thorough analysis of research studies that linked pedagogy to reading improvement.

Pressley (2000) recommended the following strategies for teachers:

- Teach decoding, with an emphasis on morphology.
- Teach the use of context cues and monitoring meaning.
- Teach vocabulary.
- Encourage extensive reading.
- Encourage students to ask their own 'Why?' questions of a text.
- Teach self-regulated comprehension strategies, for example:
  - prior knowledge activation
  - question generation
  - construction of mental images during reading
  - summarisation.
- Analyse into story grammar components.
- Encourage reciprocal teaching.
- Model strategies and scaffolding for independence.
- Encourage transactional strategies instruction.

In my view, this list is both comprehensive and exemplary, and clearly one could devote an entire book to describing in detail how a teacher might implement Pressley's strategies. *Understanding Reading Development* (Harrison, 2004) is my attempt to do this, but the emphasis in the present chapter is on skills and strategies for the 21st century, and the ways in which new technologies can support learners in achieving them, so we shall not dwell on all the skills in detail.

Traditionally, the verb 'reading' has been assumed by default to mean 'reading a book', but this is changing, as children are more likely to be reading electronic texts of various types, and this fact has two massively significant implications:

- first, the 21st century text itself is more likely to be multimedia, and to be a composite of text, images, graphics, sound files and mini-applications
- second, navigating from one section of a document to another will not be a matter of turning a page—it will require understanding of the architecture of an information source, and this may not be at all transparent, and is unlikely to be linear.

Parents and teachers are only just beginning to take in the implications of these changes to reading, and perhaps the reason for this is that clicking on a computer mouse seems such a trivial thing to do. But in reality, we need to think of a child who is about to click on an Internet link as a person standing inside a room that has 25 billion doors, and having absolutely no idea of where a door may lead or what is behind it.

In the main sections of this chapter, I wish to describe some ways of using new technology, each of which can support readers, and particularly weaker readers, in ways that would hardly have been imagined 30 years ago. I have divided the approaches into four groups, and each group represents a stage on a continuum that moves from a reader working on a standalone computer (with only an optional connection to the Internet) to a reader forming an active part of an Internet community of readers and writers. Each stage involves supporting readers in using technology, but the support moves from helping to develop fluency in beginning reading through to knowledge creation, and then broadens out still further to text production and finally to supporting communities of readers and writers.

These stages can be linked to a concept that has been much debated in the United Kingdom recently, namely the use of Web 2.0 tools. The phrase 'Web 2.0' is generally used in education to refer to some of the more creative and interactive uses of the intelligent Internet; for example creating blogs, wikis, podcasts and participating in online communities. Web 2.0 tools, and the social networking associated with them (for example using programs such as Facebook or Bebo) are massively popular with teenagers in the United Kingdom, though they have yet to make any discernable impact on the curriculum in most schools (Crook & Harrison, 2008). I argue that as the reader moves along the continuum from no Internet use to using Web 2.0 tools, changes in the nature of the student's activity (from passive learner to active creator of content) and changes in the teacher's role (from pedagogue to learning guide) are also implied.

What types of software can support the teaching of literacy and the development of literacy?

Supporting beginning reading—Internet optional

Here, I describe two types of program that support the earlier stages of reading: Soliloquy's *Reading Assistant* and Broderbund's *Living Books*. Both are aimed at beginning or younger readers, but each could also be used with older readers, or with those who have not made a confident start in reading, though in different ways. Software to develop reading is ever changing, but there are important aspects of both these programs that are worthy of attention, even if these products evolve or mutate into other products.

Soliloquy's *Reading Assistant* is a program that performs right at the current limits of speech recognition. It uses some very sophisticated software to perform one of the most valuable jobs a teacher (or parent) can do, namely to 'listen' to a child read and to offer a range of supportive feedback. All beginning readers benefit from practice in reading aloud to a supportive listener, one who will pause, prompt and praise as necessary, and the software does all this and more. The child chooses an electronic book from a library, dons a headset and then makes a voice recording to teach the computer something about his or her vocal characteristics. Then the child begins to read aloud into the microphone and the *Reading Assistant* kicks in. First, as the child reads, the text turns from black to grey, signalling that the computer has logged the reader's progress with every word. As reading progresses, if the reader pauses for longer than 3 or 4 seconds, the computer will supply the next word, and if the program judges that there are too many prompts, the child will be asked to re-read the whole sentence, to ensure fluency development.

But the software is cleverer than this. If the child misreads a word, the computer detects the misreading, highlights and remembers the word, and offers the correct pronunciation. A reader can click on any word to hear it pronounced, and can look up any difficult words in the computer's glossary. The child can call up a recording of his or her own reading, or can ask to hear the whole passage being read. After the reading is complete, all the information is stored so that the teacher is able to call up a recording of the reading. The teachers get a colour-coded version of the text that shows where errors, hesitations or prompts were made (see Fig. 5.1 for a screen shot of the record of an individual child's reading). Over time, the computer generates a very comprehensive picture of the reading proficiency of each child, and *Reading Assistant* knows a great deal about how each child's fluency is developing.

Teachers' reactions to this software are often highly polarised, and range from enthusiastic approval to outright rejection. I have met teachers who are incredibly keen on this software, for three reasons. First, because the program does a job no teacher can do, namely to provide, for a number of children simultaneously, sensitive, immediate and individual support for those who need help in developing reading fluency. Second, the program stores detailed information on each child's reading that can be retrospectively interrogated by the teacher, session-by-session.

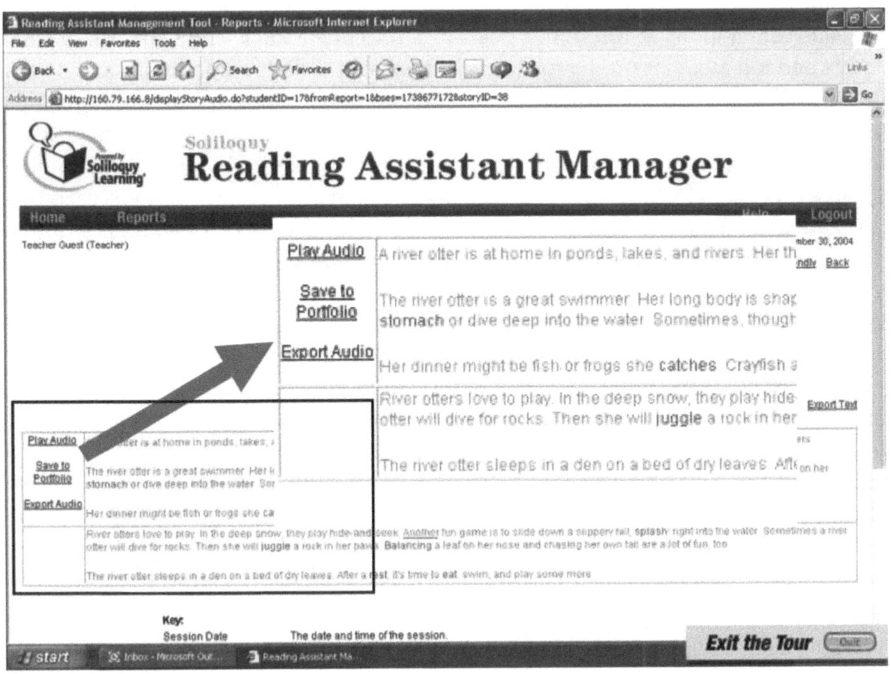

Fig. 5.1 Screenshot from Soliloquy's *Reading assistant*, showing enlarged section of the report for the teacher (© Soliloquy, reproduced with permission)

Third, the information on reading fluency is updated cumulatively over time, so that not only the teacher, but also every child can see a graph of their development, reinforcing a sense of progression and thereby helping to develop the confidence that is so important in a beginning reader or for those struggling with reading.

Naturally, some teachers (and some professors of reading) not only find this apparent replacing of the teacher to be repugnant, many are rightly suspicious about the capability of the speech-recognition part of the program to judge accurately whether a child is pronouncing a word correctly. The truth is that speech-recognition engines are far from perfect, and they still have problems with regional accents. The response from the software manufacturers is that their program is not obsessively accurate, and it permits the child to make small errors in pronunciation without interrupting the flow of reading. The other point to make, of course, is that the cautious teachers are entirely correct—the computer does not displace the teacher, whose encouragement and care can be immeasurably important in helping a child to read. The program is a support for, not an alternative to, the teacher.

Another program that received very mixed reception in schools when it first appeared was Broderbund's *Living Books.* The program presents on a CD-ROM stories that are already available in book form, together with a number of additional features. The computer screen presents pages that look very much like those in the printed book, but when the reader clicks on a 'hot spot', the text is read aloud, word-by-word, sentence-by-sentence or page-by-page. The most exciting part of the program for the child comes from the dozens of additional hot spots on the screen that cause animations to run: fence posts turn into piano keys, crabs run out of sand castles and nip a bather on the toe, cans of beans on a shelf come to life and do the Can-Can. Figure 5.2 shows a screenshot from the *Living Book* version of Dr Seuss's *The Cat in the Hat.* When this page runs, the text is read aloud, with each word highlighted as it is read. The child can then click on any individual word to hear it re-read, but in addition there are perhaps 30 or so animations on the screen that can be activated by mouse clicks: the child can play tennis or basketball with the toys on the floor, the plant bursts into flower; clicking on the rain-spotted window produces successively greater floods (complete with boats and fog horn) and, best of all, clicking the picture above the goldfish runs an animation of an angler who gets pulled into the water by the fish.

Living Books is the oldest software described in this chapter (the company began marketing this product in the early 1990s). I include it in order to talk about the relationship between motivation, interaction and learning. During the 1990s, I talked with many teachers about this software, which is still available today, about its attractiveness and impact. Some teachers were very much against having it in their school. They saw children crowding around the screen, screaming with laughter as they triggered animations and apparently paid no attention at all to the text. 'The children just play—there is no reading at all', was what one head teacher told me. And guess what, that's what the children said, too. But this does not mean that the children were not learning.

Of course, it was the case that the animations distracted the children from the story, but the animations also encouraged the children to play inside the story. Had

Fig. 5.2 Screenshot from the *Living books CD-ROM, The Cat in the Hat* (© The Learning Company)

the software been cleverly crafted such that the 'play' did not begin until after the page had been read, and each word highlighted as it was read? Of course it had. The program had been designed by cognitive scientists who knew about the psychology of both learning and play. Recently, two researchers from Israel used *Living Books* software in an experiment to teach English to Grade 1 and 2 immigrant children in Israel who spoke only Spanish and Hebrew, whose home language was Spanish, and who had not watched any television in English (Eshet-Alkalai & Chajut, 2007). The children were not yet being taught English, and they had previously encountered the Latin alphabet only in Spanish. The children were adamant that they only 'played' with the software over the 2 months of the study. But after 2 months, the children recognised and pronounced nearly 70% of the words from the story, even though the words were presented typed on a page, and in a different type style. The researchers concluded that there had been massive and effort-free 'incidental learning' from the children's activity in this playful multimedia environment.

Teachers who used *Living Books* in classes in England told me that the software made the children keen to read all the original books, and that they were just as keen to show them and each other differences between the text of the printed book and the CD-ROM version as they had been to share information on newly discovered animations. The moral of this story? After 40 years as a teacher I have come to a stark realisation: nearly ALL learning is 'incidental'!

For learning to occur, a number of things need to be present, but attention, motivation and rehearsal are three of the most important. It can also be very valuable to have some kind of reward. In the case of *Reading Assistant* that reward might be feedback (both immediate, through prompts and long-term, through progress graphs). In the case of *Living Books*, it is the discovery of new animations or new parts of the book. Where there is high motivation it is certainly sensible for the teacher to consider what can be learned from children about why it was so high. Children, and particularly those with learning difficulties, can be quick to decide that something is 'boring'. But often their use of that term is not about the learning resource itself—'boring' is often to do with authority and control: if children feel they are in control of something (the mouse, their learning, the screen), then they are much more likely to be motivated and ready to learn, especially those experiencing difficulties with reading.

Supporting knowledge creation: the Internet as knowledge source

In this section I discuss approaches to using the Internet to extend and develop reading.

UDL Editions from CAST (the Center for Applied Special Technology) are not so much books as hypermedia resources that may be used to support and develop comprehension and reading-strategy skills. This organisation uses what the developers call a Universal Design for Learning, a framework for designing curricula that enables all individuals to gain knowledge, skills and enthusiasm for learning. The *UDL Editions* are a wonderful example of this.

UDL Editions offers a web-based environment that begins with the reader selecting a text for online study that ranges in difficulty from a simple folk tale to a Shakespeare sonnet or an Edgar Allen Poe horror story. The webpage then offers the reader one of three levels of support: maximum, moderate or minimum, and according to which one the reader selects, provides the reader with a range of study guides, glossaries and support tools. In the case of the horror story *The Tell Tale Heart* for example, the resources include links to webpages on mental health, a YouTube video of the celebrated cartoon version narrated by James Mason, glossaries of all unusual words and prompts to stop and think. In the case of the Shakespeare sonnet XVIII, *Shall I compare thee to a summer's day?*, resource links include readings, a song and web resources on love poetry (see Fig. 5.3, for a screenshot of the Shakespeare study guide). As the reader delves deeper into the poem, they meet 'Stop and Think' questions, and if these are too tricky, the reader can click on one of three animated characters (or 'Coaches') who will give a strong hint leading to some possible answers. Finally, but very importantly, there is a floating 'textHELP toolbar' that the reader can drag to a convenient place on the page, and which contains highlighting tools and a button to trigger synthetic speech vocalisation of any text, with instant translation into Spanish if that is needed for Hispanic students.

The variety of tools on offer in the *UDL Editions* is interesting. The sheer range and depth of study opportunities here leads us back to the theme of user control

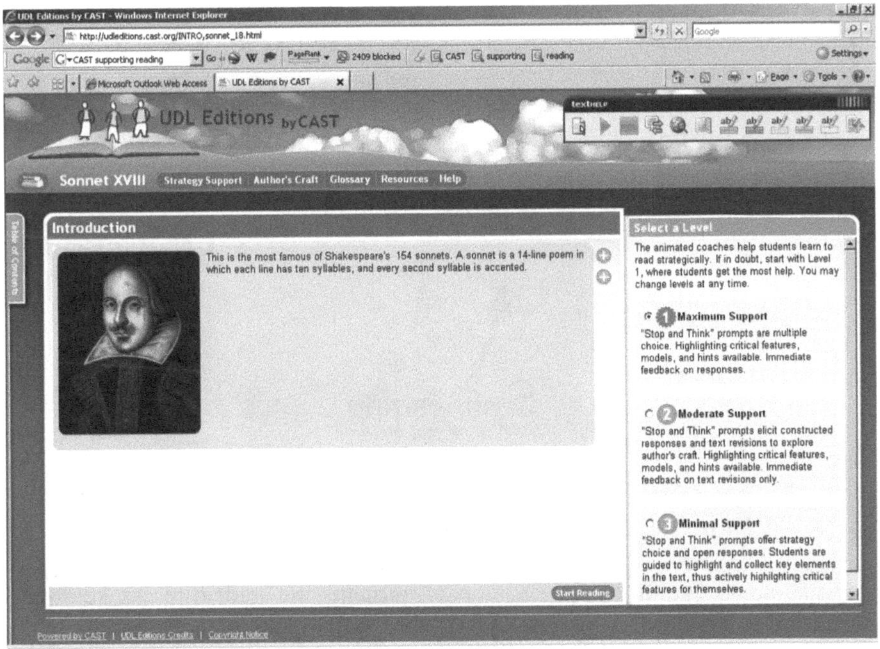

Fig. 5.3 Screenshot of the *UDL editions* guide to Shakespeare's Sonnet XVIII (© Center for Applied Special Technology, reproduced with permission)

and autonomy. What is impressive is the way in which such a full set of tools can facilitate and support independent reading with even quite insecure readers, and can therefore provide an experience of dealing with challenging classic texts that might otherwise tend to be seen as inaccessible to all but the most able.

The number of Internet-based programs to support reading is increasing rapidly, but perhaps the greatest challenge faced by weaker readers is how best to manage the Internet itself, and how to develop the skills and critical abilities necessary to use its vast stores of information effectively. For school students (and their teachers), one of the best-researched approaches is the QUEST model of Internet inquiry (Eagleton & Dobler, 2006). As the authors explain, the QUEST model (see Fig. 5.4) was based on a great deal of classroom trialling. The five stages of the QUEST model are complex and each one is treated at chapter length. Each stage is also a concept well-understood within developmental psychology as linking to deep processing and a potentially rich set of learning potentials:

Q—Questioning
U—Understanding
E—Evaluating
S—Synthesising
T—Transforming

Fig. 5.4 The QUEST model of Internet inquiry (Source: Eagleton & Dobler, 2006, reproduced with permission)

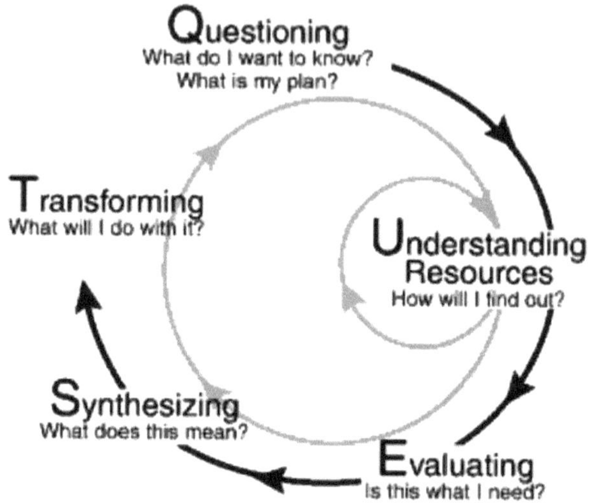

Questioning, within the QUEST model, prompts the student to ask key questions, such as 'What do I want to know?' and 'What is my plan for finding out?', but Eagleton and Dobler (2006) also make it plain that these overarching questions need to be revisited at least twice before the research cycle is complete. This phase also emphasises the importance of student choice in research. This point is not stressed in some skills training courses that simply ask children to carry out a search with a teacher-selected question, with few or no points of choice for the reader, and a bounded (often intranet rather than Internet-based) set of links that may be interrogated for information.

The Understanding phase is also rich and complex, with tasks that focus around the question 'How will I find answers to my questions?' Here, the emphasis is on acquiring skills of navigation, with critical interrogation of URLs and domain names, and a sensitisation to the special features of different web browsers. I particularly applaud the authors' stress on the need to develop strategies for generating good search phrases. They cite with approval the paper by Guinee, Eagleton, and Hall (2003), which suggests that it can be valuable to teach students to develop a two-word <topic + focus> strategy in planning Internet search phrases. This relatively simple tip could provide a massively useful key to unlock many of the resources of the Internet for young researchers.

It is interesting that Eagleton and Dobler (2006) place Evaluation at number three in a six-element cycle. For many teachers, evaluation is seen not only as the most cognitively demanding, but also as the final stage in the research and composition process. But for these authors evaluation has to come early, because it is at the point where the reader comes up against the material that a web search has brought to the screen that the skills of evaluation are most critically necessary. The key question is 'Is this the information I need?', but the question could equally have been formulated as 'Can I trust this site, and can I trust the information that is presented to me?' and it is the most challenging of all to answer, because even adults can be fooled. For younger students, however, distractions such as advertisements and the malware that often comes with them can also be a pressing concern. Again, the advice given by Eagleton and Dobler seems commonsense and simple: think about a site's truthfulness and think about its usefulness. These concepts are relatively straightforward for younger researchers to at least begin to consider, but behind them are really important issues of ontology and the authority of knowledge, and they indicate an approach to evaluation that is direct, while also profound.

The next QUEST element to consider is Synthesising. It is in the synthesising stage that ideas need to be prioritised, inferences drawn and tested, and data brought together from a range of media sources. Here again, are some massively difficult and under-researched areas, notably how we integrate information from multiple resources that present data in visual, numerical and textual modes, but at least we begin to address the problem when we recognise that there is a problem.

The final element in the QUEST model is Transforming; this is the area in which, as Eagleton and Dobler (2006) put it, information gets turned into knowledge. Information on its own is not knowledge: information needs to be selected, manipulated, restructured, reorganised and represented in order to become knowledge. The authors also remind us that in this information age, the products of Internet enquiry are not necessarily going to be written text. The product might be an essay, but equally well it might be a multimodal presentation, a poem or a movie. It might be presented in its final form, or it might be offered in a web area as a palimpsest to be written upon and added to by others. In study skills courses, many students in the United States are taught to take notes or make summaries, but even this difficult skill will only begin to address the possibilities that confront a 21st century author, who will be not only an author, but also a composer, editor, graphic designer and movie producer.

What is most important about the QUEST model, however, is that what appears to be a dauntingly challenging skill set to acquire becomes less intimidating when it is broken down into small steps, and into a set of procedures that can be activated by the QUEST acronym, then used iteratively as students become more confident and more autonomous researchers. The QUEST model has much to commend it, both because it has been used successfully with the full range of student abilities (including those with learning difficulties) in terms of reading and background knowledge, but also because it can provide very robust scaffolding for a teacher who is using the model to support their role as mediator of information technologies for learning.

Supporting text production: mind maps, concept maps and writing

I want to say a little about mind maps, because so many teachers find them useful, and because it is instructive to learn why so many psychologists are extremely cautious about mind maps. Teachers like mind maps because they offer children an opportunity to reflect on the organisation of information in non-linear ways. Only a small proportion of children are able to plan a piece of writing in their head; most children (and most adults for that matter) find it incredibly difficult to reorganise information in their head, but begin to be able to do it if they can put at least part of what they are thinking down on paper, and see how ideas relate to each other. Many of our ideas are not organised spatially; they are not even organised in a linear way; they are just a series of loose associations that pop into our consciousness. This is why psychologists are so cautious about mind maps—they may not be representations of any deep structure or conceptual ordering that is fixed in our 'mind', but rather a somewhat arbitrary set of loose associations that happened to have been produced under a particular set of circumstances on a particular day. To talk about a mind map as if it has a structural relationship with the architecture of knowledge in the brain might therefore be inaccurate on two counts: because it might fail to represent any internalised semantic or conceptual organisation of knowledge, and because it might be much more arbitrary than it appears.

But mind maps can be extremely valuable. This is because, even if a mind map is somewhat arbitrary, and might have come out differently on a different day, and even if it is little more than a shopping list presented in the form of a spider diagram, a mind map presents an opportunity for another person to construct a representation of knowledge that is transparent, and therefore open to discussion, debate and challenge. It presents the world in 2D semantic space and not in a sentence. Its nodes are words, and its links are semantic relationships, not verbs, and this makes it very powerful, because the other person looking at the map can supply their own verbs and construct their own grammar to link the nodes. This lack of specificity and closure brings an openness and provisionality that is attractive and inviting—and it makes the mind map an ideal tool for the construction of social networks of knowledge that characterise Web 2.0.

Many people are confused by the use of the terms 'mind map' and 'concept map'. The term 'mind map' is often used to refer to a tree or root-structured verbal map. By contrast, 'concept map' is generally interpreted as a rather freer representation, one that does not even necessarily use word labels at all. Notwithstanding the reservations expressed above about making too many assumptions about the relationship between a person's maps and their understanding of a topic, concept maps have been used in research as well as teaching, to investigate depth of understanding, and to explore how a person's representation of the world can change over time.

One of the most widely used concept mapping tools available over the Internet is the freeware Cmap program (IHMC, 2009), which not only permits users to construct concept maps on their own computer, it also enables them to share them on servers anywhere on the Internet, and even to edit them collaboratively across continents in real time. This gives concept mapping real power, and moves it into Web 2.0 territory, since such use empowers users to construct knowledge collaboratively, and to share and critique that knowledge.

The ImpaCT2 project in England, which investigated the relationship between information technology in the school curriculum and school attainment (Somekh et al., 2002), was set up before Cmap became widely known, but it used concept maps to explore in a systematic way how children's understanding of computers

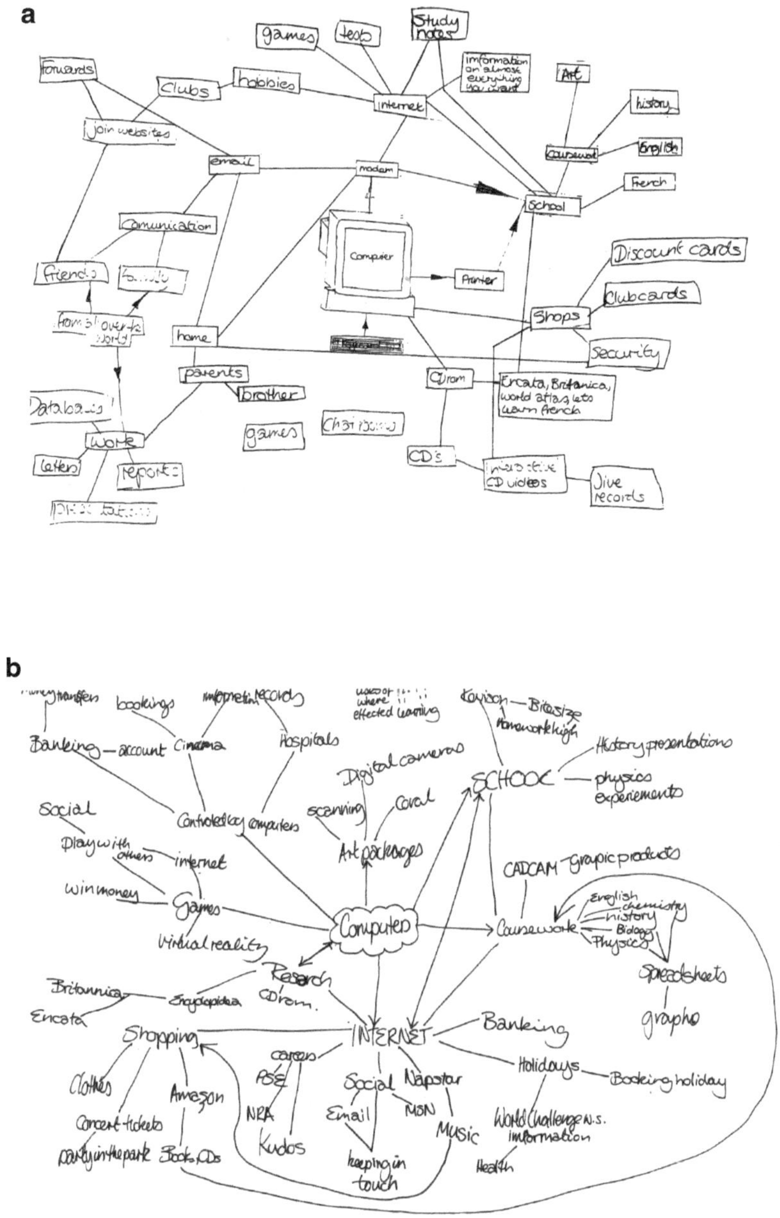

Fig. 5.5 Gail's concept map of computers in today's world (a) age 15; (b) age 16

in today's world developed over the course of a year. In this national project, over 2000 students drew concept maps, and the research team evaluated them qualitatively and quantitatively. Figure 5.5a, b show two concept maps drawn by the same student, Gail, a year apart, in response to the invitation to draw a map showing her understanding of 'computers in today's world'.

The two concept maps show interesting emerging patterns in Gail's ways of representing computer technologies. Her first map, drawn when she was 15 years old, is carefully crafted, with a central image of a computer, and ruled lines connecting nodes. There are three main zones of use: school, home and shops, and two spheres of thinking: e-mail and the Internet. A year later, Gail's map is drawn freehand, with text nodes only, and she does not put boxes round the text. There is a lot more information in the later map: the number of nodes increases from 43 to 62, and the number of links from 102 to 140. The range of zones of use is extended, with more community and commercial nodes (hospitals, cinemas, banking and holidays are added), and the range of technological equipment referred to is also broadened out, which in the later map includes digital cameras, scanners and CAD (Computer-Aided Design). One general trend is clear: for Gail, the concept of ubiquitous computing has arrived. New technologies now permeate every zone of life, and do so in ways that multiply the number of devices that carry information, and multiply the complexity of communication channels that interconnect them. Not every young person would produce maps as complex and rich as these, but the world represented in Gail's maps is one that is potentially available to every child in technologically advanced nations. The challenge before us as teachers is to make children aware of how best to access and utilise those resources to improve their lives and the lives of others.

I mentioned that psychologists urge caution against over-interpretation from concept maps. This is a very proper concern, but in the case of the concept maps generated by the ImpaCT2 project, generalisations are defensible, simply because pairs of maps representing a year's growth in understanding were elicited from over 2 000 young people. It is my belief that, as graphical and indeed multimodal forms of communication become more prevalent in children's lives, and in the lives of their teachers, concept maps will come to be used more and more as valid representations of the outcomes of learning.

Does producing a mind map or a concept map help children to improve their writing? Elaine Cockburn (2004) has been looking systematically at the best ways to help children structure their writing by using concept maps to help in the planning stage of writing, and she has shared some interesting findings:

- Many younger children will, if left to themselves, only produce one kind of concept map: a spider or star diagram, with a central node and a series of relatively independent 'legs' or links; this is essentially a kind of list structure, albeit a 2D one.
- A spider or star diagram may help a writer to bring ideas to the surface, but it may not necessarily contribute much to organising those ideas.
- A concept map may be useful in planning writing, but may be unhelpful if the map does not match the structure of the planned composition; if the writing is chronological, for example (as in a biography), a 2D mind map may be less useful than a 1D timeline.

- For young writers (Cockburn's authors were between 9 and 11 years old), the relationship between concept map and final text structure may not be congruent: some of Cockburn's students produced rich and complex concept maps but then went on to produce a list-like or poorly organised piece of writing; others produced a poor concept map with few links around a single node, but went on to author a complex and well-structured piece of writing.
- Teaching children to draw concept maps may aid their writing, but it is more likely to do so if the potential relationship between a concept map and the subsequent writing is made explicit and modelled, rather than left to establish the connection itself by a sort of cognitive osmosis.

In my view, therefore, mind maps and concept maps have immense potential for all students, and particularly for those with learning difficulties. They seem likely to become a more significant element as our use of new technologies puts greater emphasis on graphical and visual communication, and as collaborative construction of these representations becomes embedded in our literacy practices.

Supporting Web 2.0 communities of readers: RealeBooks

As Charles Crook's (2008) review showed, the vast majority of young people who enter the Internet communities are currently consumers rather than producers. In this final section of the chapter dealing with software, I wish to spotlight one type of Web 2.0 activity that is much more centrally involved with production as well as consumption: *RealeBooks*. RealeBooks are little books, generally written by children that are not only printed, but also published and made part of a literacy community across the Internet.

RealeBooks (pronounced 'really books'; see <www.Realebooks.com>) were begun by Mark Condon, a middle-school teacher in the United States who wanted to motivate children by helping them to become authors of real printed books that they could hold in their hand and take home to their parents. The first RealeBooks software was standalone, and cleverly worked out how to print pages so that, with just one paper cut and one staple, a book was produced that had pages printed back to back and numbered correctly. The much more ambitious web-based software that enabled every reader to become a member of a worldwide community of readers and authors came later, and with it the potential of archiving community texts and preserving community languages.

RealeBooks are incredibly attractive for teachers of children with limited literacy achievement, because they make the goal of becoming a published author accessible to every child, regardless of how modest their literacy skills are, provided that their words and ideas can be turned into print. Like many good ideas, RealeBooks look simple, but are underpinned by software that is more complex than it appears, and by a philosophy that is deeper than might be imagined at first glance. All that is needed to produce a RealeBook is a digital camera to take photographs, a computer and a printer, plus an author. Many RealeBooks have been produced by teachers who

a

b

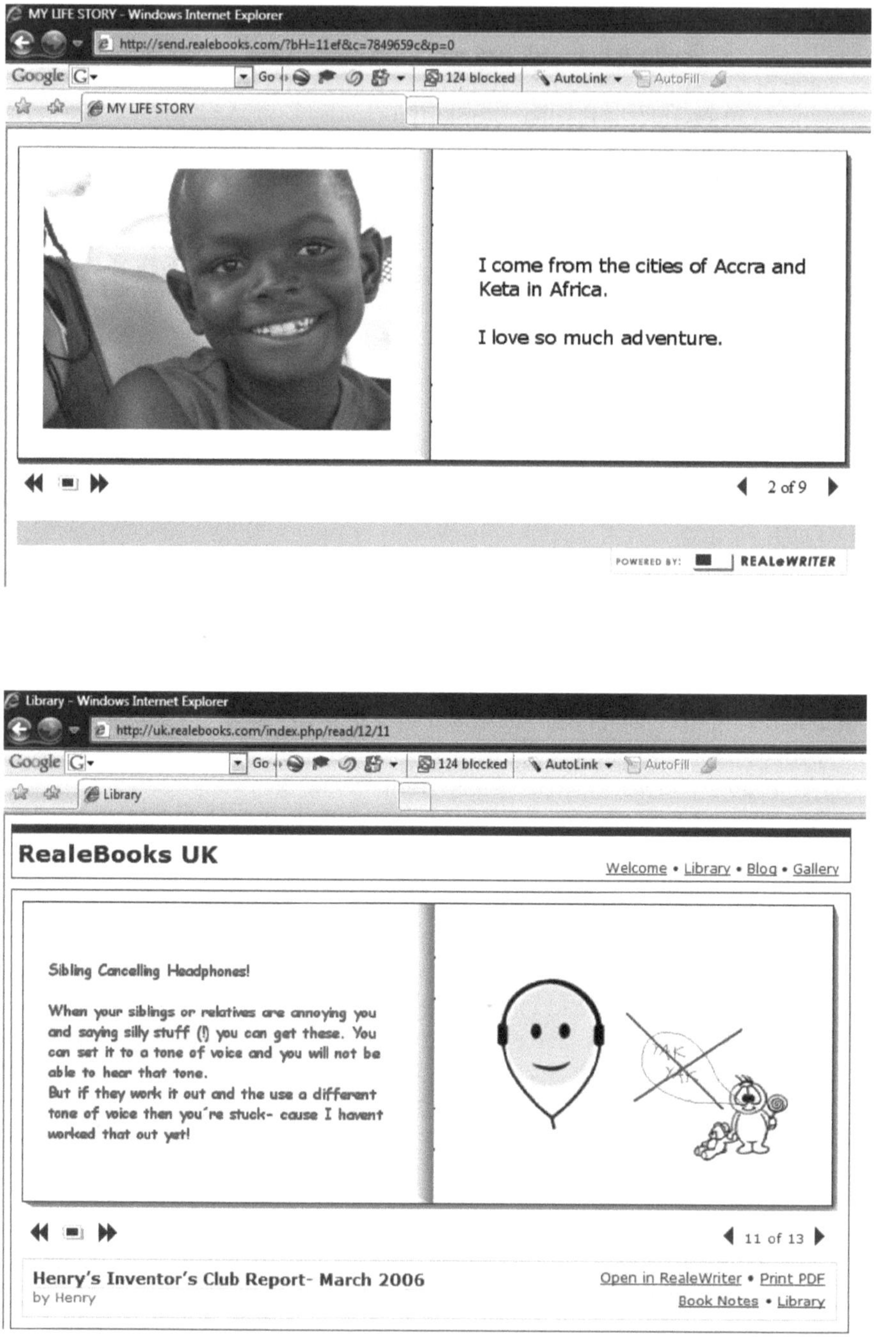

Fig. 5.6 (a) A page from a Realebook produced in Ghana, West Africa, (b) A page from a Realebook produced as part of the Innoed project, showing one of Henry's many inventions: sibling-cancelling headphones (© RealeBooks, <www.Realebooks.com>, reproduced with permission)

have put into print the words of a child, but many have been produced by children themselves. Sometimes the child-authored books have been part of a project, in local history or geography. Other projects have paired older children with younger ones, and the older child has written a book to order that was requested or even composed by a younger child.

As a visit to <www.realebooks.com> will demonstrate, many RealeBooks have been produced in community languages. Over 200 books have been produced and published online as part of a project in New Mexico with the Tse'ii'ahi' Community School, one goal of which is to help preserve Navajo ways and the Navajo language in the United States. Many RealeBooks have been used to build bridges between the languages of Spanish and English that are spoken in so many United States schools. RealeBooks have also been a pivotal aspect of a number of international literacy projects. Figure 5.6a shows a page from a RealeBook that was written and printed in Ghana as part of an international initiative. RealeBooks are at the core of a World Bank project that aims to produce and distribute a million books written in each of the nine indigenous languages of the Republic of South Africa. RealeBooks cost less than one United States dollar to produce, and they make publishing in community languages with a very modest print run economically feasible.

As a final example of the part RealeBooks can play in building an international literacy community, Fig. 5.6b shows a page from a book produced as part of the Innovation Education or Innoed project (<www.innoed.is>), a Europe-wide project originally begun in Iceland (where Innovation Education is part of the national curriculum), which had as its goal the encouragement of young people to design, develop and envisage the commercial potential of inventions that might benefit humankind. One of the Innoed projects that ran in a primary school in England used RealeBooks to record the children's inventions, and also extended the international element by twinning with an elementary school in Texas, in the United States. The children in each school set the other school three innovation challenges, and they were asked to come up with design solutions. The challenges included developing a tool that would enable a small child to walk a big dog, and developing a technological solution to the perennial problem of quickly locating 25 pencils at the end of a lesson. The Sibling-Cancelling Headphones shown in Fig. 5.6b are an example of the ingenuity and wit that 10-year-old children brought to this project. RealeBooks began as a standalone, computer-based publishing project, but has gradually become a Web 2.0 tool, as children, teachers and parents have used it not only to publish, but also to build literacy communities using the Internet.

How do new technologies change and extend the teacher's role?

Leu and Kinzer (2003), in a landmark chapter on teachers and new media, argued that far from becoming marginalised as new literacies become more important, the teacher's role becomes more significant

> It seems certain that Internet resources will increase, not decrease, the central role that teachers play in orchestrating learning experiences for students as literacy instruction converges with Internet technologies. Teachers will be challenged to thoughtfully guide students' learning within information environments that are richer and more complex than traditional print media, presenting richer and more complex learning opportunities for themselves and their students. (p. 30)

I share this view, but while it suggests an exciting future for teachers as well as students, it also presents massive challenges for individual teachers as well as for the whole educational system, for three reasons. First, we know that teacher change occurs slowly and gradually, and that if it is rushed, its results may not be enduring. Second, we know that even if teacher change is slow, it is often more rapid than change in the curriculum and in the pedagogical and assessment structures that frame that curriculum. Third, we know that students spend three or four times longer per day on a computer at home than on a computer at school, and so radical change that will impact learning by presenting richer and more complex learning opportunities outside as well as inside the classroom will only come about if technology is used to construct new types of scaffolding for learning that extends beyond the school walls.

This topic is important enough to merit much fuller consideration than is possible here, but I want to at least discuss briefly four ways in which the literacy specialist's role is likely to change over the next few years, and some ways in which those changes might be supported. I want to suggest that in the future, teachers who are specialists in literacy are likely to be

- more knowledgeable and active in the area of reading development
- more knowledgeable about the Internet information architecture
- information managers rather than managers of student behaviour
- well-supported in their professional development by ICT literacy coaches.

Most teachers of beginning reading are already much more knowledgeable about reading and the reading process than was the case 20 years ago. Rather than being dependent upon a district-level adviser for a second- or third-hand understanding of the processes of learning to read, teachers have been able to take advantage of much more research-informed, pre-service and inservice training, and know a great deal about the development of preschool literacy and the importance of both good books and the systematic teaching of phonics to help children get a good start in reading. But fewer teachers are as knowledgeable about reading development beyond the early stages, and there are two reasons for this. The first is that reading development is still a relatively new field: it is only comparatively recently that specialists have come to realise that nearly all readers will benefit from additional help in learning how better to interrogate texts using the full range of skills in Michael Pressley's (2000) list given at the start of this chapter. The second reason is that the higher-level skills needed are changing: traditional reading comprehension exercises will do little or nothing to help a reader become a more critical reader of the results of a Google search. A richer and more elaborated set of skills will be needed for this, and teachers will need to learn how to support the development of those skills. This in turn will require them to be more knowledgeable about the architecture of information on the Internet.

I have characterised the Internet as a room with 25 billion doors, and unlike a traditional library in which the result of a search is access to a printed text, the result returned from an Internet search could be a webpage, a document, an image, a movie, a sound file or a map of further potential information sources (see, for example Kartoo.com or Grokker.com). If teachers are to support their students' learning, they are coming to understand that it is not enough to say 'Research Hinduism for your homework', because the skills needed by their students are more sophisticated than those necessary to search an encyclopaedia. This, in turn, suggests that teachers will not only become much more expert in interrogating web resources in the ways discussed in the section earlier in this chapter on the QUEST model, but they will also become skilled at setting up tasks and task environments in which their students can acquire these skills.

This leads directly to my third prediction, which is that teachers will gradually become information managers rather than managers of student behaviour, and I am aware that this is my most contentious prediction, because it would seem to imply a massive shift in what most teachers believe they are doing, most of the time. But I would wish to suggest that this change is already occurring. Ten years ago, I was involved in an evaluation of Integrated Learning Systems in the United Kingdom (Wood, 1998). The systems that the evaluation teams investigated were ostensibly managing the students' learning in mathematics and reading, but of course the teacher's role in ensuring that the children got on the computers, and saw their activity as worthwhile, was immense. As one teacher said to me, 'If ILS works, thank a teacher'. It was the teacher, too, whose role it was to look at the data stored on the system by the ILS software, and to see that good use was made of it. What I am suggesting here is that this monitoring role of the teacher is likely to increase, as online student activity becomes more fully integrated into the curriculum, and as e-assessment and teacher feedback on student activity become a daily part of the job. And this will inevitably mean an increase in the role of the teacher as information manager, and a corresponding reduction in the role of classroom behaviour manager.

My final point is that teachers will not be making this journey alone. Most of us learn in two ways: independently (often by trial and error, but with occasional assistance from the Help screen), and from or with others. What this suggests is the importance of building in opportunities for such informal and unstructured learning for hard-pressed teachers in and beyond their workplace, and this is where an ICT literacy coach could be really valuable. An ICT literacy coach would be an expert (at least in relative terms), but one who was willing to come into the classroom of a teacher and work alongside, supporting and sharing knowledge with students as well as the teacher. Those of us who belong to ICT in education e-mail lists see potentially distressing postings every day reporting computer suites that are empty and underused because teachers are too battered by assessment and curriculum imperatives to become better informed about how ICT could improve their teaching and the lives of their students. If we are able to locate, fund and employ literacy coaches who could work alongside their colleagues and model new approaches, while avoiding being either patronising or an additional burden, then, even within our currently coercive educational systems, literacy learning with new technologies could really begin to change. I hope that the ideas in this chapter make some contribution to the challenge of indicating the direction that such change might take.

Essential next questions

There needs to be a better match between theory, research, policy and practice. The questions that follow try to suggest not only some approaches that would provide us with much-needed answers, but would enable some better connections to be made between what we know about how children learn with ICT, and how this understanding can be turned into better informed and even more effective teaching.

The questions for research are ambitious and incredibly challenging, but we need to push developers to move the field forward in ways that make full use of the massive processing capacity of tomorrow's computers. Questions for research are as follows:

- As computer speech-recognition improves, can computers do a better job of listening to children read, and giving them (and their teachers) appropriate feedback?
- Can researchers find more attractive and educationally sound ways of encouraging reading using personalised ICT tools that are matched to the individual needs of every reader?
- Can researchers create tools that will evaluate and give feedback on a reader's critical Internet literacy skills?

However good the software and hardware, and however skilled the teacher, ICT in schools will not really take off until policy makers put new technologies at the heart of learning, rather than on its periphery. Questions for policy makers are as follows:

- Can policy makers square the circle and make Internet access both universal and safe?
- Can policy makers rid themselves of the assessment chains that lead schools to waste money on test-preparation software instead of using ICT to support real learning?
- Can policy makers give serious attention to the need for ICT coaches to model good innovative practice in classrooms of every teacher?

Now that teachers are becoming digital natives rather than digital immigrants, we are reaching the point when they can begin to bring about real changes in learners' lives using ICT. Questions for practitioners are as follows:

- Can teachers become confident users of ICT, finding new roles as guides, coaches and mentors, rather than as pedagogues and law-makers, as the emphasis shifts from schooling to learning?

- Can teachers take on the massively important job of teaching children critical Internet literacy?
- Can teachers find ways of encouraging the creative potential of Web 2.0 approaches within the regular school curriculum?

References

Cockburn, E. (2004, September 9–12). *Can 8–10 year old children use mind maps to help in narrative writing?* Paper presentation, EARLI SIG2 'Text and graphics comprehension'. Valencia: University of Valencia.

Crook, C. (2008). *Web 2.0 technologies for learning at KS3 and KS4: The current landscape*. Retrieved January 2, 2009, from http://partners.becta.org.uk/index.php?section=rh&&catcode=_re_rp_02&rid=15878

Crook, C., & Harrison, C. (2008). *Web 2.0 technologies for learning at Key Stages 3 and 4: Summary report*. Coventry: Becta. Retrieved September 29, 2008, from http://schools.becta.org.uk/upload-dir/downloads/page_documents/research/web2_ks34_summary.doc

Dwyer, B., & Harrison, C. (2008). 'There's no rabbits on the internet': Scaffolding the development of effective search strategies for struggling readers during internet inquiry. In Y. Kim & V. J. Risko (Eds.), *57th Yearbook of the National Reading Conference* (pp. 18–202). Oak Creek, WI: National Reading Conference, Inc.

Eagleton, M. B., & Dobler, E. (2006). *Reading the web: Strategies for internet inquiry*. New York: Guildford.

Eshet-Alkalai, Y., & Chajut, E. (2007). Living Books: The incidental bonus of playing with multimedia. *Journal of Educational Multimedia & Hypermedia, 16*(4), 377–388.

Guinee, K., Eagleton, M. B., & Hall, T. E. (2003). Adolescents' internet search strategies: Drawing upon familiar cognitive paradigms when accessing electronic information sources. *Journal of Educational Computing Research, 29*(3), 363–374.

Harrison, C. (2004). *Understanding reading development*. London: Sage.

IHMC (2009). *IHMC Cmap tools web site*. Retrieved May, 5, 2009, from http://cmap.ihmc.us/conceptmap.html

Leu, D. J., & Kinzer, C. K. (2003). Towards a theoretical framework for new literacies on the internet: central principles. In J. C. Richards & M. J. McKenna (Eds.), *Integrating multiple literacies in K–8 classrooms: Cases, commentaries and practical applications* (pp. 18–38). Mahwah, NJ: Lawrence Erlbaum Associates.

Pressley, M. (2000). What should comprehension instruction be the instruction of? In M. Kamil, P. B. Mosenthal, P. D. Pearson, & R. Barr (Eds.), *Handbook of reading research: Volume III* (pp. 545–562). Mahwah, NJ: Lawrence Erlbaum Associates.

Somekh, B., Lewin, C., Mavers, D., Fisher, T., Harrison, C., Haw, K., et al. (2002) *ImpaCT2: Pupils' and teachers' perceptions of ICT in the home, school and community*. Coventry: Becta. ICT in Schools Research and Evaluation Series – No.9. Retrieved 18 January, 2010, from http://partners.becta.org.uk/page_documents/research/ImpaCT2_strand_2_report.pdf

Stanovich, K. (1980). Toward an interactive-compensatory model of individual differences in the development of reading fluency. *Reading Research Quarterly, 16*, 32–71.

Wood, D. (1998). *The UK ILS evaluations: Final report*. Coventry: British Educational Communications and Technology Agency.

Chapter 6
Essential Provisions for Quality Learning Support: Connecting Literacy, Numeracy and Learning Needs

Peta Colbert

Internationally, an increasing accountability focused on achievement outcomes has heightened the need for schools to reconsider the effectiveness of their policies and practices, especially those concerning support for students with learning difficulties. In Australia, national testing occurs in literacy and numeracy across years 3, 5, 7 and 9, but each sector and system within each state and territory currently employs different literacy and numeracy curricula and assessment programs—a challenge being addressed by the Australian Curriculum, Assessment and Reporting Authority with the implementation of the country's first national curriculum in 2011. Recently, the Australian government has also moved to consider the approaches used to support students with learning difficulties in schools. As each school system and sector again employs a combination of augmented programs targeting different student groups, the government moved not to regulate this delicate type of literacy and numeracy support, but rather to build a strong evidence base from which to promote informed teacher and school choice of a program that would make a difference for students with learning difficulties. Through the national program, titled *Effective Teaching and Learning Practices for Students with Learning Difficulties Initiative*, government funding was provided in order to build a picture of available support provisions at classroom and school level in the states and territories, and to determine the effectiveness of those practices. This funding was allocated to strategic projects designed to increase knowledge and understanding of how to enhance the literacy and numeracy development of students with difficulties in learning in the early and middle years of schooling.

P. Colbert (✉)
Griffith University, Brisbane, QLD, Australia
e-mail: p.colbert@griffith.edu.au

Some parts of this chapter were drawn from Volume 1 and Volume 4 of the InLaN project report (Wyatt-Smith et al., 2007). Interested readers should visit the case-study reports (Volume 4) and the synthesis findings (Volume 1) for more detailed information on the sites.

C. Wyatt-Smith et al. (eds.), *Multiple Perspectives on Difficulties in Learning Literacy and Numeracy*, DOI 10.1007/978-1-4020-8864-3_6,

One such project was the Interventions in Literacy and Numeracy (InLaN) project[1] designed by Griffith University researchers to identify those effective teaching and learning practices that led to measurably improved outcomes in literacy and numeracy for students experiencing difficulties in learning in the primary and middle years of schooling in Queensland schools. The project, which was the inspiration for this book, comprised four related inquiries and used a range of methodological perspectives. The final report provided the Australian government with a comprehensive overview of practices in use for supporting students with learning difficulties in Queensland schools.[2]

In this chapter I present the key features of effective learning support provision, as demonstrated in 14 case-study schools, and identify the importance of these in supporting the literacy and numeracy development of all students, particularly those with difficulties in learning. The chapter is presented in four parts. The first presents the theoretical framing for the discussion and turns to definitional matters relating to the category, 'students with difficulties in learning'. The second outlines the methodological approach adopted in the project and case-study sites. The third part presents the effective practices observed in the case-study schools and the fourth considers essential next questions that might guide future research.

Part 1: theoretical framing and definitions

Sociocultural perspective and learning support

The 'intellectual development of children is inherently involved with participation in sociocultural activities' (Rogoff & Chavajay, 1995, p. 871). Finding its foundation in Vygotsky's work, sociocultural theory considers social interactions and cultural context as vital elements influencing student learning. The interaction of the student in the classroom, that is the teaching–learning environment, is central to this theory (Vygotsky, 1987), since it is 'impossible to separate the learning competencies and problems of individual children from the contexts in which they live and function' (Keogh, Gallimore, & Weisner, 1997, p. 107). All interactions within this teaching and learning environment are influenced by the cultural perspectives of the teachers, students and the organisational element guiding operations—the school. This is the framework guiding the discussion of features of support provision. It is recognised

---

[1]The InLaN project was a joint enterprise of Professors Claire Wyatt-Smith and John Elkins of Griffith University, Education Queensland (EQ), Queensland Catholic Education Commission (QCEC) and Independent Schools Queensland (ISQ).

[2]The outcome of this project is a four-volume report, titled *Changing the Nature of Support Provision. Students with Learning Difficulties: Interventions in Literacy and Numeracy Project* (InLaN), available for download from the Department of Education and Training's publication website: http://www.dest.gov.au/sectors/school_education/publications_resources/profiles/effective_teaching_learning_practices_stud_learn_difficult.htm#authors

that the interactions and culture of the players are intertwined, influencing each other inextricably.

As social interactions are dependent upon schooling policies and frameworks, and the people involved, with each representing varied cultural experiences and approaches, it follows that one approach will certainly not fit all. Critical to the identification of effective practices and approaches for learning support is the notion that what influences a student's learning and achievement in one setting may be different in other settings (that is learning domains/disciplines and teachers) for that student, other students in the classroom, other students in the school and other schools entirely. Individualising practices and approaches provides an appropriate solution. In addition, it is entirely practical, both financially and logistically, to use a range of practices and approaches to learning support.

Coming out of research examining effective inclusive schooling in the United States and the United Kingdom, Rouse and Florian (1996) found that 'effective inclusive schools are diverse problem solving organisations' (p. 71). Furthermore, they recognised that there 'are different ways of achieving this goal that must reflect the cultural and social context' (p. 71). More than a decade later, and this view has changed little when considering inclusive schooling or learning support provision in varied contexts—with both currently being used in schools in Australia (Anderson, Bourke, & Carrington, 2007; Forlin, 2005; Klassen & Georgiou, 2007). Again, the most appropriate approach to learning support is entirely contextual. The ways in which the social and cultural elements of students, teachers, the school and parents work together vary, with strategies and approaches catering to context. Hence, the framing of practices suited to supporting the particular needs of students with learning difficulties should begin with close attention to the distinctive features of the school's cultural and social contexts and students.

With full appreciation of the nature of the school's context, attention can be targeted to the range of features that may have an impact on provision of support and school-level decisions, as explained in the next section.

Effective learning support provision

Many facets of how learning support is enacted in a school represent operationally based, school-level decisions made in consideration of system or sector priorities, district support, local needs, availability of specialist services, funding, and the training beliefs and assumptions of those involved. Effective learning support provision has many features; however, in this chapter, discussion centres on features falling within three categories of school operation. Each category can be thought of as being operationally based, in that each represents significant decision-making opportunities for influencing a school's support provision. The categories, *leadership*, *networks* and *support* are represented in Fig. 6.1. The balance of effective learning support represents the context-specific balance of the three categories, which ultimately assures positive outcomes for the student and their family.

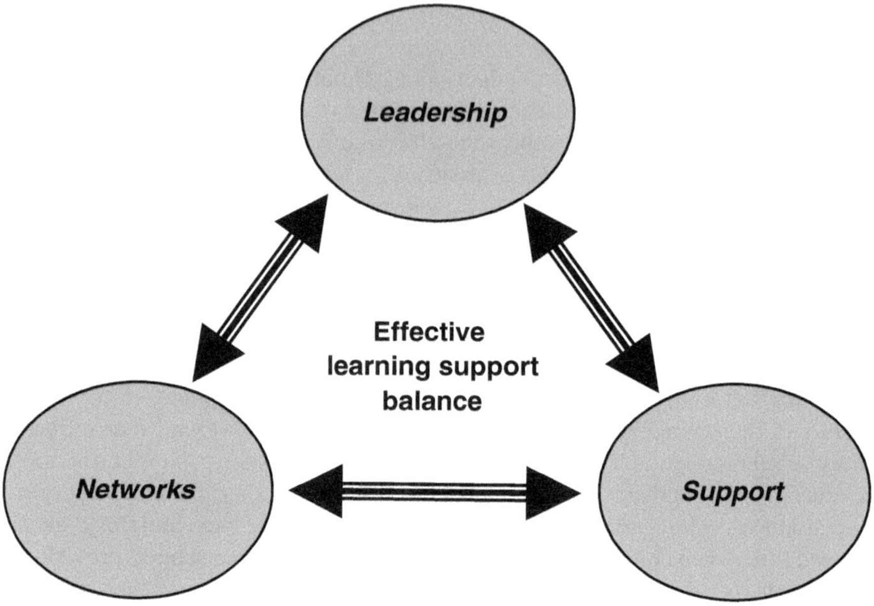

Fig. 6.1 Operational-based categories of support provision

The role of leadership, networks and support

As an overview, the first category, *leadership*, is crucial in providing the impetus for learning support and for strategic decision-making about how support will be enacted in the school. Second, the implementation of learning support plans within a school relies on teams—*networks* of people. Finally, decisions have to be made about the how the students needing *support* will be identified, how the support will be provided, who will provide the support and what support programs are to be implemented. The decisions inherent in each of these categories are contextually dependent, with many choices limited by human and financial factors.

Let's now consider the theoretical base for each of these categories. Following this discussion, I examine findings from the case studies that inform this model.

Leadership

Several Australian research projects highlighted the role of school principals in ensuring appropriate curriculum development, with one project focusing on numeracy stating that 'strong leadership is needed to elevate individual practice to shared practice and to make "good" practice an object of inquiry' (Commonwealth of Australia, 2004, p. 162, concluded from work by Hiebert, Gallimore, & Stigler, 2002). While Gunn concurred, she also noted the role of specialist teachers in strengthening leadership for shared, effective practices: 'Effective schools need strong leadership both from the principal and ... specialist teacher with consistent and collaborative implementation of policies and resources including support in ongoing observation, modelling and feedback among colleagues' (Gunn, in Wyatt-Smith, Elkins, Colbert, Gunn, & Muspratt, 2007, Vol. 2, p. 162). Principals are shown to be instrumental in developing a visionary culture within a school, and for ensuring that staff work together towards the vision (Geijsel, Sleegers, Stoel, & Krügern, 2009; Graczewski, Knudson, & Holtzman, 2009). As students proceed through the years of schooling, it is imperative that the support offered across the years be coordinated in approach, irrespective of whether support is individually planned and delivered (Chard & Linan-Thompson, 2008) or designed for small group and cohort intervention. Hence, strong leadership at the school and unit levels is essential in initiating the thrust of the vision, actualising its form of learning support and fostering its coherence and effective coordination across the years of schooling.

Networks

Improved outcomes from schooling can be fostered through networking within and among schools (Hargreaves, 1999; Mercier Smith, Fien, Basaraba, & Travers, 2009; Rhodes & Beneicke, 2002). Specialist staff, classroom teachers, parents and family members have vital roles to play in supporting students with learning difficulties. The strength in creating an effective network lies in the clarity of communication between members, and the shared recognition of the network's identified gatherer and disseminator of information, as frequently as possible. Also, key is the need for regular evidence-based reviews of progress, shared decision-making and shared understanding of any changes in the support program.

Support

Provision of learning support involves the targeted instruction of students in a variety of contexts that may include withdrawal from the classroom, in-class support with a specialist teacher, through to adapted programs implemented by the classroom teacher. Targeted instruction is also characterised in a number of ways, including use of an active learning approach, modification of a classroom program and/or delivery of a specific program, in addition to the classroom program. Typically, learning support encompasses one or more of these characteristics. How learning support is enacted in schools varies relative to the decisions inherent in the points above. What varies less so in schools in Australia is the personnel responsible for delivery of the learning support, including learning support teachers,[3] classroom teachers, teacher aides, specialists, parents and students.

The essential role of effective communication

An underlying element of these three categories of effective support provision is the high level of communication implicit within and across their function. For example, leadership can only have an impact if there is communication of the message and directives to members of the networks and other staff within the school. Similarly, communication is central to the operation of the networks and how support is enacted. All three categories require a high level of communication in order to operate in a way that assures the effectiveness of each. In a school, communication mechanisms range from formal mechanisms, such as policy messages, scheduling or adaptations to classroom programs, to more informal mechanisms, such as 'catch ups' in the staff room or chats between teachers and parents at the end of the school day. Regardless of the mechanism, effective communication is a central component in ensuring cohesiveness of the support provision offered in a school, within and across all three categories.

Before moving on to outline the case-study methodology, let's consider a definition of who the students are who have difficulties in learning.

Students with difficulties in learning

In Australia, the preferred term to refer to those students not achieving in classroom programs to the same degree as their peers is 'learning difficulties'. The term 'learning disability' refers to a more specific group of students who have, or are expected to have, persistent problems with learning over an extended period of time (Elkins, 2002). Additional terms used in Australia include 'at-risk', 'at educational risk', 'special needs', 'needing support' and 'needing higher than normal levels of support' (Louden et al., 2000; National Health and Medical Research Council, 1990). These terms are used by schools to identify students for whom targeted support is required, and across the states these terms are used in site-specific ways (Louden et al., 2000). However, the determination of which students are to receive support is a school-based, operational decision, with many schools providing support to many more students than any definition stringently applied. The support itself, as detailed later, is linked primarily to literacy and numeracy programs.

---

[3] In Australia, the Queensland Department of Education and Training uses the title Support Teacher: Learning Difficulties. Other titles include Learning Support Teacher, Support Teacher, Special Needs Teacher, Support Teacher—Inclusive Education and Head of Department—Learning Support. The title Learning Support Teacher (LST) is used in this chapter.

In this chapter, the term 'learning difficulties' is most commonly used to refer to the students identified as achieving at a level lower than their peers. InLaN project's design applied the second part of the Department of Education, Science and Training (DEST, 2002), definition, which states:

- students with 'a disability', as defined in the *States Grants (Primary and Secondary Education Assistance) Act 2000*, which affects their ability to learn (for example those with cerebral palsy, autism)
- other students identified as having significant difficulties in acquiring literacy and numeracy skills due to factors that are intrinsic to the individual, other than social, cultural or environmental factors (for example students with dyslexia or attention deficit disorders).

In some of the case-study sites presented in Part 3 of this chapter, the above definition was used as a guide in identifying students, though it was not treated as a stand-alone criterion to determine support. Wyatt-Smith and Elkins found in a survey of Queensland principals that

> almost 60% of responding principals do not use the DEST definition of learning difficulties, or a similar definition to determine which students receive support. The surveys showed that there is no consensus about policies or practices for determining those students with learning difficulties eligible for learning support teacher assistance. (Wyatt-Smith et al., 2007, p. 54)

This finding is not uncommon. MacMillan, Gresham, and Bocian (1998) observed a similar finding relating to the identification of students with learning disabilities in the United States.

What is changing, however, is the prevalence of students identified for support or, more specifically, the estimates by schools of the number of students they support. Surveys administered nationally in Australia over the past few decades have identified the prevalence rate as ranging from 6 to 20% (Andrews, Elkins, Berry, & Burge, 1979; Louden et al., 2000) with states varying, often greatly, in terms of their identified prevalence rates. In the same survey of principals referred to above, prevalence rates for literacy were higher than for numeracy and higher for boys (29% literacy, 22.5% numeracy) in both categories than girls (22% literacy, 20.3% numeracy). Further, Wyatt-Smith et al. (2007) noted that more students were identified than were being supported.

A distinction needs to be made at this point regarding the identification of students with learning difficulties in order to meet funding criteria. There are the students who are identified using statewide testing, and other assessments whereby the school provides support for the student using formal funding allocations. Then there are the students supported by schools where a need has been determined, but where, however, the severity of the learning difficulty does not meet system or sector funding criteria. Funding for learning provision for these students rests with the school. When the students are considered in relation to the prevalence of learning difficulties, the prevalence rates rise again—considerably in some cases. The support for students with learning difficulties and the funding implications are revisited again later in this chapter.

Part 2: methodological approach and sites

Case-study methodology

The InLaN project applied a multi-layered theoretical approach, with the case studies representing only one of the four inquiries of the project. Through the case studies, an overview of learning support practices being used in schools was obtained. The aim was to identify effective practices being used by schools to support students with difficulty in learning in various contexts. A further underlying aim was to provide information useful to sectors, schools and teachers, in order to promote better understanding of the nature of the learning difficulties experienced by students, and to enable greater effectiveness in provision of support.

Schools were selected on the basis of how they met the selection criteria, which were designed to identify schools with effective literacy and numeracy learning support programs for students with learning difficulties. The criteria included achievement data on statewide, cohort-based literacy and numeracy test programs, and system or sector advice as to school outcomes related to effective provision of support.

The final set of 14 case-study schools represented 10 Education Queensland (EQ) state schools, two independent schools in the Independent Schools Queensland (ISQ) sector and two Catholic schools in the Queensland Catholic Education Commission (QCEC) sector. Data collection for the 14 case studies was undertaken onsite by one to three researchers, depending on school size, during 2005 and 2006. In brief, the case-study data sets comprised: interviews with the principal (and other administrative staff), together with a sample of class teachers across Years 1–7, other specialists (learning support teachers, guidance officers, etc.), parents and students in Years 3–7. In addition, school documents and classroom artifacts were collected, reproduced and analysed; the former included documentation relating to the school's literacy and numeracy programs. Finally, observation records of relevant instructional practices were made, including whole-class, small group and withdrawal practices, as appropriate.

Cross-case analysis leading to identification of key features

A reporting framework was used to document all facets of the school, with a particular focus on the learning support system. A strength of applying the framework was that it allowed identification of common practices and themes, and variation of support provision across sites. From this, a cross-case analysis was undertaken.

This led to the development of evidence-based accounts of practice to inform school and classroom practice, policy directions and research on learning difficulties. This chapter highlights findings from the cross-case analysis that point to key features of effective support provision for students with difficulties in learning literacy and numeracy.

Part 3: key features of effective support provision

The cross-case analysis revealed a number of features that characterised the case-study school sites. These features are presented as representative of the three categories identified earlier: *leadership, networks* and *support*.

As the case-study schools represent sites where effectiveness has been evidenced, the features can be considered 'best existence proofs' (Pressley, 2002, p. 397) that point to practices that are effective in supporting students with learning difficulties. All 14 sites employed strategies and engaged in practices falling within all three categories, though with varying emphasises. That is, effective provision does not involve one feature within a single category (for example establishing a network of support staff); it is characterised by strategies and practices in place in a school that range across all three categories of operation. There is no approach that fits all schools, as the social and cultural aspects in schools differ and an approach must be catered to context to ensure that teachers and students are best placed to take advantage of the teaching–learning environment created. Across the sites, all three categories were evidenced to be equally important in ensuring sustained effective support provision in a school.

Leadership

The importance of the role of leadership cannot be overstated, especially as it relates to whole-school planning for learning support provision in schools in order to ensure coherence and alignment with other school priorities. School leadership was central to strategic management in the case of the allocation of sufficient resources for staff and students, ensuring appropriate professional development, and management of physical space. Leadership was also important in cultivating collaboration between staff, which is essential in ensuring effective support provision.

Leadership drives support initiatives

In the school sites, members of the *leadership team*, most notably the principal, were key in driving learning support initiatives at the school level. In many sites, the learning support teacher was a member of the school's leadership team and worked closely with the principal in determining the school's approach to learning support. Devolved leadership practices were also evident at some sites. In these cases, the learning support teacher/s assumed a leadership function in relation to the implementation and monitoring of support initiatives. Where this occurred, the learning support teacher/s were supported by management, and it was clearly communicated to, and understood by, staff in the school that they were speaking with the support of the leadership team.

Strategic decision-making assists in managing student diversity

In the sites, one of the most widely practised techniques adopted for *managing student diversity* in classrooms was ability-based grouping of students for targeted curricular and pedagogical delivery. As implied, targeted curricula refers to groupings for discipline-specific instruction. That is, across all sites, most ability-based groups were formed for reading, writing and/or numeracy. For other curriculum areas, students remained in their usual multi-ability-level classrooms. At all 14 school sites, ability-level grouping was a practised strategy for managing diversity, though implementation varied across sites. We now consider some of these variations and the features of these structures.

Ability-level groups were formed within class, across year-level cohorts, and across year levels. The group could be located within the classroom or another site.

- *Within-class grouping*—practised at many sites, with classroom teachers supported by a team teacher, learning support teacher, teacher aide or volunteer. Different ability-level groups would most often work on the same material, though with adapted expectations based on the group's level. In some site observations, groups were working rotationally with, for example, one group on the computer, one group playing games and one group reading, with all having an overall focus on literacy. At Beecham Primary, a whole-part-whole approach used a combination of whole-class instruction at the beginning and end of the period, with the middle section focused on adapted programs targeted toward ability-level.
- *Across year-level cohorts*—at several sites, students were allocated to ability-level groups across the entire cohort at specific times of the day.[4] Students would form their group, which might necessitate a move to another classroom, and also could mean a different classroom teacher for that period of instruction. The two Catholic Education sites (Camdon Primary and Cloakton Primary) both practised this approach for literacy and numeracy, and more formally termed the groups Journey Groups, which in practice was closely linked to a secondary school model.
- *Across-year levels*—this form of grouping was not as common, due to the difficulties of timetabling, though the practice was in evidence. Due to the multi-age organisational structure of the two Catholic Education schools mentioned above,

---

[4] Interested readers should see Chelton Primary and Fulton Primary site reports in *Volume 4: Case Studies* (Wyatt-Smith et al., 2007).

> they both loosely fell into this category as well. Canville Primary used this approach, though being a very small school it was easily managed by the teaching principal, and Burgess Primary used this approach across Years 2–4 for Guided Reading. This practice was more easily adopted for students in year levels near to each other, but when used across many year levels, downward movement of students was limited due to the negative connotations associated with that placement.

The ability-level grouping structure was also used for targeted instruction in a *withdrawal context*, where the student's usual classroom may have been working on similar curricula, but the student met with like-ability students for a focused lesson directed by support personnel. An example of this is provided in Sandville Primary, where the focus student, Helen, was withdrawn from class to join a small group of students from other classes in the same year level for a numeracy support lesson with a teacher aide. Other sites also practised this approach, with students sometimes joining across year levels to meet as a group. An ultimate aim of the timing of the withdrawal lesson was to align it with the classroom program, which was not always organisationally possible across year levels.

Group membership was determined through periodic, school-based assessment, usually performed at the beginning of the year, the classroom teacher's or the support personnel's assessment of the student's 'readiness to proceed' (as noted by Fulton Primary). These assessments were usually related to observation of performance, work product or level attainment. The placement of the student into a new group would occur after consultation between the classroom teachers and support personnel, and on occasion with the parent or carer.

Support staff often worked with the 'lowest' ability-level group, though this varied. At some sites, the classroom teacher, often when groups were formed within the classroom, worked with the 'lowest' ability-level group, while a teacher aide would work with other groups.

Strategic decision-making assists in managing physical space and learning support image

Managing *physical space* was recognised as a leadership decision-making challenge that was important in ensuring the cohesiveness of all aspects of the school's teaching and learning. However, at the school sites, extremes existed as I demonstrate here. One large school found that it had no physical space available to take students withdrawn from the class for a learning support lesson. Support staff worked with students outside the classroom—in the hallway or in lunch areas. At the other extreme was Chelton Primary. Though not often a situation afforded to school leaders, it was a newer school where consideration for managing teaching spaces and learning support was given during the planning and construction of the school. The room chosen to be the learning support room was located in the middle of the primary school's building cluster. In addition to the location of the learning support room, all classrooms in the school were purpose built. With 18 classes and 21 available classroom teachers, the extra spaces enabled the Junior Primary year-level classes to be split into three groups at times throughout the day, thus reducing the student–teacher ratio for particular teaching events. The learning support teacher joined each year-level for one period per week, and at other times members of the learning support team joined and/or taught in the classroom. The extra space available in this built environment facilitated other staff joining the classroom to become part of the normal weekly teaching regime. The effect of this arrangement was a reduction in the negative connotations associated with the learning support staff targeting particular students.

At Chelton Primary, in addition to the benefits a purpose-built site provided, strategic decisions were made to specifically reduce any *negative association* with students entering the learning support room and/or receiving assistance from any of the learning support staff. For instance, the room:

- held computers with the Internet access that all students could use during breaks
- housed video and entertainment equipment, and other resources and materials accessed by classrooms throughout the day
- was used by the school's Reading Club, where students worked with volunteers and items that students could buy based on points earned in the Reading Club were displayed in one corner of the room.

The room was a hive of activity, with students continually visiting for a variety of reasons that did not always require them to speak specifically with any learning support staff member. Another initiative in the school that reduced the negative connotations of working with the learning support staff centred around an event in the school called Discovery Day,[5] primarily organised by the support staff.

Networks

As within any organisation, the forming of networks and collaboration among staff is a key factor assuring success. Formalised networks (that is committees) and less formalised networks (support personnel) worked together at the case-study sites, with the central aim of identifying, planning, assessing and evaluating support programs. Effective sites incorporated a significant array of networks, with communication between members and across networks required for success.

[5]Discovery Day was run over one whole day for all students in Years 4–6. With assistance of students in Year 10, teachers and volunteers, the day included numerous activities whereby students could select electives based on what they would like to discover (for example, digital photography, robotics, pottery, orienteering, first aid, computing and 'weird science'). The only requirement of students' choice was that they did something new or different from their range of experiences.

Membership varied for the network's purpose and, importantly, included specialists and parents—and students in some cases.

Consultative committees form for decision-making

With the exception of two schools, key support decisions were made through a formalised committee process. In the main, membership in these *committees*[6] was fixed and common across sites, and usually included the principal, deputy principal, curriculum coordinator, learning support teacher/s, guidance officer and other specialist school staff. Flexible membership was held by classroom teachers, allied health professionals, liaison officers and parents. The case-by-case discussion for these meetings determined the committee's full membership for each meeting.

Regular meetings were scheduled at most sites, once per week or fortnightly, depending on school and student characteristics. In the main, the agenda of meetings included discussion of student cases, whether presenting as new cases or cases for review, the latter involving discussion of whole-school support plans and initiatives. In most sites, there were clear processes guiding referral of student cases to the committee, with the responsibility for initiating these most often being with the classroom teacher. Several sites[7] formalised the referral process or intervention plan using flowcharts.

At all sites, *decision-making in relation to support provision* for individual students was 'point-in-time' *evidence-based.* A range of sources of information was considered by committee members during the decision process. Common sources of information were work samples, classroom teacher assessments and observations, results in school-based and statewide testing programs, and learning support teacher and specialist assessments, if available. At all sites, the classroom teacher consulted parents prior to the final decision to refer the student to the committee. Committee decisions regarding support provision, with details on follow-up or cyclic review recorded as part of a formalised register. Each site maintained a register system in print and/or digital format (see Beecham Primary).

The creation of teams for *curricular and pedagogical planning* also enhanced teacher ownership of local support practices. The teams, comprising classroom teachers and often the curriculum coordinator, were responsible for planning and implementing classroom curricula, ensuring that students' individual programs were incorporated.

---

[6]The committee name varied across sites ranging from Special Needs Committee, Student Support Team, Special Needs Intervention Committee, Special Needs Assessment Committee and Special Needs Team.

[7]Interested readers should see Beecham Primary, Chelton Primary and Shepton Primary in the InLaN report, *Volume 4: Case Studies* (Wyatt-Smith et al., 2007).

Networks of support personnel plan, implement and assess support

Intervention was most effectively provided through a strong network of personnel. In sites, key network members were learning support teachers, classroom teachers, teacher aides, volunteers and parents/carers. In many of the schools, the network also included specialist teachers and allied health professionals. The formalised concept of the network existed in several schools. For example, Cloakton Primary viewed the Intervention Network as the third layer of support in the school's learning support approach. Across all sites, it was recognised that a group of individuals had agency and responsibility for providing support to students, not just one stand-alone figure.

High levels of dedication and enthusiasm were evident in members of the support network, most notably the learning support teachers and teacher aides in some sites. They were *key members* often recognised by principals and classroom teachers as the drivers of support who also led local additional professional development. In relation to the value of networks, the case studies brought to light different understandings about the role of the learning support teachers, especially in relation to the classroom teacher and teacher aids. Overall, there was a clear move away from withdrawal as the primary mode of intervention support. This was consistent with the move in some schools to regard the classroom teacher as the 'coordinator of intervention' (see Fulton Primary). Where this model was emerging, the stance taken was that the classroom teacher spent most of the day with the student and was the person primarily responsible for implementing or overseeing adapted programs. Additionally, they were responsible for continuing assessment and were best placed to coordinate support and monitor improvement.

At many sites, teacher aides were viewed as pivotal to the success of effective intervention and were highly valued members of the teaching community (see Cloakton Primary). At some sites, teacher aides received targeted professional development that enabled them to administer particular small group and individual literacy and numeracy intervention programs (see Sandville Primary). Shepton Primary provides a further example of a school with an early intervention focus (Preschool to Year 2) that relied heavily on having trained teacher aides in classrooms, 4 days per week, full-time. There was clear and growing acknowledgement of the value of teacher aides' experience and training, with some sites recognising their role as trained paraprofessionals (see Canville Primary, Hillside Primary and Shepton Primary).

The matter of how networks functioned in the local contexts was directly tied to the availability of *specialist support and allied health professionals*. Where specialist support was available, the overall effectiveness of support program initiatives was improved.

Support

Features of the support category include identification of the students to receive support, alignment of support delivery with literacy and numeracy teaching and the use of a range of approaches and programs for support delivery. How support delivery is enacted in practice is often the result of whole-school planning and policy decisions. Sites with effective support provision had clear methods for identifying the students for whom to provide support, worked within current whole-school literacy and numeracy programs to allow support to a wide definitional group of students and to reduce disruption to classroom programs, and incorporated a range of approaches and strategies as a part of a student's learning support program. Further, effective sites used evidence-based approaches to determine support provision and for continuing assessment of student progress and evaluation of program and/or approach.

Scope of provision: who are the students?

At some sites, the DEST (2002) learning difficulties definition was used as a guide to identifying students, though it was not treated as a stand-alone criterion to determine support. Evidence-based decision-making approaches were used in identification, including school-based and statewide testing, classroom teacher observations and, at times, through expressed parental and student concern. However, it was clear at many sites that many more students were supported than classically fit any definition, as illustrated by the following comments from teachers at the sites

> Our programs support all students' learning, not just those with learning difficulties.
> All students are targeted for support—all have various needs.
> Any student experiencing difficulty with their learning.

Essentially, it was evident that assistance was provided to a wide range of students and, in some cases, to all students with varying needs. Specific prevalence figures, detailed for some schools, also supported the notion that more students were being supported in schools than previously reported in the literature. Rates of specific assistance to students varied from 9.5% in a very small school, to 20.4% in a medium-sized school to 77.5% (literacy only) in a remote school.

Evidence-based, decision-making approaches were used for identification at the sites. Students were identified through school-based and statewide testing (emphasis on early identification through the Year 2 Diagnostic Net), classroom teacher observations and, at times, through expressed parental and student concern. The trend of methods used to identify students with learning difficulties has not altered significantly in Queensland since the project by Louden et al. (2000).

It was also clear that sites provided support for students from other categories of disadvantage under the banner of learning difficulties, in order to access support networks already in place. Richardson Primary, for example, classified its entire student body as comprising those for whom English was a second or subsequent language, and estimated that 80% of students had hearing difficulties. At another site, students who fell short of disability funding categories but who were achieving at lower levels than their peers accessed support through the learning difficulties programs.

Literacy, numeracy and the curriculum: central to support

Organisational, whole-school management approaches were given high priority across sites as a means to ensure all students had focused access to literacy and numeracy learning. In line with this, nearly all sites adopted the notion of literacy and numeracy blocks, these being set periods of time dedicated to each domain, with variation across sites in terms of how this was managed. For example:

- Cloakton Primary used what was termed Dedicated Learning Sessions, whereby students would break into multiple year-level ability-based groups (Journey Groups) and move to new settings with the possibility of new teachers for set times of the day, where the focus was on literacy (reading and writing—1 hour each) and mathematics (1 hour). Five days per week, students formed their Journey Groups, with each stage of schooling breaking at different times of the day. The movement of the groups has already been discussed above.
- Fulton Primary broke into year-level, ability-based groups for Guided Reading at different times during the morning, for 30 minutes, 4 days per week.
- Shepton Primary's students broke into a 2-hour literacy block, 4 days per week, with a 1-hour focus on reading and the rest of the period focusing on various literacy-related activities.

Primacy was given to support for students with learning difficulties in aspects of literacy and numeracy over other Key Learning Areas. In many cases, this was purely a strategic management decision, often necessitated by the timetable in whole-school programs coordinated around key aspects of literacy and numeracy teaching. In some cases, students had modified classroom programs in addition to targeted learning support and, in these cases, modified programs did exist for these students during lessons in other discipline areas.

Support comprises a range of approaches and programs

The focus on *early intervention*[8] was the dominant approach by schools in providing overall support to students. Considered 'a prevention approach', early intervention was viewed as optimum for reducing learning support in the longer term. As such, it was valued as a deliberate prevention method for staving off the onset of difficulties students may experience. This approach to support was most commonly provided through the formative years of primary education, being preschool/preparatory to Year 3. Support programs and initiatives were directed at entire cohorts at the outset, with additional support targeted at the conclusion of the scheduled period of support to small groups or individual students who were determined through testing or observation as requiring continuing support.

---

[8]Schools with formalised early years intervention programs included Burgess Primary, Chelton Primary, Lakeside Primary and Shepton Primary, though many of the other sites also practised this approach to support.

Support programs and initiatives often were *layered,* in that there would not be a reliance on one particular approach in supporting students across these early years. Across sites, oracy and phonemic awareness were given primacy in the formative years through the implementation of school-developed programs constructed to meet the particular needs of the student population and a mix of commercially available programs (for example, *Letterland*, *Jolly Phonics* and *Smart Words*, among others).

In addition to early intervention approaches, most sites also implemented a *suite of programs* across the school, often done so to address financial and human constraints that limited the number of students who could be supported. In implementing a suite of programs, strategic management of these restraints ensured a higher level of support to a greater number of students. Across the 14 case-study sites, there was no typical approach[9] to the support of all students or the learning support approach for students with learning difficulties. At each site, programs

Table 6.1 Lakeside Primary's support program showing early intervention and whole-school plans

<table>
<tr><th></th><th>Year 1</th><th>Year 2</th><th>Year 3</th><th>Year 4</th><th>Year 5</th><th>Year 6</th><th>Year 7</th></tr>
<tr><td rowspan="5">Early Years Intervention Program</td><td colspan="2">Gross Motor Skills Program</td><td></td><td></td><td></td><td></td><td></td></tr>
<tr><td colspan="2">Inclass support for At Risk students with teacher aide or teacher assistance</td><td></td><td></td><td></td><td></td><td></td></tr>
<tr><td colspan="2">Early Literacy and Numeracy Program (all students)</td><td>Early Literacy and Numeracy Program (students identified in Net)</td><td></td><td></td><td></td><td></td></tr>
<tr><td colspan="2">Introduction to the New Basics syllabus</td><td></td><td></td><td></td><td></td><td></td></tr>
<tr><td colspan="2">Team teaching</td><td></td><td></td><td></td><td></td><td></td></tr>
<tr><td rowspan="6">Whole-school approach</td><td></td><td></td><td colspan="5">Fully implemented New Basics syllabus</td></tr>
<tr><td colspan="7">Philosophy in Schools</td></tr>
<tr><td colspan="7">Environmental Education in context</td></tr>
<tr><td colspan="7">Mathematical problem-solving</td></tr>
<tr><td colspan="7">Multiage classroom organisation</td></tr>
<tr><td colspan="7">Withdrawn, individualised or small group support as required</td></tr>
</table>

[9]For an overview of some of the varieties in approaches see *Part B: learning support overview* of Burgess Primary, Chelton Primary, Camdon Primary and Lakeside Primary case reports in *Volume 4: Case Studies* (Wyatt-Smith et al., 2007).

Table 6.2 Burgess Primary's support program

Preschool	Year 1	Year 2	Year 3	Year 4	Year 5	Year 6
Binyi: Wuku linking to Phonemic Awareness Program - KPAK						
Number patterning						
	Guided Reading in class ability-based groups	Guided Reading in ability-based groups across year levels				
		Reading Recovery				
			Buddy Reading – Years 3 and 4 students read to Year 1 students			
	Other commercialised programs as required					
Other individualised or small group withdrawn support as required						

Table 6.3 Camdon Primary's support program

	Year 1	Year 2	Year 3	Year 4	Year 5	Year 6	Year 7
Whole-school approaches	Home Class organisation for Integrated Units with adaptation for individual need						
	Dedicated Literacy Session in ability-based groups with DL Literacy Teacher						
	Dedicated Numeracy Session in ability-based groups with DL Numeracy Teacher						
		Smart Words Spelling Program					
	Multiage classroom organisation						
	Withdrawn, individualised or small group support as required						

and approaches were selected on the basis of local context considerations, including: student population characteristics and the aforementioned human and financial resources available. It was also common to see a mix of commercial and site-developed approaches in use across all year levels. Three site-specific examples, Tables 6.1, 6.2 and 6.3, show the variability in the delivery and range of programs across year levels. These three site examples also exemplify early intervention, whole-school and targeted approaches to support program planning.

The use of *withdrawal and in-class support* also varied across sites. Withdrawal remained widely practised across many sites, with some schools reporting strong

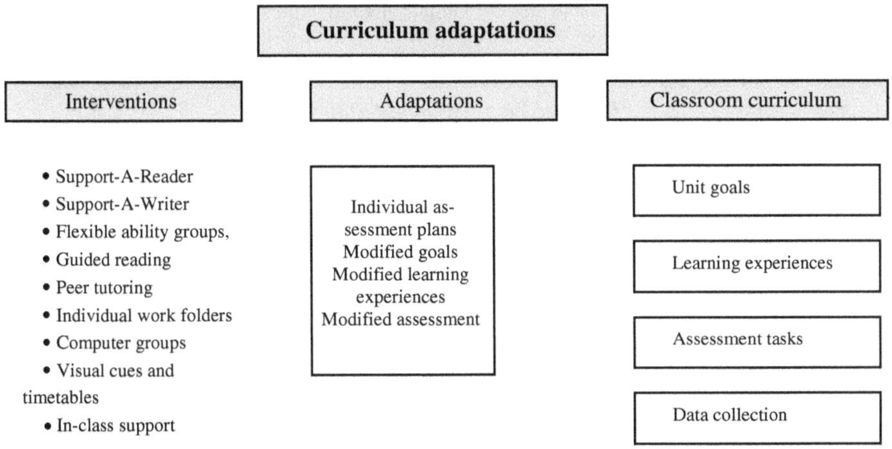

Fig. 6.2 Curriculum adaptations for students with learning difficulties

support for withdrawal as being the critical intervention for some students. This was specifically the case where students were identified as 'not contributing' in whole-class settings and where social intimidation was indicated. In the main, however, schools were making a move towards an in-class support model as the primary form of intervention.

Within *whole-class provision*, the focus on curricula adaptations to support the diverse range of students in the classroom was paramount, and the professional autonomy over planning facilitated the adoption of whole-school and learning support initiatives (see Fig. 6.2 for an example of a curriculum adaptation plan). A key strategy that sites were using in making the move to whole-class provision, and this was a trend across the 14 sites, was the identification and implementation of like-ability groupings. This was valued as a mechanism for managing diversity, specifically pertaining to diverse abilities.

Assessment and evaluation inform support provision

As there was no common or set list of recommended assessment materials or techniques, sites used a wide variety of school-based assessment protocols and strategies, in addition to statewide testing initiatives. In some sites, whole-school assessment programs were formalised[10] with set timetables for administration of testing and use of data for tracking purposes. While the priority of generating point-in-time assessment evidence to substantiate a judgment was paramount, far less attention was given to profiling individual and cohort performance over time.

[10] Interested readers should see Beecham Primary, Chelton Primary and Cloakton Primary for examples.

Table 6.4 Chelton Primary's whole-school identification process

Year level	Assessment item
Preparatory	*Brigance Prep Screening Test* (pre-entry)
	PIPS Screening Test (once in school)
Year 1	*PM Benchmarks* [a]
	Year 2 Diagnostic Net (August)
Year 2	*PM Benchmarks*
	Curriculum-based Math Pre Test
Year 3	*PM Benchmarks*
	Curriculum-based Math Pre Test
	Year 3 Aspects of Literacy and Numeracy Tests (August)
Year 4	*South Australian Spelling Test* [Westwood, 1999]
	Group Reading Test II—Sentence Completion—Form C
	Curriculum-based Math Pre Test
	Ravens Progressive Standardised Matrices (August)
Year 5	*South Australian Spelling Test*
	Group Reading Test II—Sentence Completion—Form D
	Curriculum-based Math Pre Test
	Year 5 Aspects of Literacy and Numeracy Tests (August)
Year 6	SO1 Testing
	South Australian Spelling Test
	Group Reading Test II—Context Comprehension—Form X
	Curriculum-based Math Placement Tests
Year 7	*South Australian Spelling Test*
	Group Reading Test II—Sentence Completion—Form Y
	Curriculum-based Math Placement Tests
	Year 7 Aspects of Literacy and Numeracy Tests (August)

[a] *PM Benchmarks* are resources used to assess students' abilities in aspects of literacy.

Several sites had in place a formalised *whole-school identification or monitoring* process (see Tables 6.4 and 6.5) with administration of testing at scheduled times of the year. Assessment protocols were, in most cases, administered by the classroom teachers, with consultation of results with the learning support teacher and other relevant personnel. Of note in the school-based assessment programs was the variety of testing techniques and programs used across all years. Several sites had introduced screening at the end of preschool (approximately 4–5 years old) beginning of Preparatory, which is consistent with the noticeable movement in the sites toward early intervention. Site-specific examples of formalised whole-school identification and monitoring programs follow.

Two sites were longitudinally tracking cohorts for evaluation of the effectiveness of interventions or for informing decision-making. The learning support teachers at Beecham Primary performed cohort analyses using data from assessments implemented by the classroom teacher as a part of the Whole-School Assessment and Monitoring Program. The classroom teachers were responsible for performing the class analysis, with advice provided by the learning support teachers as

Table 6.5 Beecham's whole-school literacy assessment and monitoring timetable

Year level	Term 1	Term 4
Year 1	PM Benchmark SPIOL—Oral Language Letter ID	PM Benchmark SPIOL—Oral Language Letter ID Canberra Word Test Hearing & Recording Sounds Writing sample
Year 2	PM Benchmark SPIOL—Oral Language Letter ID Hearing & Recording Sounds SA Spelling Test	PM Benchmark SPIOL—Oral Language Letter ID Hearing & Recording Sounds SA Spelling Test Burt Word Recognition
Year 3	PM Benchmark Letter ID Hearing & Recording Sounds SA Spelling Test	PM Benchmark Letter ID Hearing & Recording Sounds SA Spelling Test Burt Word Recognition Writing sample
Years 4–7	PM Benchmark SA Spelling	PM Benchmark SA Spelling Test Burt Word Recognition Writing sample

needed. There was a strong focus on partnerships and working together to understand assessment data, so that programs were ensuring positive outcomes for students.

In the main across sites, there were limited, easily accessible individual intervention histories for students, though some schools had implemented processes and methods for doing so as shown here with Sandville Primary and in Fig. 6.3 with Shepton Primary.

Sandville Primary had in place a variety of processes for recording and tracking individual interventions, including:

- a referral register
- an appraisement register
- an archive register
- an access database—recorded and tracked all students in the areas of mathematics, reading and comprehension. Using this database, the school was able to make queries to identify at-risk students (for example, Stanine 3 and under)
- a tracking register—monitored Year 1 students who had received intervention, Year 2 pre-net intervention students and Year 3 students who had received intervention before appraisement
- intervention register—once a student had been identified, an intervention register was used to track the support provided to individual students.

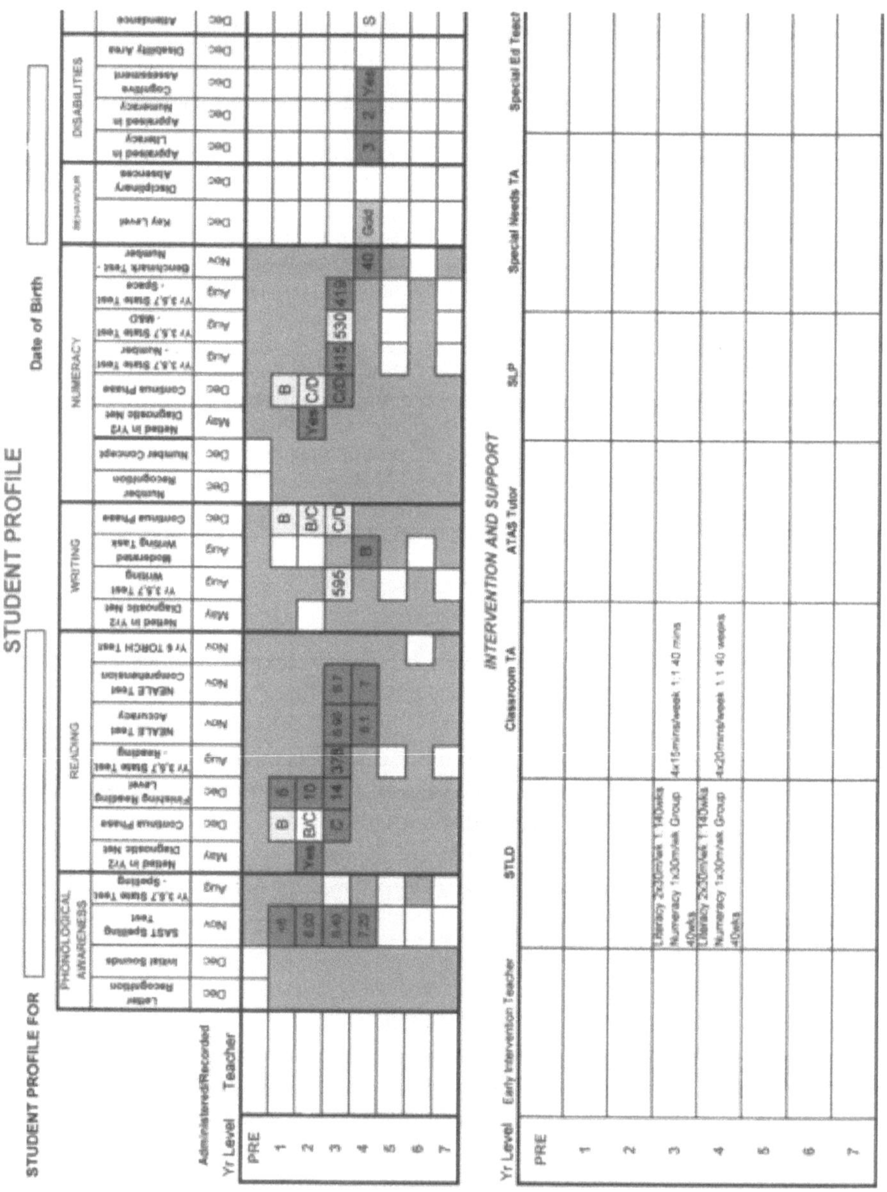

Fig. 6.3 Shepton Primary—student profile for tracking student results and intervention measure

The challenge for schools

As shown throughout this chapter, there are many contextually specific features that led to effective provision for all students, including those with difficulties in learning. Revisiting Fig. 6.1, shown in Fig. 6.4, the operational categories now include some of the features discussed in this chapter.

Each of the features represents to readers a decision point useful for analysing current practice. The emphasis placed on any one is a locally enacted decision with what works for one school not necessarily being the most appropriate selection for another. Every school, teacher and student presents a social and contextual mix that requires site-specific decisions to ensure that support provision is effective—measured ultimately by student improvement in learning. What is clear is that these features are shown to be effective in the case-study sites, so emphasis on many, if not all, of the features across the three categories, regardless of emphases is one way forward to work toward effective provision. Additionally, the importance of ensuring effective communication is maintained using a range of both informal and formal mechanisms within and across each of the categories cannot be understated.

The good news is that there are schools who are achieving positive outcomes for students with difficulties in learning and, most notable about these schools, is that the approach to support is data-driven with evidence-based decision-making leading

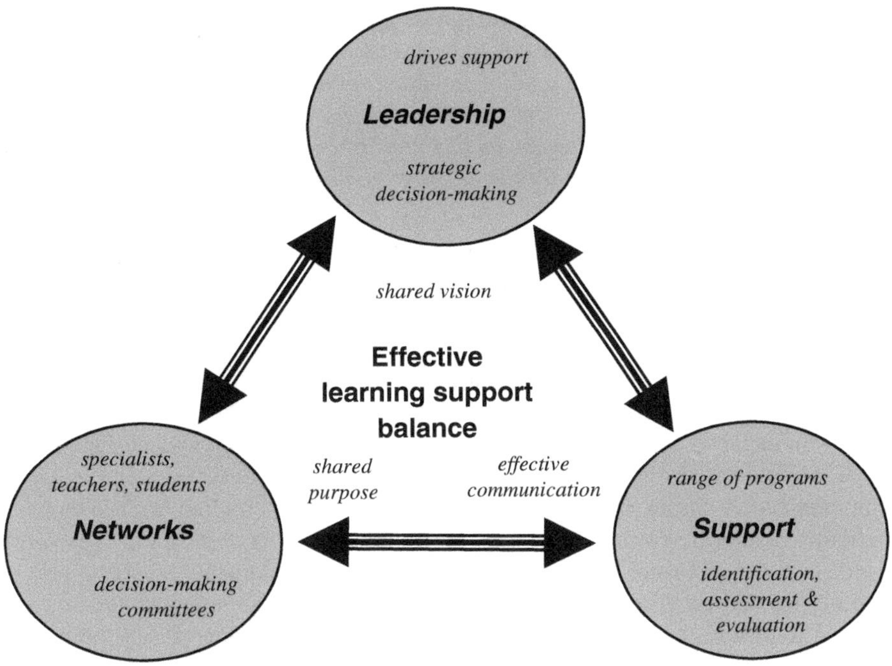

Fig. 6.4 Operational-based categories of support provision revisited

to alterations of individual and whole-school plans. The balance that fits the school, and how learning support is enacted in site-specific ways, will always remain the challenge for schools.

Essential next questions

Student agency

Except for several very specific teaching practices observed at the studied sites, student agency in learning support was not given priority. Yet, it was central to how learning, teaching and assessment routinely occurred. With a focus on students, the care of staff (expressed and observed), support program delivery and informal tracking of progress were evident; however, formalised mechanisms to ensure students 'had a say' in voicing their learning needs and interests and the help they received is less certain. In short, learning support provision was done to students who, in turn, were the recipients of the support. Wyatt-Smith and Gunn (2009) highlighted Sadler's (1989, 1998) work when looking at a student's role in the use of assessment standards. Specifically, they identified that his work on 'formative assessment "provided a model for a teaching–learning–assessment nexus that shows how improvement follows when students are inducted into assessment knowledge and expertise"' (p. 93). Underlying this, as Wyatt-Smith and Gunn pointed out, is the teacher's 'critical ability and willingness to facilitate students' transition from feedback to self-monitoring. For this to occur, the teacher must already possess the knowledge of what constitutes quality and must value opportunities for sharing this knowledge' (p. 93). Essential next questions for research include: *What are effective features of learning environments in which students are taught how to engage in self-assessment?* and *What are the elements required to ensure teacher capacity to create this environment?*

Parent, student, school partnerships

As Epstein (2001) noted, 'when parents, teachers, students and others view one another as partners in education, a caring community forms around students and begins its work' (p. 403). In the studied sites there was variability in how often parents were included in learning support decisions, though it was evident that parents were informed of decisions and some discussions with parents revealed the use of communication diaries and suggestions for helping their children with homework at home. Further development of the parent, student, school partnership or network needs to occur to ensure the promotion of two-way communication and support of approaches and programs. It is important that the networks envelop students and parents to ensure the forming of valuable partnerships, for the school, home and community. An essential next question for research relates to systematic *mapping of existing networks to identify effective network practices that promote supportive relationships and improved student outcomes.*

Longitudinal tracking

Across sites there was variability in the comprehensiveness of data analysis, with very few sites longitudinally tracking cohorts for evaluation of the effectiveness of interventions or to inform decision-making. The critical issue for all the studied sites related to access to, and interpretation of, reported large-scale testing data, knowledge of how to manage the data and time to complete the analyses. Some sites did not keep electronic records, which also hampered these efforts. Another concern for schools was the practice of tracking intervention histories and ways in which to provide an overall 'picture' of a student's achievement and strengths and weaknesses. There was certainly significant variability across sites in the collection, management and integration of student records by paper and electronic means. Essential next research questions include: *How can schools and systems work together to develop effective methods for electronically integrating achievement recording (school-based and statewide) with intervention histories? This focus reflects the understanding that such methods are a priority for determining the effectiveness of interventions over time.* Additionally, *training in the use of such systems is a priority for relevant school personnel, with access by teachers being the ultimate aim.*

References

Anderson, C. J. K., Klassen, R. M., & Georgiou, G. K. (2007). Inclusion in Australia: What teachers say they need and what school psychologists can offer. *School Psychology International, 28*(2), 131–147.

Andrews, R. J., Elkins, J., Berry, P. B., & Burge, J. A. (1979). *A survey of special education in Australia.* St Lucia, Queensland: Schonell Educational Research Centre.

Bourke, P., & Carrington, S. (2007). Inclusion education reform: Implications for teacher aides. *The Australasian Journal of Special Education, 31*(1), 15–24.

Chard, D., & Linan-Thompson, S. (2008). Introduction to the special series on systemic, multitier instructional models: Emerging research on factors that support prevention of reading difficulties. *Journal of Learning Disabilities, 41*(2), 99–100.

Commonwealth of Australia. (2004). *Researching numeracy teaching approaches in primary schools.* Canberra: Author.

DEST (Department of Education, Science and Training). (2002). *Request for proposal for funding under the effective teaching and learning practices for students with learning difficulties initiative.* Canberra: Author.

Elkins, J. (2002). Learning difficulties/disabilities in literacy. *Australian Journal of Language and Literacy, 25*(3), 11–18.

Epstein, J. (2001). *School, family, and community partnerships: Preparing educators and improving schools* (pp. 403, 590). Boulder, CO: Westview Press.

Forlin, C. (2005). Moving onwards and upwards: Inclusive education in secondary schools. *Australian Journal of Special Education, 29*(2), 87–91.

Geijsel, F. P., Sleegers, P. J. C., Stoel, R. D., & Krügern, M. L. (2009). The effect of teacher psychological and school organizational and leadership factors on teachers' professional learning in Dutch schools. *The Elementary School Journal, 109*(4), 406–427.

Graczewski, C., Knudson, J., & Holtzman, D. (2009). Instructional leadership in practice: What does it look like, and what influence does it have? *Journal of Education for Students Placed at Risk, 14*(1), 72–96.

Hargreaves, D. H. (1999). The knowledge creating school. *British Journal of Educational Studies, 47*, 122–144.

Hiebert, J., Gallimore, R., & Stigler, J. W. (2002). A knowledge base for the teaching profession: What would it look like and how can we get one? *Educational Researcher, 31*(5), 3–15.

Keogh, B. K., Gallimore, R., & Weisner, T. (1997). A sociocultural perspective on learning and learning disabilities. *Learning Disabilities Research and Practice, 12*(2), 107–113.

Louden, W., Chan, L., Elkins, J., Greaves, D., House, H., Milton, M., et al. (2000). *Mapping the territory—primary students with learning difficulties: Literacy and numeracy*. Canberra: Department of Education, Training and Youth Affairs.

MacMillan, D. L., Gresham, F. M., & Bocian, K. M. (1998). Discrepancy between definitions of learning disabilities and school practices. *Journal of Learning Disabilities, 31*(4), 314–326.

Mercier Smith, J. L., Fien, H., Basaraba, D., & Travers, P. (2009). Planning, evaluating, and improving tiers of support in beginning reading. *Teaching Exceptional Children, 41*(5), 16–22.

National Health and Medical Research Council. (1990). *Learning difficulties in children and adolescents*. Canberra: Australian Government Publishing Service.

Pressley, M. (2002). Conclusion: Improving comprehension instruction: A path for the future. In C. C. Block, L. B. Gambrell, & M. Pressley (Eds), *Improving comprehension instruction: Rethinking research, theory and classroom practice* (pp. 385–399). Hoboken, NJ: Wiley.

Rhodes, C., & Beneicke, S. (2002). Coaching, mentoring and peer-networking: Challenges for the management of teacher professional development in schools. *Journal of In-Service Education, 28*(2), 297–310.

Rogoff, B., & Chavajay, P. (1995). What's become of research on the cultural basis of cognitive development. *American Psychologist, 50*(10), 859–877.

Rouse, M., & Florian, L. (1996). Effective inclusive schools: A study in two countries. *Cambridge Journal of Education, 26*(1), 71–85.

Sadler, D. R. (1989). Formative assessment and the design of instructional systems. *Instructional Science, 18*, 119–144.

Sadler, D. R. (1998). Formative assessment: Revisiting the territory. *Assessment in Education, 5*(1), 77–85.

Vygotsky, L. S. (1987). *The collected works of L. S. Vygotsky, Volume 1: Problems of general psychology*. New York: Plenum Press.

Westwood, P. (1999). *Spelling: Approaches to teaching and assessment*. Camberwell: ACER Press.

Wyatt-Smith, C. M., Elkins, J., Colbert, P., Gunn, S. J., & Muspratt, A. (2007). *Changing the nature of support provision - Students with learning difficulties: Interventions in literacy and numeracy project (InLaN)*. Canberra: Department of Education, Science and Training, Australian Government. Available from www.dest.gov.au/sectors/school_education/publications_resources/profiles/effective_teaching_learning_practices_stud_learn_difficult.htm#authors

Wyatt-Smith, C. M., & Gunn, S. (2009). Towards theorising assessment as critical inquiry. In C. Wyatt-Smith & J. Cumming (Eds), *Educational assessment in the 21st century: Connecting theory and practice* (pp. 83–102). Dordrecht: Springer.

Chapter 7
'Reading' the Home and Reading in School: Framing Deficit Constructions as Learning Difficulties in Singapore English Classrooms

Anneliese Kramer-Dahl and Dennis Kwek

Introduction

> Half of the students in the Normal (Technical) stream leave their classroom every day not understanding their lessons. They spend less than an hour a day studying. One in four said they had difficulty studying because their English was poor. At home, they speak mainly Mandarin or dialect. Poor study skills and habits are common. These students have trouble concentrating and listening in class, so they also have difficulty remembering what has been taught. They also manage their time poorly (Tan, 1996).

The above excerpt is from a newspaper article published in the *Straits Times*, Singapore's English broadsheet, and reflects the popular perception about what makes a 'Normal (Technical)' (NT) student, so named because students in Singapore are 'streamed'—labelled into categories so that they can be assigned in groups into different kinds of classes. NT students are classified as the 'slowest learners' in Singapore's secondary-schooling system and, as the excerpt shows, are often assumed to have learning difficulties. In a highly disciplined society such as Singapore, where educational failure tends to have far-reaching consequences, the education system is quick to implement measures such as streaming to help 'slower' students. Subsequent failure on the part of streamed students to do well in class often leads to an attribution of blame by teachers and policy makers: blame on the student's innate intelligence, family background or personality. Each of these, however, assumes that these students have a *deficiency outside the school* that is *primarily responsible* for the learning difficulties they have *in school*. Such a way of reasoning is, in and of itself, not necessarily 'wrong'. Students from the lower streams may be inundated with very real material problems that impact upon their abilities to perform in school. The issue with deficit beliefs and thinking is the insidious way in which they seek to provide the obvious, often visible or measurable, reasons for learning difficulties while pointing the arrow of blame away from the methods and materials we use in classrooms that give rise to them in the first place, and, ultimately, from the invisible but broader structural configurations and

A. Kramer-Dahl (✉)
National Institute of Education, Nanyang Technological University, Singapore, Singapore
e-mail: anneliese.kramer@nie.edu.sg

C. Wyatt-Smith et al. (eds.), *Multiple Perspectives on Difficulties in Learning Literacy and Numeracy*, DOI 10.1007/978-1-4020-8864-3_7,

ideological practices that perpetuate the conditions for school failure and educational inequality. Unless brought to the surface, interrogated and replaced with alternative frameworks of viewing students and their academic difficulties, the circularity and persistence of deficit constructions will continue to create challenges to innovations and reforms that seek to improve the lives of teachers and students alike.

This chapter provides case studies of two secondary English teachers in Singapore as they participated in a professional development project to build their repertoires of reading and teaching reading. We examine the perceptions these teachers have of their students' learning difficulties, and how these frame their discourses about student (in)abilities and affect the way they construct and enact their curriculum. Our analysis is informed by theories on deficit thinking and the larger debates on class, ethnicity and normative views on young people, family and schooling (for example Comber, 1998; Valencia, 1997). Through our exploration of deficit constructions of students' learning difficulties, we hope to extend current research on deficit discourses by discussing how teachers' assumptions about their students' reading habits and family backgrounds constrain what is offered in their classrooms as well as severely delimit what can be achieved in professional development attempts to help them broaden their offerings. Ultimately, we aim to explore teachers' ways of accounting for learning difficulties through probing the relationship between teachers' beliefs about 'what their students bring from home' (Freebody & Baker, 2003, p. 237)—beliefs by no means invented by the teachers but part of the larger public common sense—and how these inform what comes to count as reading instruction in school.

The chapter is structured as follows. First, we offer a review of current research on deficit thinking. Second, we briefly outline the educational context of Singapore; in particular, the practice of streaming. Third, we describe the research context, the methodology and selection of data. Fourth, we present a multi-layered analysis of teachers' deficit beliefs, using categories that recurred throughout the duration of the project, with the proviso that these categories are interlinked in theory. Finally, we close with a discussion of essential questions that can drive further research on deficit thinking.

Review of deficit thinking

The way teachers teach, and the strategies they employ in classrooms, do not emerge from a vacuum. Even though pre-service and inservice education helps to equip teachers with numerous approaches to prepare them for diverse student populations, more often than not such knowledge is itself shaped by deficit assumptions (Gutierrez & Orellana, 2006), and then in turn gets filtered through teachers' perceptions and expectations about the nature of learners and learning. Moreover, variations of the same discourses circulating in educational policies, popular media and social networks further impede upon and reinforce teachers' belief systems. Over time, teachers become inducted into particular discourses that constitute certain students as deficient in their ability to learn in school. Such student deficits are often attributed to a range of factors—from low socioeconomic status, lack of access to resources, dysfunctional family background, lack of motivation to study and language difficulties, to innate intelligence and poor genes (Comber, 1998; Valencia, 1997). Yet, these habits of 'naming' students and their difficulties become particularly counter-productive, if not dangerous, when held by teachers, who are, after all, the most important variable in making a difference for students (Comber & Kamler, 2004).

To elaborate, deficit thinking constructs a *logic of practice* (Johnston & Hayes, 2008) that attributes failure or learning difficulties in the following manner. First, deficit discourses construct homogeneous stereotypes of student groups, drawn from the causes of their learning difficulties (for example intelligence, family, socioeconomic background). Blame for these attributes is then levelled at individual students and their families. Learning difficulties therefore become *individualised*, and the solution is often *neighbourhooded* (Freebody & Welch, 1993); that is confined to the families and individuals. These discourses in turn prevent those who are thus constituted from recognising that the constructions are not natural but made. Instead, these students are coerced into accepting the validity of such constructed representations of themselves as lacking attributes and cultural resources, which a presumed 'mainstream' student possesses. Further, they are expected to view their learning difficulties as the result of personal inability and lack of effort, rather than their position within the social hierarchy and other structural factors. Finally, any alternative explanations for deficit constructs are rejected if they do not fit into existing belief systems, furthering the continued resilience of such constructs against attempts to problematise them. Unfortunately, deficit constructs have considerable impact on the expectations and beliefs of policy makers, researchers and teachers, through limiting what can be accomplished. For example, Bartolome (1994) argues that any pedagogy that subscribes to a deficit-driven belief system that renders groups of students disadvantaged and in need of 'fixing' is doomed to be ineffective. In this way, a vicious cycle, starting and ending with deficit thinking, perpetuates educational failure and learning difficulties. What gets obfuscated in this logic of practice is the ways in which schools and the macropolitics of educational policies are structured to prevent learning and exacerbate the unequal distribution of educational resources. This shift in attention away from structural and institutional practices prevents a critical examination of the circulation of deficit constructions among teachers, and the impact of these upon teaching practices in the classroom.

In order to provide a deeper understanding of how deficit constructions operate in Singapore's context, we describe in the next section the country's education system and the place the policy of streaming occupies within it.

Singapore's education system

Singapore is a compact, highly urbanised, nation-state in Southeast Asia with a population comprising multi-racial, multilingual and multi-religious citizens. Over 75% of the population is of Chinese, Malay (15%), Indian (7%) and other (3%)

background, and the country's official languages are English, Mandarin, Malay and Tamil. English is the dominant language of administration, business and education. Since independence in 1965, education has continuously been a key policy focus, having significant governmental investment. Realising the importance of producing an efficient education system, the government initiated a series of reforms designed not only to update the system, but to ensure that schooling embodies a meritocratic ideal—that every child has a chance to succeed in school and society so long as they put in an effort. In the late 1970s, in the midst of aggressive economic growth, the government implemented a massive restructuring of the school system through 'streaming', or ability grouping, students at both primary-school and secondary-school levels according to their learning abilities. This was to curb student failure and streamline students' educational experiences (Ministry of Education (MOE, 1979)). Through a series of high-stakes examinations that test for academic performance, streaming subsequently hailed a major success in ensuring near-universal primary and secondary education for all Singaporean children.

To be more specific, all students undergo a 6-year compulsory primary school education, during which a streaming examination is conducted at the end of Grade 4 to determine which of the two major ability streams students will be channelled into. A glance at the streaming criteria shows how the education system promotes English and mathematics (taught in English) as important subjects, and thus favours certain groups of students. It is therefore not surprising to find students from non-English-speaking homes over-represented in the lower streams, especially those of minority Malay ethnicity, given the importance placed by their families on retaining their mother tongue as the language of the home (Barr & Low, 2005).

After primary school, higher-stream students are usually progressed onto the academic track—the 'Express' (E) stream—for their secondary school education, and, eventually, universities. However, primary school students who are placed in the lowest stream typically end up in the 'Normal (Academic)' (NA) and 'Normal (Technical)' (NT) secondary streams, and subsequently progress to the polytechnics or vocational institutes that provide training for a range of manual, service and technical jobs. In 2005, about 40% (17,000) of the secondary-school cohort were in these Normal streams (Ng, 2005). Most of these students were over-represented in what are locally referred to as 'neighbourhood schools'—so named because they are located in mostly underprivileged, working-class neighbourhoods and receive students from these areas. With few exceptions, the neighbourhood schools tend to be poorly positioned in the national school league table, a ranking and appraisal system that bands schools annually according to academic performance.

The practice of streaming is, of course, not unique to Singapore. Indeed, it has been vociferously debated in educational circles outside Singapore (for example Gamoran, 1992; Oakes, 2005), but far more reluctantly within Singapore itself (Singapore Government, 2003). On the one hand, streaming rests on the premise that children have different inborn abilities, which means that the supposed benefit of streaming stems largely from teachers tailoring their instruction for a homogeneous group of students studying at the same pace. On the other hand, despite the ideal of optimal learning for all, streaming has turned out more often than not to be unfair and harmful, especially to lower-stream students. One of the major reasons given is that streaming fails to lead to long-term equitable outcomes, since the students' particular stream status affects teachers' perception of their abilities, which in turn leads to their qualitatively different instructional treatment (Caughlan & Kelly, 2004; Cazden, 2001). Because teachers 'teach to the average level of students' ability in their classrooms' (Lleras, 2008, p. 890), the streaming and school-ranking systems, taken together, have the unintended effect that teachers—particularly in neighbourhood schools—often perceive their students' learning difficulties as an aggregate at the school level. This *tacit streaming* is salient when teachers compare their students with students from other schools; in Singapore it is common to hear teachers speak of how their E students are equivalent to NA students in what they perceive to be a better school. Streaming, therefore, creates a highly stratified system of schooling, consequently manifesting a situation whereby perceived student learning difficulties have yielded common consequences for stakeholders. For teachers, streaming encourages a deficit-based pedagogy and curriculum for students in the lower streams, frequently no more than a caricature of what those in the higher streams receive. For students and parents, there is increasing pressure to perform well during high-stakes examinations, in order to avoid streaming downwards. Some critics argue that, given its introduction at such an early age and its 'intrinsic anticipatory nature', streaming has been 'directly though not solely responsible for the creation of the current Singapore school system's pressure-cooker emphasis on grades and examination' (Barr & Skrbis, 2008, p. 114).

The research context

Introducing the research project

The data and findings reported in this chapter are derived from a 2-year professional development project that focused on building teachers' pedagogical capacities in secondary English language teaching (Kwek, Albright, & Kramer-Dahl, 2007). Started in 2006, the project was premised on the broad findings that Singapore English language teachers had limited capacity to generate local, school-based curriculum that was coherent and responsive to students' needs (Luke, 2005). Furthermore, despite the secondary English syllabus calling for higher-order work with texts, teachers tended to shy away from setting intellectually rich or cognitively demanding curriculum and classroom tasks, resulting in a reduced syllabus.

The focus of the project did not include an investigation into the nature of teachers' deficit constructions. Over the project's duration, however, teachers' reductive constructions of their students consistently undermined the professional development work, providing ready-made ways of accounting for their pedagogical difficulties and for their narrow and shallow curricular offerings, and severely limiting our attempts to get them to envision alternative, more open-ended notions of literacy in general and reading in particular. It was the persistence and durability, even after our repeated interruptions, of these deficit ways of thinking and talking that compelled us to revisit our project data and seek to understand the relationship between classroom interactions, teachers' beliefs and the reproduction of educational inequity.

In the following sections, we provide brief descriptions of the school and the teachers to help understand the background and context of the subsequent analysis.

Introducing Bukit Secondary School

Bukit Secondary School (BSS) is situated in a neighbourhood that is one of Singapore's poorest, comprising largely working-class residents in a 'mature' estate. Given its catchment area, it is no surprise that BSS offers mainly NA and NT classes. In 2006, it was ranked as Band 3 (with 4 being lowest) on the national school league table, a stark contrast to most independently funded and government-aided schools that consistently make Bands 1 and 2 rankings. Although official figures were unavailable, our observations indicated that at least half of the school's students were from ethnic minority backgrounds, and almost all of them come from non-English-speaking homes.

Because BSS's annual O-level English language examination results have been relatively weak, it encounters far less pressure to maintain annual league table standings than many other schools. Subsequently, the school decided it could afford to take the risk to initiate a school-based, teacher-designed English language arts curriculum in 2005, a year prior to our project. The Language Arts Programme (LAP) was designed for the lower secondary level, with the aims of making English language learning more interesting and engaging for its students and de-emphasising dense curriculum coverage and high-stakes examinations in favour of the injection of some project-based learning. The principal, herself a former English teacher, told us that she and the teachers believed that because their students, given their 'special background and their more aesthetic leanings', were underperforming in their English language examinations, there was 'no harm in trying something different' (Interview, 16 October 2006).

Methodology

Black (2007) argues that in order to explore institutionalised classroom practices and how ideological and sociocultural influences are manifested in classroom talk, there is a need for a multi-layered methodological framework that can enable an analysis of how teachers' beliefs and intentions impact upon the micro-level of classroom practice, how they are connected to broader ideological and cultural discourses at the macro-level, and ultimately how they 'reproduce the unequal distribution of educational capital' (p. 22).

The framework draws upon multiple data sources in order to interpret the findings. In our case, each source sheds light upon teachers' deficit constructions of their students' learning difficulties. We limit our analysis to two teachers—described below—who represent different inflections of how teachers' deficit beliefs impact upon classroom pedagogy. The layers of analysis are as follows:

1. The first layer presents a classroom discourse examination of the teachers' reading lessons. It considers the nature of the pedagogy enacted by the teachers and what the teachers validate as valuable knowledge in the English language classroom.
2. The second layer presents dominant themes that emerged from data from teacher interviews and reading circles that illuminate the two teachers' beliefs about their students' learning difficulties, their family backgrounds and the teachers' role in BSS.
3. The third layer examines policy, social and institutional discourses that serve to circulate normative understandings of students' difficulties to learn in school. Specifically, we investigate how teachers' deficit beliefs are aligned with broader discourses about particular groups of students with learning difficulties, and how these are recontextualised into a highly limited, and limiting, pedagogy and curriculum.

Introducing the two teachers: Mrs Tan and Mrs Chan

Mrs Tan, who had a master's degree, was the English language department head, and, at the beginning of our project, had been teaching for 7 years at BSS. Aged in her early 30s, she came from a middle-class family. She had spearheaded the LAP innovation, conceptualising its overall goals and structure, and ensuring that the program conformed to the national English language syllabus. As head, she was always concerned about the English academic results of BSS students.

Mrs Chan was a teacher in her late 30s who had taught at BSS for over 3 years. A university graduate and from a middle-class background, she taught English across all secondary levels. She was also a teacher counsellor for many of the school's lower secondary students. These two teachers were chosen because they represented different inflections—one psychological and the other moral, bordering on religious—of the deficit discourses that are shared among the teachers. As we will see below, at one extreme was Mrs Tan, with her strong beliefs about her students' cognitive and linguistic deficiencies, which impeded successful learning, and at the other extreme was Mrs Chan, with her equally strong insistence that her students, as products of largely 'problematic' families, needed their school and its teachers to provide the missing moral anchor.

A multi-layered analysis of teachers' deficit constructions

We begin the analysis with the first layer that examines the classroom interactions of the two teachers, before proceeding to the second layer on teacher beliefs and the third layer on policy, social and institutional discourses. It should be emphasised that, though presented below in a linear way, these layers have to be viewed, in a Chinese-box fashion, as interdependent and nested within each other, hence allowing for intricate moves forwards and backwards to be made between the different analytical foci.

Layer 1: analysing classroom interaction

Mrs Tan's comprehension lesson

The following excerpt is taken from a comprehension lesson taught by Mrs Tan to a secondary 2 NA class. The texts used were a video clip on the Hiroshima bombing, with an accompanying transcript. At first glance, this lesson seemed to attempt to raise students' meta-awareness of how to categorise comprehension questions and gain a better understanding of how to answer them. Mrs Tan started the lesson by asking students to write down a taxonomy of comprehension questions. According to her taxonomy, there were three possible comprehension question types: Knowledge (K) questions have answers directly extractable from the text; Inferential (I) questions have answers that are inferred from the text; 'In your Own words' (O) questions are extractable from the text but must have their key words modified.

After watching the video clip, Mrs Tan turned the students' attention to the accompanying comprehension worksheet.

The excerpt (Fig. 7.1) highlights a few points. First, what was being taught here was not unfamiliar to the students in Mrs Tan's class, as can be seen from the quick response by one of the students (Turn 137), the eager volunteering of responses by several other students in the background and her later acknowledgement that students should be familiar with K questions from primary school (Turn 150). Second, in response to an alternative answer, 'engulfed' (Turn 141), Mrs Tan is quick to discount the student's contribution as invalid, not because he was unable to understand the particular passage, but because he failed to comply with her terms for proper engagement with the text, which was to display a particular, narrowly conceived kind of comprehension. Such lack of compliance by students like S2 convinces her that despite many years of exposure her students are not yet 'ready' to move on to more cognitively and linguistically demanding questions, and that her highly repetitive, uni-dimensional question posing is necessary to ensure that students 'get it right because they keep failing to get the basics right' (Interview, 16 October 2006).

Overall, the pattern that emerged in the reading lessons we observed was the same: Framing them as rehearsals for examinations, Mrs Tan would, in recitation script, walk through lists of literal short-answer questions, usually discrete and independent of one another. Selected mainly for their ability to provide examination practice, the texts were treated less as meaning-making objects than as 'geological sites from which words and phrases must be laboriously quarried' (Dombey, cited in Dadds, 1999). Correct answers were therefore more a matter of adhering to technical procedures than of understanding.

136	T	Okay. Question two. Hiroshima can be said to be enveloped in fire. Pick up the phrase from which…
137	S1	'Totally covered.'
138	T	Exactly. It says 'pick'. That means it's from the passage. A 'K' question. What's the phrase?
139	S1	'Totally covered in fire.'
140	T	Okay, if you put down 'totally covered in fire' or if you say 'in fire', that's fine.
141	S2	'Engulfed.'
142	T	'Engulfed' is a good word. But is it found in the passage?
143	S3	No.
144	T	No. Did he read his question carefully?
145	Ss	No.
146	T	Okay. He is treating it as a what question?
147	S4	'O'.
148	T	In-your-own-words question. Right? If you don't read your question carefully, you will interpret as a different question type. When they say you get it out from the text, you get it out from the text. Unless they ask you, 'Hiroshima can be said to be enveloped in fire. In your own words explain what the author means by that'. Then he will be correct. Okay?
149	Ss	Yes.
150	T	In primary school, you are very used to K questions. You just lift it up straight. Now in secondary school, we have a few more question types. We have O, I. First passage we keep things simple for you. K question. But if you don't read carefully, sometimes you answer in the wrong way. It's never unclear.

Fig. 7.1 Mrs Tan's comprehension lesson (22 August, 2006)

Mrs Chan's Romeo and Juliet lesson

The following excerpt (Fig. 7.2) is from a lesson from a Poetry and Drama unit on the play *Romeo and Juliet.* According to the lesson objectives, the teacher, Mrs Chan (C), was supposed to discuss with her secondary 1 English students how characterisation is accomplished in plays. However, after a couple of minutes, she left the curriculum text and objectives behind to venture into an 'object lesson' of a moral point about what she perceived as her students' deficit in the area of character assessment.

In this question-and-answer routine, Mrs Chan wanted to make a point of discussion and subsequent moral commentary what she seemed to take as a student deficiency, namely their predilection to make hasty judgments about a person's character. Through elicitations, she moved the class away from their offers of superficial attributes towards the answer she had in mind: that a person's impression of another, often hastily formed on the basis of surface characteristics, should be open to revision upon hearing other people's opinions. In our conversation after the lesson, Mrs Chan referred to her students' moral and experiential location, and her out-of-class observations of their hasty judgments of the character of others as reasons for deviating from the curriculum. At the same time, it also gave her the

141	C	How do you tell a person's character? You look at your friends around you, your parents, or anybody else. Now, all of you came from different primary schools. On the first day of secondary school, you meet each other for the very first time in your life. As you mingle, you realise there are certain character traits of this classmate. So how do you tell what kind of person your friend is?
142	Ss	Actions. Attitude. Appearance. The way they talk!
143	C	Yes. Not only the way they talk, but the words used. If you come in, you hear Jason use the F word all the time, you know what kind of person he is already, right? But if he speaks nicely, gently, you also know what kind of person he is, right?
144	S6	Background of a person.
145	C	Background of a person? How? When you see a person, is the background written on the person? I live ... I come from a ... But this one not so easy, *lah*. Not so easy to know unless you spend enough time with the person. There's one more. Can you think?
146	S2	Start with what letter?
147	C	Okay, if I tell you, 'John is such a lazy boy, never does his homework on time'.
148	S7	Gossip.
149	S1	Impression.
150	C	The impression of John from ...
151	S1	Another person.
152	C	Me! So you also get to know people's character from ...
153	S2	Another person.
154	C	The impressions you gathered from other people! But someone might say, 'No, *lah*. John changed. He's such a good boy, used to be a terrible student! Now you see him, he's the top student in school!'
155	Ss	Wah.
156	C	Impression change again, right? So you gather information not from one source, but from various sources to find out a person's character, okay?

Fig. 7.2 Mrs Chan's Romeo and Juliet lesson (24 August 2006)

chance to weave in examples, albeit half-way humorously, that singled out certain students and their misbehaviour (Turns 147 and 152)—and make it part of her lesson on morality. In her view, because students like the ones in BSS lacked support structures in their homes to inculcate such civic dispositions and moral values, the logical extension was that she had to take compensatory measures within her own language classrooms.

In Bernsteinian terms, these lessons from the two teachers, despite overtly privileging different kinds of knowledge, were nevertheless similarly strongly framed, as the teachers decided what knowledge was to be transmitted and students were given few opportunities to deviate from the sequence and selection of the knowledge (Bernstein, 1990). As for the knowledge on offer, particularly the ways with which language and texts are being co-constructed, these lessons appeared relatively shallow and to dwell on the familiar. In Mrs Tan's lessons, the focus was on basic literacy skills through the recycling of what was taught in primary school, with most of her reading instruction driven by the examination agenda. In Mrs Chan's lessons, the language or literacy element was often barely evident, with the text on hand

and the curriculum topic serving largely as openers for extended talk about cultural and moral values. Texts were therefore treated as distant objects around which ritualised, almost mechanical language and exercises were set, with lessons drawing attention away from text and language, towards commonsense knowledge about morals, manners and civic sensibility (Hunter, 1994).

Layer 2: analysing teachers' beliefs

In order to understand why the teachers taught the way they did, and why they privileged certain kinds of knowledge over others, it is important to unpack the beliefs and values they held. This section analyses teachers' beliefs through their participation in the reading-circle discussions, and the occasional interview. In examining the reading circle and interview data, two recurrent themes emerged, which shed light on the teachers' deficit beliefs. The themes—*learning difficulties* and *moral and cultural ballasts*—are not isolated clusters of meaning, but are interlinked with one another in how teachers perceived their students. For explication purposes, they are separated below.

Teachers' beliefs about learning difficulties

In our interview with her, Mrs Tan expressed the belief that her students' English learning difficulties were attributable to the lack of opportunities to use the language, especially at home: 'They speak their mother tongue at home, and in school, too. There's a lack of English exposure at home and in school, even during English lessons' (Interview, 16 October 2006). The students' unwillingness to use English *anywhere* is thus strongly implicated in their lack of language and literacy potential, and exacerbated by their socioeconomic history. Their limited experiences with English language at home are not just seen as causing their poor academic achievement, but for Mrs Tan, as justification for the drive to get her teachers to focus language and literacy instruction on procedural knowledge, thus reiterating the bottom-up and readiness themes that we earlier heard her voice in her comprehension lessons.

When we brought in activities to introduce the teachers to multiple readings and to illustrate the insights their students could gain about textual constructedness by reading pairs of texts against each other, rather than engaging with how these texts could be effectively used in their classrooms, Mrs Tan pre-empted further exploration by insisting that BSS students would not understand them, that 'they do not have the language capacity'. Even our insistence that despite their potential for rich interactions the texts were rather simple linguistically, and that possible scaffolds or 'enabling strategies' (Langer, 2001) could be easily used to help the students make sense of them, yielded nothing but a vehement rejection by the teachers. The students' lack of preparedness—'Our students are grappling with basic things. We talk about interpretation but it boils down to the fact that they are hampered by their

65	C	One test used a *Readers' Digest* recount of an incident regarding this special needs child. I know Mrs Tan included it because she thought the students could identify with it. But surprisingly they didn't show better understanding. They just can't understand the language!
66	A	They watch quite a bit of Chinese drama, don't they? They learn history and they seem to be able to question a lot about it ...
67	C	But they don't question! They don't learn! When we brought them to watch the play, Blythe Spirit, one teacher brought her kids, and they laughed at moments when the rest of the audience didn't laugh. We're not sure how much they got out of it.
68	T	We're always worried. This year's mid-year paper, I chose the *Readers' Digest* passage about why certain students had difficulties reading. It says that it all stems from a few factors, reading habits from young, the family, the activities that you engage in, that allows you to engage in more literary activities. It was interesting because when I read it, I realised that many problems mentioned in it actually applied to our kids.
69	C	So it stems from their history when they were just a child.

Fig. 7.3 Reading circle, 18 May 2007

language and word level' (reading circle, 30 May 2007)—meant that aiming for such 'higher-order skills' as interpretation was at best a lofty ideal and at worst beyond students' intellectual potential. Hence, drawing on back-to-basics and literacy-as-lockstepped-processes ways of reasoning, Mrs Tan kept invoking the familiar deficit and readiness arguments when justifying her reluctance to attempt to give such texts and activities a try.

Similarly, another recurrent explanation for the students' inability to engage with texts was the impact of their poor reading habits at home on learning. These habits, so the teachers believed, 'stemmed from their history when they were just a child' (Turn 69, below) and hence were beyond school to break. Some teachers doubted that their students read anything out-of-school at all, while others argued that their restrictive reading and viewing choices did not help to improve their English learning. We conducted student interviews to ascertain their out-of-school literacy activities, and showed the teachers evidence of the rich literacy lives the students had. In response, the teachers swiftly shifted the issue from the texts the students read to *how* they read them. The excerpt above (Fig. 7.3) from a reading-circle discussion between Mrs Chan (C), Mrs Tan (T) and one of us (A) demonstrates this point.

As we can see, the official account from the *Readers' Digest* is mobilised here in support of the teachers' view that it is ultimately 'the family' and its failure in inculcating school-related 'reading habits from young' (Turn 68) that are to be held responsible for the students' low English proficiency and reading level. Perhaps more insidiously, by bringing this official, 'authoritative' explanation as reading material into their classroom, in the hope of getting 'the students to identify with it' (Turn 65), the teachers seemed to consider it a gesture of psychological support to convince their students of the validity of its reasoning and thus to educate them about their family's role in their own lack of educability.

The category of 'learning difficulties' and its attendant deficit beliefs allow us to re-examine Mrs Tan's lesson by introducing another interpretive layer to the classroom interaction. We see how these deficit beliefs have compelled her to delimit the texts for her students to what she considers linguistically and structurally simple ones. They have also severely limited the range of textual activities that are put on offer to her students, and led her to view a pedagogy of basic procedural skills as the only way to ameliorate her students' literacy problems. All of these are, significantly, solutions that 'remain within the existing framework of the school, calling for few structural changes' (Gunn, Forrest, & Freebody, 1995, p. 181) in its ways of doing things.

In Mrs Chan's case, we notice similar ways of accounting and locating for the students' lack of English literacy achievement in their histories and family problems—poverty, dysfunctionality, illiteracy, etc. Her foremost 'solution', then, the introduction of character-building skills, suggests that for her a shift of priorities from a language teacher towards a moral and pastoral educator is warranted. Since her students' language difficulties appear near insurmountable, and given their limited academic prospects, to provide them with life skills and a sound moral compass seems more important for their survival.

Teachers' beliefs about their role as moral and cultural ballast

Though most readily voiced by Mrs Chan, we found that all BSS teachers held strong beliefs that their students needed to be protected and insulated from the kinds of unsavoury values and ideals they were exposed to at home and in their neighbourhood. In our interviews, many teachers argued that their students' backgrounds were very different from those of other schools. Subsequently, the issue of tacit streaming arose during a reading circle, during which BSS teachers began comparing their students with those from another working-class neighbourhood school. While similarly ranked in the national league tables, the teachers pointed to differences in locality and student catchment area that led to different student socioeconomic backgrounds, with theirs faring 'more poorly'. For example, Mrs Chan insisted that their 'kinds of students are more likely to have single parents in BSS . . . majority of [broken families and divorce] cases here are extremely unpleasant. My student just told me about his father who is single, mentally unstable and had an accident' (reading circle, 26 April 2007). The teachers readily highlighted how different their school was, encapsulated in the attributes given to the families that 'their kinds of students' came from. One is reminded of Freebody and Baker's (2003) observations in a similar context, which led them to conclude: 'Professionals are expert at producing "kinds of kids" to fit any proposal about the character of their school population, and can propose and counter-propose endlessly' (p. 237).

Mrs Chan's foregrounding of her students' 'family problems and complications which they have to endure from a very young age' led her, as we have shown earlier, to reframe her teaching role foremost in pastoral, almost religious terms: 'Teaching is very challenging but it's the only salvation for kids like ours' (Interview, 16

43	T	As an institution, our selection of texts is very important. As far as the texts go, we go with promoting universal values. Certain values, we try not to venture into because we have to be careful about the selection of texts. So universal values, yes.
44	A	What is the problem with this text?
45	T	There is no right or wrong. But if you talk about universal values like friendships, filial piety, it is something that ALL would agree regardless of culture or whatever. But this type of text is a very 'iffy' thing. It's very difficult to comment and when students hear the text, all of them have personal experiences with their own parents, they enter in at a different entry point and we come in from our point and tell them, you should tolerate or it's my value that you should tolerate.
46	A	Some people would argue that individual autonomy is a universal value.
47	T	In Asia, no. It's more the family and the group.
49	A	That's the interesting thing, because the social group may be considered a universal value from another position. It's very difficult to argue about universal values in some ways.
50	T	That's what I mean. Certain values cannot be argued—filial piety is something that you do not argue about regardless of culture. In this story, the wife stayed, didn't get a divorce. Her value system is to tolerate and live with him. But if I open up to discussion I wouldn't venture into this area. You're opening up a Pandora's Box.

Fig. 7.4 Reading circle, 26 April 2007

October 2006). Like other teachers, Mrs Chan considered it one of her main teaching responsibilities to act as '*moral ballast*', which can protect her students from the 'realities' of home. Furthermore, there was the firm belief that given their students' lack of exposure to preferred cultural and literacy activities at home and their likely lack of such activities in their future, they needed to be given exposure, even if only for a moment, to highly valued world knowledge. Hence the inclusion, however, disparate from the actual curriculum, of the highly abridged *Romeo and Juliet*, and the one-off visit to the national theatre to see *Blythe Spirit*, both seen as opportunities for the students to glimpse at a cultural 'otherness' (Freiberg & Freebody, 1995) to which they likely will never have access.

In addition, the 'salvation' notion brought up by Mrs Chan tended to occur as a dominant pedagogic principle throughout their lessons, in which personal values and virtues were imposed upon the students. For example, the excerpt above (Fig. 7.4) is from a reading-circle discussion that occurred after 'Story of an Hour', a short story by Kate Chopin, had been introduced to the teachers. The story is about interrogating women's traditional role in marriage. Here, Mrs Tan opens a discussion around her teachers' responsibility to uphold 'universal values'.

Therefore, the teachers' dual belief that the students would struggle with linguistically and structurally challenging texts because of their learning difficulties, and that they should be shielded from 'unsafe', 'iffy' texts because of the need for the school to inculcate the strong moral values their unstable families neglected to provide them with, meant that the texts that were ultimately brought into the classroom were severely limited. Furthermore, the strong framing and classification practiced by teachers when teaching these select texts meant that students were not given opportunities to interrogate or offer alternative readings of them. The ubiquity and forcefulness of the teachers' deficit beliefs about their students' learning difficulties, and of their understandings of their own role, in compensation for home deficiencies, as moral and cultural ballasts, suggests their status as 'cultural models' (Black, 2007), which manifest themselves in the broader policy, social and institutional discourses of schooling, and education. It is to an examination of these wider discourses that we now turn.

Layer 3: aligning teachers' beliefs with policy, social and institutional discourses

This section highlights how policy, social and institutional discourses in Singapore create conditions for the production, interpretation and circulation of teachers' deficit constructions about student learning difficulties and teachers' role as moral and cultural ballasts.

Learning difficulties

The theme of 'learning difficulties' is premised on just under 30 years of enculturation into the belief that children should be streamed and taught 'at different rates according to their capacity to absorb learning' (Goh, 2004, pp. 197–8). The measure of a student's academic ability is then dictated by this learning pace, while the impact of structural factors like ethnic and socioeconomic influences becomes marginalised. Streaming and its attendant assumptions, enshrined in policies such as the Goh Report (Goh, 2004) and in subsequent reforms, therefore emphasises a student's ability to learn as a matter of *individual cognitive ability*. This shift towards intelligence as a measure of academic success is partly political, due to the need to decouple intelligence from ethnicity and socioeconomic status for the sake of social cohesion. It can be argued that the official policy discourse therefore abrogates the responsibility of addressing systemic causes of learning difficulties to something that is beyond the scope of policy makers—innate intelligence. Streaming can be seen as an attempt to mitigate learning difficulties by homogenising groups of students to facilitate learning. In doing so, policies over the past few decades have centred on reducing content, pacing and introducing vocational curriculum for lower-stream students, while providing higher-order thinking and more advanced academic content for students with 'gifted' abilities. The former is exemplified in a newspaper article, which stated that lower-stream students are provided with 'a watered-down syllabus for subjects such as English and mathematics, while taking classes in technical studies and computer skills' (Lee, 2004).

Further forms of stratification have also been created due to streaming. The official curriculum is often designed based on a stratified understanding of learning. For example, the new Singapore English Language Syllabus for Primary and Secondary schools (MOE, 2009) is premised upon three student types of increasing language proficiency, ranging from those who are able to attain only a foundational understanding of the language, to increasing language competence and finally the minority elites who can master the language at an internationally comparable level. This inherently assumes a hierarchy of literacies (Tan & McWilliam, 2009), whereby basic competencies are required prior to more advanced ones, and the latter reserved only for a select few, a view debunked by many literacy educators (for example Paris, 2005). Other curricular stratifications are visible in the prioritisation of academic content between neighbourhood schools, and top-performing schools with predominantly E classes. Citing examination results by Normal students, which showed that they performed better in technical subjects or subjects less abstract in nature (see Nirmala, 1997), teachers in neighbourhood schools were often quick to adjust lessons so that they are 'not purely academic' (Koh, 2003). The NT syllabus was revised in 2004 to offer more practice-oriented learning, with character building and aesthetics, rather than differentiated academic content.

Furthermore, streaming has reified certain stereotypes of students and the roles they are destined for in society. A stratified society was articulated as early as 1966 by then Prime Minister Lee Kuan Yew, when he argued that the education system needed to produce a 'pyramidal structure' comprising 'top leaders', 'good executives' and a 'well-disciplined and highly civic-conscious broad mass' (Lee, 1966, as cited in Tan, 2008). Ultimately, these stratifications have an impact on teachers' beliefs about their students' learning difficulties. In revisiting Mrs Tan's lesson, for example, we can see how these wider discourses informed her beliefs that the students, being naturally 'slow' learners and destined for the 'broad mass', needed a 'watered-down' curriculum and a highly framed pedagogy that prevented their access to more critical, higher-order and independent thinking. At the same time, though, the deficit thinking held by the teachers is more nuanced than that held by policy makers, who would view low intelligence as the main hurdle to student learning. These teachers, working in a neighbourhood school, have a cultural model that incorporates their strong beliefs about their students' family backgrounds. In other words, in their daily lives, the teachers have made visible what has been left invisible in official policy discourse—the impact of social, and linked to it, ethnic background, on student learning. However, their deficit thinking nevertheless blinds them from asking themselves the crucial question of 'what schooling requires, and even presupposes, of [students]' (Freebody & Baker, 2003, p. 237).

Moral and cultural ballasts

The teachers' roles as moral and cultural ballasts are also a consequence of the policy and social discourses that have circulated in Singapore regarding students with learning difficulties. Media reports often paint such students as those who 'couldn't follow lessons' and are 'always lost in class' (Ng, 2008), and NT students more specifically as those who 'came from troubled families' and have problems ranging from 'financial woes, frequently quarrelling parents, and parents who are unable to supervise or guide their children' (Tan, 2005). Deficit discourses about Normal students are common in newspapers that portray them as troublesome and prone to illicit activities. Against the background of such stock portrayals, it is no surprise to find neighbourhood schools like BSS contend that while it would be an uphill task to achieve school-wide academic success, their major role is one of inculcating student dispositions and skills that could meet the 'broad mass' demands of Singapore's economy. BSS's mission statement, for example, emphasises strong moral and character development and, forgoing academic excellence, it has chosen to focus on sports, aesthetics and extra-curricular activities. Other neighbourhood schools have taken similar paths and have been showcased in the media for their 'innovative practices', such as organising field trips to Paris to select art students or buying discarded pianos to give them access to music—access that they would not have at home. The ubiquity of such larger social discourses about these students' backgrounds, their lack of access to values (high), culture and family stability, helps to explain why Mrs Chan, as evident from her repeated insertion of moral lessons into the English classroom, and from her reasoning about this decision, and why Mrs Tan, as evident from her addition of bits of (high) culture events and texts into the students' school life, have made the curricular choices they have.

Essential next questions

Our intention in writing this chapter is not to blame the individual teachers and hold their actions responsible for the production of educational inequality. As Freebody and Baker (2003) point out, teachers are 'not inventing the discourse, merely replaying it with local variations' (p. 236). It is precisely because it is too easy to blame, and much harder to examine how there are multi-faceted structural factors, and intersecting layers of ideologies, cultural beliefs, institutional practices and social norms that constitute an elaborate web of legitimation that continues to perpetuate deficit constructions in school and society. In extending the research on deficit thinking, two areas are in need of particular attention—understanding this web of legitimation, and disrupting it so as to open up a space for educational policy makers, researchers, teachers, parents and students to step away from the deficit cycle. Some of the essential questions that emerge from these, then, are as follows.

Why do deficit beliefs persist, and who perpetuates them?

We have already alluded to the need to move away from the game of blaming students, parents and teachers, and to recognise structural factors at play that circulate deficit beliefs. There is, therefore, a need to understand how such beliefs are circulated at the policy-making level, and the nature of the evidence base that policy makers draw upon that perpetuates particular deficit beliefs. Similarly for curriculum planning that is done at the national level, there is a need to problematise assumptions embedded within various curricula that circulate narrowly construed forms of student learning, or do not afford critical spaces to examine structural factors that contribute to disadvantage or inequality.

What 'deficiencies' matter to educators?

This question, raised by Freebody and Baker (2003, p. 233), seeks to unpack the nature of deficits, particularly those attributed to the home. We acknowledged earlier in the chapter that there are material realities that students encounter daily that impact upon their educational experiences, yet there is a need to ask exactly which 'deficiencies' in the home are important to educators for the specific purpose of *redressing* educational inequity in schools. Following through this question requires opening up issues of class and cultural differences, ethnicity and socioeconomic disparities, with the recognition that educational achievement is currently defined by the dominant (middle to upper) classes. For countries such as Singapore, where social cohesion is a national imperative and discussions of race and ethnicity are deemed seditious, there is a need to consider reflexively how avoiding discussions of such structural inequality factors may not only make deficits impervious to any alternative discourses, but ultimately exacerbate social discord in the long-term.

How can teachers break the deficit cycle?

We recognise that any attempt to disrupt deficit beliefs will not be a simple task. Luke and Goldstein (2006) suggest that the first step to disrupting the deficit cycle is to encourage teachers to 'learn ways of "performing" pedagogy that display and value other cultures—that they become intercultural actors whose knowledge and competences can be recognised by students' (p. 4). To become 'intercultural actors', teachers need to be able to look at their own 'interactional genres' (Lefstein, 2008)—patterns of teacher–student interactions that frame participants' expectations of classroom talk—and change them. An example of a successful attempt to disrupt deficit thinking and improve teaching is an intervention project by Comber and Kamler (2005). Central to their 'turn-around pedagogies' is the concerted effort to bring teachers directly into the students' lifeworlds, to become 'intercultural actors' able and willing to traverse class and cultural divides. Of course, given their durability, established beliefs and interaction patterns are difficult to shift, and our recent professional development experience has shown us that attempts to do so pose a host of often unanticipated challenges, including our own propensity to make similar deficit assumptions about the teachers as they struggle to disrupt their reductive ways of thinking about their students. Nevertheless, in this chapter we hope to have taken a step in the right direction towards a deeper understanding of the complex nature of deficit constructions, and the issues and dangers associated with them when they become the primary resource that teachers have available to interpret their students' learning difficulties.

References

Barr, M., & Low, J. (2005). Assimilation as multiracialism: The case of Singapore's Malays. *Asian Ethnicity*, *6*(3), 161–182.

Barr, M. D., & Skrbis, Z. (2008). *Constructing Singapore: Elitism, ethnicity and the nation-building project*. Copenhagen: NIAS Press.

Bartolome, L. I. (1994). Beyond the methods fetish: Toward a humanizing pedagogy. *Harvard Educational Review*, *64*(2), 173–194.

Bernstein, B. (1990). *Class, codes and control, vol. IV: The structuring of pedagogic discourse*. London: Routledge.

Black, L. (2007). Analysing cultural models in social-cultural discourse analysis. *International Journal of Educational Research*, *46*, 20–30.

Caughlan, S., & Kelly, S. (2004). Bridging methodological gaps: Instructional and institutional effects of tracking in two English classes. *Research in the Teaching of English*, *39*(1), 20–62.

Cazden, C. B. (2001). *Classroom discourse: The language of teaching and learning* (2nd ed.). Portsmouth, NH: Heinemann.

Comber, B. (1998). Problematising 'background': (Re)constructing categories in educational research. *Australian Educational Researcher*, *25*(3), 1–21.

Comber, B., & Kamler, B. (2004). Getting out of deficit: Pedagogies of reconnection. *Teaching Education*, *15*(3), 293–310.

Comber, B., & Kamler, B. (2005). *Turn-around pedagogies: Literacy interventions for at-risk students*. Newtown: Primary English Teaching Association.

Dadds, M. (1999). Teachers' values and the literacy hour. *Cambridge Journal of Education*, *29*(1), 7–19.

Freebody, P., & Baker, C. (2003). Categories and accounts in literacy research and education: Changing targets. In M. Anstey & G. Bull. (Eds.), *The literacy lexicon* (2nd ed., pp. 225–240). Sydney: Prentice Hall.

Freebody, P., & Welch, A. R. (1993). Individualization and domestication in current literacy debates in Australia. In P. Freebody & A. R. Welch (Eds.), *Knowledge, culture & power: International perspectives on literacy as policy and practice* (pp. 209–232). London: Falmer Press.

Freiberg, J., & Freebody, P. (1995). Analysing literacy events in classrooms and homes: Conversation analytic approaches. In P. Freebody, C. Ludwig, & S. Gunn (Eds.), *Everyday literacy practices in and out of school in low socio-economic urban communities* (Report to the Australian Commonwealth Government of Employment, Education and Training). Brisbane: Griffith University.

Gamoran, A. (1992). Is ability grouping equitable? *Educational Leadership*, *50*(2), 11–17.

Goh, K. S. (2004). *Wealth of east Asian nations*. Singapore: Marshall Cavendish Academic.

Gunn, S., Forrest, T., & Freebody, P. (1995). Perspectives on poverty, schooling and literacy: The view of participants. In P. Freebody, C. Ludwig, & S. Gunn (Eds.), *Everyday literacy practices in and out of school in low socio-economic urban communities* (Report to the Australian Commonwealth Government of Employment, Education and Training). Brisbane: Griffith University.

Gutierrez, K. D., & Orellana, M. F. (2006). The 'problem' of English learners: Constructing genres of difference. *Research in the Teaching of English*, *40*(4), 502–507.

Hunter, I. (1994). *Rethinking the school*. Sydney: Allen & Unwin.

Johnston, K., & Hayes, D. (2008). 'This is as good as it gets': Classroom lessons and learning in challenging circumstances. *Australian Journal of Language and Literacy*, *31*(2), 109–127.

Koh, Y. (2003, October 13). Turning a normal (tech) kid from this. *The Straits Times*. Retrieved October 12, 2008, from http://www.lexis-nexis.com

Kwek, D., Albright, J., & Kramer-Dahl, A. (2007). Building teachers' creative capabilities in Singapore's English classrooms: A way of contesting pedagogical instrumentality. *Literacy*, *41*(2), 71–78.

Langer, J. A. (2001). Beating the odds: Teaching middle and high school students to read and write well. *American Educational Research Journal, 38*(4), 837–880.
Lee, L. (2004, March 22). Normal (tech) doesn't mean the N. *The Straits Times*. Retrieved October 12, 2008, from http://www.lexis-nexis.com
Lefstein, A. (2008). Changing classroom practice through the English national literacy strategy: A micro-interactional perspective. *American Educational Research Journal, 45*(3), 701–737.
Lleras, C. (2008). Race, racial concentration, and the dynamics of educational inequality across urban and suburban schools. *American Educational Research Journal, 45*(4), pp. 886–912.
Luke, A. (2005). *CRPP intervention plan: Moving from the core to pedagogic change*. Unpublished technical report, Centre for Research in Pedagogy and Practice. Singapore: Centre for Research in Pedagogy and Practice.
Luke, A., & Goldstein, T. (2006). Building intercultural capital: A response to Rogers, Marshall, and Tyson. *Reading Research Quarterly, 41*(2), 202–224. Online supplement.
MOE. (1979). *Report on the Ministry of Education 1978*. Singapore: Author.
MOE. (2009). *English language syllabus 2010 (primary & secondary)*. Singapore: Author.
Ng, J. (2005, December 20). Better N(T) students get a shot at O levels. *The Straits Times*. Retrieved October 12, 2008, from http://www.lexis-nexis.com
Ng, J. (2008, August 29). Pitching it right to help students cope. *The Straits* Times. Retrieved October 12, 2008, from http://www.lexis-nexis.com
Nirmala, M. (1997, December 6). First batch of N (tech) students can enter ITE. *The Straits* Times, p. 61.
Oakes, J. (2005). *Keeping track: How schools structure inequality* (2nd ed). New Haven, CT: Yale University Press.
Paris, S. G. (2005). Reinterpreting the development of reading skills. *Reading Research Quarterly, 40*(2), 184–202.
Singapore Government. (2003). *Report on Remaking Singapore*. Retrieved July 30, 2005, from http://www.remakingsingapore.gov.sg
Tan, D. (1996, November 28). The woes of normal (technical) students. *The Straits Times*, p. H39.
Tan, T. (2005, April 1). Fewer students are dropping out of school. *The Straits Times*. Retrieved October 12, 2008, from http://www.lexis-nexis.com
Tan, J. (2008). Managing diversity: The Singapore experience. In G. Wan (Ed.), *The education of diverse student populations: A global perspective* (pp. 159–181). Dordrecht: Springer.
Tan, J. P.-L., & McWilliam, E. (2009). From literacy to multiliteracies; Diverse learners and pedagogical practice. *Pedagogies: An International Journal 4*(3), 213–225.
Valencia, R. R. (1997). Conceptualizing the notion of deficit thinking. In R. R. Valencia (Ed.), *The evolution of deficit thinking* (pp. 1–12). London: RoutledgeFalmer.

Chapter 8
Parent, Family and Community Support for Addressing Difficulties in Literacy

Janice Wearmouth and Mere Berryman

Introduction

Many students experience difficulties in literacy. In this chapter we adopt the view that school-sanctioned literacy is only one of many different literacies in people's lives (Street & Street, 1995; Street, 1997) and that, as well as school staff, there are a number of 'mediators', such as siblings, parents, carers, relatives and community members, who can support literacy development among students who experience difficulties (Gregory, 1998; 2004a, Gregory, 2004b; Gregory, Long, & Volk, 2004; Gregory, Williams, Baker, & Street, 2004). The attitude of any educational institution to the role of parents, families and community members, as prime educators of children, is therefore of great significance (Wearmouth, 2004).

The ways in which schools respond to family culture and background and the kinds of parent/family/community—school partnership arrangements that exist can serve to include or alienate the very communities, families, parents and children that schools seek to serve. Embedded within the particular discourses, approaches and strategies of schools are a variety of preconceptions about the ability and right of parents, families and/or communities, from a diversity of backgrounds and cultures, to support the literacy development of their children (Dale, 1996). In this chapter, we propose the fundamental importance of schools recognising these preconceptions and assumptions in order to negotiate more effective home—school literacy programs. A priority is to explore how schools can harness all available resources to address difficulties in literacy development effectively and help to maintain the 'broad-based' instruction (Wearmouth, Soler, & Reid, 2002; Wragg, Wragg, Haynes, & Chamberlain, 1998) required to support all children's literacy development. We discuss the importance of schools addressing issues of cultural differences between home and school, if they are to understand and respect the potential of partnerships between parents and teachers, and construct literacy learning tasks and

J. Wearmouth (✉)
University of Bedfordshire, Bedford, UK
e-mail: janice.wearmouth@beds.ac.uk

C. Wyatt-Smith et al. (eds.), *Multiple Perspectives on Difficulties in Learning Literacy and Numeracy*, DOI 10.1007/978-1-4020-8864-3_8,

contexts that recognise and affirm the literacy values and practices that are evident in different cultural communities (McNaughton, 2002).

Models of literacy acquisition: the reading process

Underlying the techniques and programs that are commonly used in home—school literacy partnerships are models of literacy acquisition, particularly of reading, and associated teaching practices that can be linked to the issue of the power relationship between home and school (Wearmouth, 2004; Wearmouth et al., 2002). As already noted in Chapter 2, there are two common, contrasting theoretical perspectives on the reading process. Each leads to a different approach to the teaching of reading to students who experience difficulty. From one, reading comprises the decoding of the alphabet, and, through this, the simple reconstruction of the author's meaning. To learn how to do this, children must go through a staged process: learn the letters of the alphabet and establish the principle of sound—symbol identification, then apply this to decode words. It implies the use of teaching methods that emphasise the mastery of phonics and word recognition, which tend to be seen as part of the role of professionals rather than parents (Wragg et al., 1998). The second perspective on reading is influenced by psycholinguistics. Reading is seen as the active *construction* of meaning.

Clearly, there is tension between the two perspectives, phonic and psycholinguistic. On the one hand, Adams (1994, p. 140) warns against encouraging beginning readers to rely too much on contextual cues to read text. On the other hand, supporters of the notion of a psycholinguistic approach argue that children's literacy emerges out of their interaction with language and their experience of the world of print around them. Competence in phonics may develop as a result of access to stories and print, rather than the other way round. Direct instruction in phonics as a precursor to reading is less important than provision of literacy rich environments.

A third, many might say more balanced (McNaughton, 2002), view of reading is the 'interactive' model (Stanovich, 2000), which suggests that readers use information simultaneously from different levels and do not necessarily begin at the graphic (bottom-up) or the context (top-down). During the development of reading skills, some readers may rely more heavily on some levels than others.

Wragg et al. (1998) comment that reading programs for parents to use with their children usually fall into the category of a psycholinguistic approach

> It is interesting to note that ... the involvement of parents in their children's learning sits more easily with meaning-based approaches to teaching reading than with those emphasising the acquisition of skills ... When reading is seen as an enjoyable shared activity, the aim of which is understanding, then parents, relatives and siblings can all join in. If reading is seen as a series of skills to be mastered, however, then teaching it is more likely to be claimed as the prerogative of the professionals ... (p. 34)

Literacy: individual skill or social practice?

Where literacy is understood to be a discrete skill based either on decoding of sound or reconstruction of the author's meaning, interventions designed to overcome the difficulties that students may experience are highly dependent on the particular understanding that a teacher or other educator may hold about the process of becoming literate and, therefore, of what might cause difficulties in literacy acquisition. However, many researchers, for example Gee (1996), Rassool (1999) and Street (1997), would argue that viewing literacy as a set of discrete skills is too simplistic and restrictive. It is more appropriate to view literacy as a practice that takes place within a social context and is linked to culture, knowledge, power and the dominant ideology of any particular societal group. If literacy is conceptualised as a social practice, then, logically, barriers to literacy may also lie in the social context, such as in the expectations, social norms, types of texts and overall pedagogy of the literacy classroom (Wearmouth, 2002). When viewing literacy in this way, issues related to whether and how schools should collaborate with students' families and communities to support literacy acquisition and address barriers that they might face become highly pertinent.

Students' culture and the school

Schools are increasingly working with students from a wider diversity of cultural groups. Students' cultures really matter in terms of the relationships and pedagogies employed by schools (Bishop & Glynn, 1999; Bishop, Berryman, Tiakiwai, & Richardson, 2003) if students from these communities are to be successful. People who share a culture may, on one level, simply have common interests and preferences, and may meet together occasionally to share activities and experiences. However, on another level, culture has a much more important and deeper meaning. In New Zealand, for example the Waitangi consultancy group (Quest Rapuara, 1992) considers that culture involves collectively held values, beliefs and practices (ways of living and working together) that provide a basis of shared meaning and understanding for members of a group or community, including the processes of decision-making and communication, the way families are structured and what they regard as important. Culture also includes collective memory and heritage that often finds expression in art, music, drama, literature, religion and social events. These collective values, beliefs and practices are learned by living within a particular group or community over a long time. Culture, therefore, is an essential dimension of personal identity and wellbeing, and strongly influences how individuals see, interact with and make sense of the world around them. It is important that, when the students' home cultures differ from the culture of their teachers or their schools, culture is not something that is put aside while responses to educational issues are identified and addressed.

As educators, '... our interactions with others are deeply affected by our everyday intuitive theorizing about how other minds work' (Bruner, 1996, p. 45). Teaching and learning occur as 'a direct reflection of the beliefs and assumptions the teacher holds about the learner' (ibid., p. 47). Teachers draw from a variety of metaphors and discourses to make sense of the experiences they have when relating to and interacting with students and their home communities (Wearmouth, Glynn, & Berryman, 2005). Embedded within different home—school partnership arrangements are presuppositions about the ability and right of families and/or local community groups from a diversity of backgrounds and cultures to support their children in schools. For example, in regards to Māori, the marginalised indigenous cultural group in New Zealand, Nash (1993) suggests that 'family resources, both material and cultural, are the key transmission mechanism of educational disadvantage, rather than the structures of the education system' (p. 124). Chapple, Jefferies, and Walker (1997) also suggest there is strong evidence that a significant and substantial cause of educational disparities is the relatively poorer family resource position of Māori. Similar teacher attributions of lower student achievement by Māori were found across the curriculum at years 9 and 10, by Bishop et al. (2003). In a United Kingdom context, also, Gregory (1996) notes that, for many years, bilingual homes were also viewed as poorer environments for children's literacy learning in English. In 1967, for example, references to bilingualism in the Plowden Report (Central Advisory Council for England, 1967) were largely negative

> The argument in the report runs as follows: 'Immigrant children' are 'deprived' and at a disadvantage in school owing to the poor education of their families. In addition, they will be 'handicapped' by their unfamiliarity with the new language and culture ... The solution to children's needs during the 1960s was, therefore, seen to lie in assimilation of immigrants through tighter control of new arrivals, through bussing children to other schools when the proportion of immigrants reached 33% and through providing a 'compensating environment' by 'enriched intellectual nourishment'. (Gregory, 1996, p. 5)

In the United Kingdom, the cause of lower literacy scores among particular students groups has often been attributed to differences in social class. For example, Davie, Butler, and Goldstein (1972), as part of the National Child Development Study, tested the reading attainment of 7-year-old children born in 1 week in 1958. Relatively poor achievement in reading was shown by 30% of the students. This poor achievement correlated strongly with a number of home factors, including social class. The children of semi-skilled, manual working-class fathers were more than twice as likely to be poor readers than those whose fathers held professional or technical posts. A survey of teachers undertaken by Morgan and Morris (1999) linked 62% of teacher responses to 'something to do with the pupil or his or her home background', when identifying reasons for student failure, while only 18% of the response statements were positioned with 'something to do with me, the teacher' (p. 68).

In their synthesis of 'best evidence', Biddulph, Biddulph, and Biddulph (2003) identified deficit, difference and empowerment/enhancement among the metaphorical positions assumed by researchers and educators in the area of community and family influence on students' learning and learning outcomes. These researchers identified the negative impact that these theories had upon teachers' abilities to engage with families and communities. Rubie-Davies, Hattie, and Hamilton (2006) also raised the important influence of teachers' expectations and judgments that adhere to different ethnic stereotypes. Stereotyping by teachers of cultural minority groups such as:

- Asian students are diligent and supported by parents who value education
- Pacific students are heavily influenced by their church
- Māori students do not value education and come from unsupportive homes

affects the expectations that teachers hold for these students and the possibility of family support for literacy learning.

Home–school relationships

The deficit view of some parents and families fails to capitalise on opportunities to draw on existing patterns of family literacy to inform children's learning (Hannon, 1999). The importance of the relationships between family processes and students' literacy progress at school has been an important focus in some countries for a considerable time now (Glynn, Wearmouth, & Berryman, 2006; Wearmouth et al., 2002). Research working with parents as literacy tutors has been described over the past 30 years in New Zealand (Glynn, Mcnaughton, Robinson, & Quinn, 1979; McNaughton, Glynn, Robinson, & Quinn, 1981) and in Australia (Houghton & Glynn, 1993). These procedures have been successfully used with reading when English has been the first language and when English has been the second language (Berryman & Glynn, 2003). The model of working with parents and communities has challenged the deficiency model of poor working-class and ethnic minority culture families (Blackledge, 2000). The former model has also challenged assumptions that successful reading tutoring is a highly contentious and solely specialist domain (Berryman, 2000).

These studies and a number of others continue to offer a strong reason for believing that many families from across the cultural and social spectrum have the potential to provide a very important additional resource in supporting the literacy acquisition of children who experience difficulties. In the United Kingdom, as Wragg et al. (1998, p. 269) note

> The Plowden Report (1967) on primary education devoted a whole chapter to the role that could be played by parents. Young and McGeeney (1968) experimented in London schools by involving parents in attending school functions, hearing their children read, and various other forms of participation. They found some improvements in reading performance compared with control schools where there was no such participation.

Hewison and Tizard's (1980) study of the reading attainment of 7-year-old, working-class children in Dagenham showed that half of the parents of working-class children who were competent readers regularly heard them read, although none of the parents had been encouraged by the school to do so. Subsequently, a number of research studies were set up to investigate the hypothesis that parental support at home for school-related literacy had a significant effect on improvement of children's reading. For example, in the Haringey Project (Tizard, Schofield, & Hewison, 1982) every child in two older (7 years) infant classes, chosen at random from two multi-racial, inner-city schools in London, was heard reading from books sent home by the class teachers for 2 years. The results indicated that students who were heard reading at home achieved a highly significant improvement in comparison with other students. There was no comparable improvement among those students who received the extra tuition in school.

More recently, the Education Review Office in New Zealand (2004) identified that the quality of relationships between the home and school is an important factor in influencing student achievement. Parents can participate more effectively when they have access to information about their children's learning (Education Review Office, 2004) and real opportunities for collaboration, focused on learning (Alton-Lee, 2003). Educators have begun to recognise the importance of the language and cultural practices of their families, including from minority cultures, as crucial for the educational and psychological wellbeing of individual students and their families, and consequently for the wellbeing of society (Bishop et al., 2003).

Parental involvement with schools, in the development of shared goals, can ensure greater benefit from schooling for their children (Wearmouth et al., 2005; Wearmouth et al., 2002). There are, however, important differences in views, not simply about the ability of families to support literacy learning as noted already, but also about the power relationship that should exist between schools and the families of students experiencing literacy difficulties. The partnership arrangements that currently exist between families and schools, and the degree of schools' responsiveness to family culture, can serve to include, or alienate, young people, in particular those who experience difficulties in literacy acquisition. Over the years there has been a great deal of discussion about which is the most appropriate way to support children's learning in ways that take account of a diversity of family and cultural backgrounds and, therefore, of ownership of family-based literacy strategies and programs.

Wragg et al.'s (1998) summary of the manner in which parents were involved in the reading development of their children in the schools surveyed during the Leverhulme Primary Improvement Project, for example, reflects an imbalance of power and assumptions of teachers as the primary, or sole, source of expert knowledge about individual children. Most schools professed to involving families in children's literacy acquisition, but beneath the surface there was, commonly, a tension and confusion over the role that parents who had been invited into the classroom as volunteer helpers should perform

> The generally positive reaction of schools to the involvement of parents is a strong foundation stone on which to build, but there should be no doubt about the gaps, misunderstandings and lack of knowledge that exist, even in schools as effective generally as the ones studied in this research ... Unwittingly perhaps, some schools may patronise their children's parents by glossing over their concerns, assuming that they are capable of very little beyond the most rudimentary, or, in the case of ethnic minorities, assuming too readily that they may not be equipped to help. (pp. 269–70)

Similar issues were highlighted by Dale (1996). The five partnership arrangements between schools and parents/carers identified by Dale in common home–school literacy programs are as follows:

- The traditional 'Expert Model', which is similar to the doctor–patient relationship. The professional is assumed to have expertise with which to decide what needs to be done. Parental involvement is not of prime importance, except to provide information.
- The 'Transplant Model', whereby parents are viewed as an untapped resource for helping to teach the child. The role of professionals is to transplant their skills and expertise to the parents, to help the parents to become teachers. Professionals maintain control over decision-making.
- The 'Empowerment Model', whereby the right of the parent-as-consumer is combined with a professional recognition that the family incorporates a social system. Families rely as much on informal networks of support—neighbours, other family members, friends—than on the formal network of professionals. In the Empowerment Model, the role of the professional is to recognise the family's own support network and empower family members to meet their own needs, with professional support.
- The 'Consumer Model', in which power is shifted from the professional-as-service-provider to the parent-as-consumer. As consumers, parents have the power to draw upon their own expertise and knowledge about their children in deciding what services they need for their child.
- The 'Negotiating Model', in which both parents and professionals have important contributions to offer. Negotiating over the differences between their perspectives is assumed to lead to the best decisions for children.

Across the world there are examples of partnerships between schools and parents/families that closely reflect Dale's (1996) analysis of the location of power.

The 'expert' model

In some settings, parents may welcome and initially require strong directives from teachers in supporting the literacy acquisition of their children, especially those who experience difficulties. Here, the parent is seen as reliant on the professional with expertise in deciding what needs to be done. The teacher-directed model is reflected in the United Kingdom by initiatives such as Topping's (1996) 'Paired Reading' program, whereby families

> ... commit themselves to an initial trial period in which they agree to do at least five minutes Paired Reading on five days each week for about eight weeks. Grandparents, siblings, friends and neighbours can be encouraged to help, but must all use the same technique—the target child is deliberately asked to quality control the tutoring they receive (Topping, 1996, p. 46).

The way in which Topping (1996, p. 48) describes the 'rules' of his Paired Reading method is clearly behaviourist in orientation: 'engineering in' behaviour which is approved of by the expert through the use of 'much verbal praise and non-verbal approval for specific reading behaviours' and 'engineering out' behaviours seen as 'undesirable'. Topping advises strong direction from teachers to parents about how the 'Paired Reading' technique should operate

> Tutors support children through difficult text by Reading Together—both members of the pair read all the words out loud together, the tutor modulating speed to match that of the child, while giving a good model of competent reading.
>
> On an easier section of text, the child may wish to read a little without support. The child signals for the tutor to stop Reading Together, by a knock or a touch. The tutor goes quiet, while continuing to monitor any errors, praise and pause for discussion. Sooner or later while Reading Alone the child will make an error which they cannot self-correct within 4 or 5 seconds. Then the tutor applies the usual correction procedure and joins back in Reading Together (p. 48).

A 'very simple' and 'ubiquitously applicable' correction procedure is 'prescribed'

> After pausing for 4 to 5 seconds to allow self-correction, the tutor just models the correct way to read the word, the child repeats it correctly and the pair carry on.

Some parents may welcome such clear directives from teachers. Others may be very concerned about involving themselves in initiatives directed by those who may understand little of the child's background.

The 'transplant model'

In New Zealand, Glynn and McNaughton's (1985) pause, prompt, praise (PPP) technique, designed for use with parents of students experiencing difficulties in literacy acquisition, may be seen as developing from a 'transplant model'. In their rationale for involving parents McNaughton, Glynn, and Robinson (1987) expressed

> ... a growing concern for parental involvement in the education of their children ... parents, while still feeling and being held responsible for their children, are becoming more and more powerless to influence their own children's development ... The parents in our research certainly felt keenly the segregation of home and school ... We felt that parents, as well as being willing and able to help their low progress children, have a right to take part in their children's schooling. (McNaughton et al., 1987, p. 4)

PPP facilitates opportunities to self-correct errors and practise strategies for problem solving. Tutors are guided in how to implement a simple but specific set of tutoring strategies: pausing to allow for self-correction, prompting to offer word meaning or for sound–symbol identification and praising to reinforce the desired reading behaviour (McNaughton et al., 1987). Careful consideration of the type of error enables prompting to focus on meaning or on the graphical features of a word. The reading material must be of interest to the learner and also at an appropriate difficulty level so the learner meets some unfamiliar words but can read enough of the text to make semantically good guesses, even if these are miscues.

The 'empowerment model'

Developments with PPP in New Zealand (Glynn, Berryman, & Glynn, 2000) have taken these procedures into other partnership arrangements within the context of the Māori community, built on the view that tutoring skills should be available to parents and carers, thus empowering them to support their own children. This development of PPP has enabled respect for cultural diversity and sensitivity for cultural differences to be taken into account as highly significant in supporting children's literacy acquisition (Atvars, Berryman, & Glynn, 1995).

> ... having that cultural togetherness, having the tutor and the child of the same culture, has, we have found, helped enormously. The other thing that is important for our tutors is they have a really good understanding of the types of stories that might interest Māori children. That's not to say that they know the stories alone that will interest Māori children but, they're able to pick up on stories with a cultural theme, that children will enjoy reading, and that children can, Māori children specifically, can relate to. The other thing that's really important about it is there is reciprocity in the learning, and by that I mean that the tutor is able to learn about the child that they're working with, and the child benefits by being able to learn about the reading process ... So from that point of view also ... it's not directive, it's collaborative. (Open University, 2002, Appendix F, pp. 212–213)

The PPP may be seen as transforming from a 'transplant model' into an 'empowerment model' (Dale, 1996) with the transfer of skills to families and community members by teachers. Additionally, this occurs in the way in which respect and responsiveness for cultural background is encapsulated through the sharing of understandings and actions that are reciprocal between school and home.

The 'consumer model'

In the United Kingdom, there are examples of individual parent-led initiatives that reflect the 'consumer' model, which have arisen out of a sense of concern that schools are not able to address the particular literacy needs of individual children. Sometimes the resources available to the school are felt to be inadequate. Sometimes there is a difference of opinion between families and schools over the root cause of children's difficulties in literacy and, therefore, what might constitute an appropriate response to address the difficulty. Heaton (1996), for example, notes that, for her, it is extremely important to feel that she has the ability and power to offer appropriate support and help to her own child, and that teachers will listen to her when she wishes to discuss the barriers to literacy learning faced by her child. Heaton is one of a number of writers (Ostler, 1991; Riddick, 1996) who have described particular techniques and practices that they, as parents of children described as 'dyslexic', have found useful both in obtaining the additional or alternative educational provision that they have felt necessary to meet their children's learning needs and in supporting their children in their day-to-day living. From a questionnaire completed by parents and carers of students identified as dyslexic, Heaton (1996) offers advice to families on how to empower themselves to draw upon their knowledge about their own children in deciding what services are required for their child. This includes:

- being prepared to have to organise a management system for, and spend time on, all the paperwork involved in making the case for recognition and assessment of the student's difficulties in literacy acquisition
- being proactive in finding out about difficulties in literacy, appropriate teaching methods and common terminology
- maintaining close liaison with the school, and every year ensuring that the student's teachers are aware of the difficulties they experience
- working out practical strategies for personal organisation
- teaching strategies for dealing with potential difficulties with the concepts of time, space and direction.

The 'negotiating model'

In a study carried out by a community and a small rural Māori immersion primary school, PPP can be interpreted as metamorphosing into a 'negotiating model' (Berryman & Glynn, 2003). Students from this school typically completed year 8 as highly competent speakers, readers and writers of Māori. However, they faced major problems on transition to mainstream secondary school, where all teaching and assessment was conducted in English and their high level of participation and competence in literacy practices in their first language were neither recognised nor incorporated into classroom pedagogical practices in their new school. Across the world, many first-language speakers face similar problems if, when entering school in another country, they are required to continue their education through the medium of another language. As a result of being assessed (in English) prior to going to this new secondary school, the students from the community in this study, who were highly literate in Māori, were identified as being among the lowest achievers, with low expectations for academic success, and offered little or no access to effective second-language teaching and learning support. Together, the students' community and primary school negotiated the introduction of reading and writing in English to their students during the final 10-week term of their Māori immersion. This included negotiating the assistance of a researcher with tribal connections to their families and teachers, and experienced in developing effective support strategies for Māori students in both Māori and English. With the support of community elders (students' grandparents), the researcher trained English reading and writing tutors to support the students, using three specific literacy strategies: PPP for individual responsive tutoring of oral reading and two writing procedures: responsive writing (Glynn, Jerram, & Tuck, 1986; Jerram, Glynn, & Tuck, 1988) and a form of structured brainstorm (Whitehead, 1993). The program was then implemented both at home and at school. Berryman and Glynn (2003) note that students achieved substantial positive English reading and writing gains on a range of different measures, including rate, accuracy and comprehension (reading) and rate, accuracy and quality (writing), and maintained or improved their reading and writing gains at follow-up assessment points, 10 and 20 weeks after they participated in the program, and independently of their original tutor support.

Family- and community-based literacy learning

Most children spend their first few years living at home with parents, siblings and other immediate and extended family members, and learn to use language to enhance social and cultural relationships with adults and peers. Many families play a major role in providing structured, responsive and supportive contexts for children's early literacy learning, also. Along with other researchers (Farver, 1993; Gregory, 1998), Volk and de Costa (2001) note that older siblings can attune their teaching to the current level of younger children's literacy development and thus act as mediators of literacy learning in ways that enhance the literacy learning of themselves and the younger siblings. Volk and de Costa (2001) conclude that the findings of this study

> ... point to the importance for teachers of looking beyond the replication of school experiences at home to the range of literacy interactions and people in children's lives and of recognising them as genuine resources for literacy learning. Interacting with network members, asking questions and listening to their perspectives on literacy, and observing them interact with children may provide teachers with a new appreciation of network members' skill as teachers. In order for these interactions to occur, teachers must find new ways to be a part of children's families and to bring parents and other significant teachers into schools along with literacy practices from homes and communities. (p. 221)

Many children are already competent literacy learners by the time they enter preschool or school. At this point, these children are becoming literate in two major socialisation settings, each of which functions to teach specific forms of expertise. However, the relationship between home and school settings can foster collaboration and interdependence or separation and disconnect, with corresponding positive or negative influences on children's progress at school.

> As teachers learn what counts as literacy at home, it would be equally important for them to make explicit what counts as literacy in their own classrooms to themselves as well as to the children and to their families. Once that is clear, they can experiment with techniques

> and materials used at home in ways that complement their own approaches. They can also help network members understand the learning potential of activities such as making lists, reading the newspaper or the Bible, and reading, telling and reciting stories. By interweaving different approaches, teachers will make it possible for children to draw on what they learn in both settings when interacting with print. (Volk & de Costa, 2001, p. 221)

In some communities, children may have acquired literacy in schools outside the regular state school system, quite unbeknown to teachers in mainstream schools. Duranti, Ochs, and Ta'ase (2004), for example, draw attention to the function of religious schools in promoting children's literacy acquisition

> For centuries, religion has promoted literacy. Jews, Muslims and Christians alike rely on written scriptures and instruct their congregations how to read passages within them ... Far more than public schools, religious schools serve as the cornerstone of literacy across nations. (p. 159)

They note how, for example, in every pastor's school in Western Samoa, initial literacy instruction has been accomplished in exactly the same way, through what is called the *Pi Tautau*, published by the Congregational Christian Church. The *Pi Tautau* is

> ... a large poster displaying the Samoan alphabet, with Arabic and Roman numerals along the bottom. Each letter is accompanied by a picture of an object beginning with that letter. As the literacy lesson begins, you sit cross-legged on the floor with the other children in front of the teacher, who is seated on a chair, holding the *Pi Tautau* on her lap. Over time you come to understand what is expected of you. Each lesson the teacher points to the picture on the top left hand corner and asks the class to collectively recite first the letters and their corresponding images, then the letters alone and finally the Arabic and Roman numerals from one to ten. (Duranti et al., 2004, p. 160)

In many places outside Western Samoa, for example in California, Samoans have formed communities based around their local church, an important element of which is the religious school, where children are introduced to the letters, words and numbers represented in the *Pi Tautau*

> The *Pi Tautau* is thus an instrument and a symbol of continuity and even ... a tangible and safe anchor for keeping the children of Samoan descent in southern California connected to the language of their parents and grandparents. (Duranti et al., 2004, p. 165)

Gregory (2004a) draws from a number of research studies of literacy practices in different communities of economically disadvantaged groups across the world, including the work of Duranti et al. (2004), as well as her own work in the United Kingdom in Spitalfields, East London, to outline 'principles and practical implications for future family literacy involvement' (Gregory, 2004a, p. 268). Among these are

> Recognise and acknowledge the variety of literacies and 'funds of knowledge' in the lives of children and their families as practices through home and community activities (Gregory, 2004a, p. 269)
>
> Understand and support the value of different mediators of literacy ... in children's literacy development (Gregory, 2004a, p. 270)

Many Bangladeshi–British children in Gregory's Spitalfields study were involved in out-of-school Qur'anic classes and/or Bengali classes of up to 30 children, wherein the teaching was very formal and the children's role was to listen, repeat, practise and be tested. Children might also read English school books informally with older siblings, where the 'tutor' might synthesise Qur'anic and school literacy practices to scaffold literacy learning. Some researchers, for example Volk and de Costa (2004), have pointed to

> the special role which may be played by older siblings in linguistic minority families where parents do not speak the language . . . and to suggest that the ways in which children learn from older siblings in the home environment may have implications for school learning. (Gregory, 1998, p. 36)

Play between young children and older siblings can initiate children into mainstream school literacy practices and the role of the teacher (Gregory, 2004a, 2004b; Williams, 2004). In the context of the Spitalfields study above, Gregory (2004b) notes features of play-teaching, whereby younger siblings were taught to listen and repeat, as in Qur'anic and Bengali classes. Older siblings taught younger siblings the content of recent classroom literacy lessons delivered by their own teachers: demonstrating what the younger sibling should do, checking up on past learning and directly instructing the younger, while using the teaching style of their community language classes outside the mainstream school. Gregory comments that where older siblings mediate the literacy learning of younger siblings through culturally relevant teaching practices, 'it is clear that teachers have found the "perfect partners"' (Gregory, 2004b, p. 104).

Home–school literacy schemes are 'most successful when teachers are knowledgeable about the strengths of their communities and encourage a syncretism of practices' (Gregory, 1998, p. 275). Gregory concludes that teachers have much to learn from some of these older siblings in building upon such finely tuned scaffolding in reading lessons.

Implications for teachers' professional development

As we have already commented in relation to the work of Gregory (1996), Volk and de Costa (2004) and Duranti et al. (2004), one important dimension of effective collaboration is the role of teachers' knowledge of students' participation in family and community literacy contexts. It raises questions about what sorts of knowledge teachers might need in order to develop and support home and school literacy learning partnerships, and how this might be acquired through teacher professional development (Wearmouth, 2004; Wearmouth et al., 2002).

McNaughton and Glynn (1998) note two complementary forms of collaboration. One involves incorporation of school-like literacy activities into family activities. Research examples include procedures for supporting oral reading, such as PPP and patterns of reading storybooks to preschool children. A second form involves incorporation of community activities into school activities, whereby schools take seriously the notion of starting from where children 'are at'

> Where children are at is to some degree where their families and communities are at. (McNaughton & Glynn, 1998, p. 5)

Both forms of collaboration imply that teachers should have generic, and also particular, knowledge about their students' communities. Incorporating school-like activities into family practices requires a generic understanding of the forms that language and literacy practices could take within these communities. For a number of reasons, a generic understanding of cultural identity is also required as follows:

- Language and literacy practices both reflect and construct cultural identities.
- Family language and literacy practices are everyday life events forming part of children's social, economic, political, cultural and historical contexts.
- Religious schools often have a crucial role to play in literacy learning and in maintaining traditional cultural practices in some communities.

McNaughton and Glynn (1998) argue that, in addition to this generic knowledge, teachers need specific knowledge about the literacy practices of particular families of students in their classes. The diversity in ways of using language within cultural groups may be as wide as, or wider than, the differences between such groups (McNaughton, 1996).

Conclusion

The kind of home/community–school partnership arrangements that exist and the ways in which schools respond to family culture and background in literacy initiatives can serve to include or alienate communities, families and their children. Proper consideration of the location of power and its significance is a very important component in the conceptualision of ways in which families might support the literacy development of their children and help to maintain the 'broad-based' instruction (Wragg et al., 1998) that is required. 'Although many factors influence outcomes, *whānau* [family] have the power to unleash or alternately diminish potential' (Durie, 2006, p. 14). The effectiveness of information sharing is dependent upon the extent to which families feel that they can influence changes in the school (Delgado-Gaitain, 1990; Wearmouth, 1996). Schools directing parents to impose school-based policies and practices upon their children can lead to resistance (Bishop et al., 2003). This may be an important consideration in cases in which those whose experience of difficulties in acquiring literacy makes them particularly vulnerable to criticism or comment by others unaware of family circumstances or cultural background.

An important challenge for literacy education in many countries is responding to diversity in equitable ways (Wilkinson, 1998). The time has come to look critically at the range of opportunities for teachers to extend their knowledge and experience, especially in working with families and communities of the students they teach

(Wearmouth, 2000). Teachers can effectively share knowledge and understanding with the parents and communities of the children they teach. When the flow of information between home and school, and the control of that information flow is reciprocal, parents and family members can both initiate action and respond to teacher-initiated actions to help their children (Wearmouth, 2004). Literacy practices in classrooms and schools will benefit when those qualities and experiences from the home are incorporated into school-based literacy activities.

Essential next questions

Conclusions such as these have implications for future research, policy and practice if there is to be improvement in the whole area of family and home support for addressing students' difficulties in literacy acquisition. Using, for example, case study methodology that focuses on the experiences of parents and families as well as students and teachers, we need to learn more about:

- how schools can most effectively and sensitively support home-based literacy programs for students where parents' and/or carers' own experience of difficulties in acquiring literacy makes them particularly vulnerable to criticism or comment by others unaware of family circumstances or cultural background
- what kinds of professional development, including courses accredited by institutions of higher education, can most effectively support teachers to develop the kind of knowledge, confidence and sensitivity they need to work with families and communities of the students they teach
- what kinds of systems can be established in schools to ensure not only that the flow of information between home and school is reciprocal but that the control of that information flow is reciprocal, so that parents and family members can both initiate action and respond to teacher-initiated actions to help their children
- how schools can ensure that the qualities and experiences from home-based literacy practices are incorporated into classroom-based literacy activities.

References

Adams, M. J. (1994). *Beginning to read: Thinking and learning about print*. Cambridge, MA: MIT Press.

Alton-Lee, A. (2003). *Quality teaching for diverse students in schooling: Best evidence synthesis*. Wellington, NZ: Ministry of Education.

Atvars, K., Berryman, M., & Glynn, T. (1995). *Delivery of pause prompt praise in two schools using culturally matched tutors* (Report to the Ministry of Education). Wellington: Ministry of Education.

Berryman, M. (2000). Helping kids in need: Changing perspectives on positive training relationships between parents and educators for the new millennium. *Kairaranga, 1*(1), 15–19.

Berryman, M., & Glynn, T. (2003). *Transition from Māori to English: A community approach*. Wellington: New Zealand Council for Education Research.

Biddulph, F., Biddulph, J., & Biddulph, C. (2003). *The complexity of community and family influences on children's achievement in New Zealand: Best Evidence Synthesis Iteration (BES)*. Wellington: Ministry of Education.

Bishop, R., Berryman, M., Tiakiwai, S., & Richardson, C. (2003). *Te Kotahitanga: The experiences of year 9 and 10 Māori students in mainstream classrooms*. Wellington: Ministry of Education.

Bishop, R., & Glynn, T. (1999). *Culture counts: Changing power relations in education*. Palmerston North, NZ: Dunmore Press.

Blackledge, A. (2000). *Literacy, power and social justice*. Stoke on Trent, Staffs: Trentham.

Bruner, J. (1996). *The culture of education*. Cambridge, MA: Harvard University Press.

Central Advisory Council for Education (England). (1967). *Children and their primary schools: A report of the central advisory council for education* (England) (The Plowden Report). London: HMSO.

Chapple, S., Jefferies, R., & Walker, R. (1997). *Māori participation and performance in education—A literature review and research program* (Report for the Ministry of Education, May). Wellington: New Zealand Institute of Economic Research.

Dale, N. (1996). *Working with families of children with special needs: Partnership and practice*. London: Routledge.

Davie, R., Butler, N., & Goldstein, H. (1972). *From birth to seven*. London: Longman.

Delgado-Gaitan, C. (1990). *Literacy for empowerment*. London: Falmer.

Duranti, A., Ochs, E., & Ta'ase, E. K. (2004). Change and tradition in literacy instruction in a Samoan American Community. In E. Gregory, S. Long, & D. Volk (Eds.), *Many pathways to literacy: Young children learning with siblings, grandparents, peers and communities* (pp. 159–170). London: RoutledgeFalmer.

Durie, M. (2006, October). *Whānau, education, and Māori potential*. Paper presented at the V Hui Taumata Mātauranga, Taupo, New Zealand.

Education Review Office (ERO). (2004). *Catering for diversity in early childhood services*. Wellington: Author.

Farver, J. A. M. (1993). Cultural differences in scaffolding pretend play: A comparison of American and Mexican-American mother-child and sibling-child pairs. In K. Macdonald (Ed.), *Parent-child play: Descriptions and implications*. Albany, NY: SUNY Press.

Gee, J. P. (1996). *Social linguistics and literacies: Ideology in discourses* (2nd ed.). London: Taylor & Francis.

Glynn, T., Berryman, M., & Glynn, V. (2000). *Reading and writing gains for Māori students in mainstream schools: Effective partnerships in the Rotorua home and school literacy project*. Paper presented at the World Congress on Reading, Auckland, New Zealand.

Glynn, T., Jerram, H., & Tuck, B. (1986). Writing as an interactive social process. *Behavioural Approaches with Children, 10*(4), 116–126.

Glynn, T., & Mcnaughton, S. (1985). The Mangere home and school remedial reading procedures: Continuing research on their effectiveness. *New Zealand Journal of Psychology, 15*(2), 66–77.

Glynn, T., Mcnaughton, S., Robinson, V., & Quinn, M. (1979). *Remedial reading at home: Helping you to help your child*. Wellington: New Zealand Council for Educational Research.

Glynn, T., Wearmouth, J., & Berryman, M. (2006). *Responsive contexts for literacy learning*. Buckingham: Open University Press.

Gregory, E. (1996). *Making sense of a new world*. London: Paul Chapman.

Gregory, E. (1998). Siblings as mediators of literacy in linguistic minority communities. *Language and Education, 12*(1), 33–54.

Gregory, E. (2004a). Bridges to literacy. In M. Nind, K. Sheehy, & K. Simmons (Eds.), *Inclusive education: Learners and learning contexts* (pp. 263–280) London: David Fulton.

Gregory, E. (2004b, July). Invisible teachers of literacy: Collusion between siblings and teachers in creating classroom cultures. In *Literacy* (pp. 97–105). Oxford: Blackwell Publishing.

Gregory, E., Long, S., & Volk, D. (2004). A sociocultural approach to learning. In E. Gregory, S. Long, & D. Volk (Eds.), *Many pathways to literacy: Young children learning with siblings, grandparents, peers and communities* (pp. 6–20) New York & London: Routledge Falmer.

Gregory, E., Williams, A., Baker, D., & Street, B. (2004). Introducing literacy to four year olds: Creating classroom cultures in three schools. *Journal of Early Literacy Learning, 4*(1), 85–107.

Hannon, P. (1999). Rhetoric and research in family literacy. *British Educational Research Journal, 26*(1), 121–138.
Heaton, P. (1996). *Parents in need.* London: Whurr.
Hewison, J., & Tizard, J. (1980). Parental involvement and reading attainment. *British Journal of Educational Psychology, 58,* 184–190.
Houghton, S., & Glynn, T. (1993). Peer tutoring of below average secondary school readers with pause prompt and praise: Successive introduction of tutoring components. *Behaviour Change, 10*(2), 75–85.
Jerram, H., Glynn, T., & Tuck, B. (1988). Responding to the message: Providing a social context for children learning to write. *Educational Psychology, 8*(1&2), 31–40.
McNaughton, S. (2002). *Meeting of minds.* Wellington: Learning Media.
McNaughton, S. (1996). Ways of parenting and cultural identity. *Culture and Psychology, 1*(2), 173–201.
McNaughton, S., & Glynn, T. (1998). *Effective collaboration: What teachers need to know about communities.* Paper presented at the annual conference of the New Zealand Council for Teacher Education, University of Waikato, Hamilton, New Zealand.
McNaughton, S., Glynn, T., & Robinson, V. (1987). *Pause, prompt and praise: Effective tutoring of remedial reading.* Birmingham: Positive Products.
McNaughton, S., Glynn, T., Robinson, V., & Quinn, M. (1981). *Parents as remedial reading tutors: Issues for home and school.* Wellington: New Zealand Council for Educational Research.
Morgan, C., & Morris, G. (1999). *Good teaching and learning. Pupils and teachers speak.* Buckingham: Open University Press.
Nash, R. (1993). *Succeeding generations: Family resources and access to education in New Zealand.* Auckland: Oxford University Press.
Ostler, C. (1991). *Dyslexia: A parents' survival guide.* Godalming: Ammonite Books.
Open University. (2002). Video programme 4, E801 *Difficulties in literacy development.* Milton Keynes, UK: Author.
Quest Rapuara. (1992). *Cultural Identity: A resource for educators: Whakamana Tangata.* (3rd Edition). Wellington: Author.
Rassool, N. (1999). Literacy: In search of a paradigm. In N. Rassool (Ed.), *Literacy for sustainable development in the age of information* (pp. 25–53). Clevedon: Multilingual Matters.
Riddick, B. (1996). *Living with Dyslexia.* London and New York: Routledge.
Rubie-Davies, C.M., Hattie, J.A.C., & Hamilton, R.J. (2006). Expecting the best for New Zealand students: Teacher expectations and academic outcomes. *British Journal of Educational Psychology, 76,* 429–444.
Stanovich, K.E. (2000) *Progress in understanding reading. Scientific foundations and new frontiers.* New York: Guilford Press.
Street, B. (1997). The implications of the 'new literacy studies' for literacy education. *English in Education, 31*(3), 45–59.
Street, B., & Street, J. (1995). The schooling of literacy. In P. Murphy, M. Selinger, J. Bourne, & M. Briggs (Eds.), *Subject learning in the primary curriculum.* London: Routledge.
Tizard, J., Schofield, W. N., & Hewison, J. (1982). Collaboration between teachers and parents in assisting children's reading. *British Journal of Educational Psychology, 52*(1), 1–15.
Topping, K. (1996). Tutoring systems for family literacy. In S. Wolfendale & K. Topping (Eds.), *Family involvement in literacy.* London: Cassell.
Volk, D., & de Costa, M. (2001). Many differing ladders, many ways to climb: Literacy events in the bilingual classroom, homes and community of three Puerto Rican kindergartners. *Journal of Early Childhood Literacy, 1*(2), 193–224.
Volk, D., & de Costa, M. (2004). Mediating networks for literacy learning. In E. Gregory, S. Long, & D. Volk (Eds.), *Many pathways to literacy: Young children learning with siblings, grandparents, peers and communities* (pp. 25–39). London: Routledge Falmer.
Wearmouth, J. (1996). Registering special needs—for what purpose? *Support for Learning, 11*(3), 118–123.

Wearmouth, J. (2000). *Co-ordinating special educational provision meeting the challenges in schools*. London: Hodder.

Wearmouth, J. (2002). Dilemmas in the choice of responses to pupils' difficulties in literacy development at multiple levels. In J. Wearmouth, J. Soler, & G. Reid (Eds.), *Addressing difficulties in literacy development: Responses at family, school, teacher and pupil levels* (pp. 8–37). London: Routledge.

Wearmouth, J. (2004). Issues in addressing children's difficulties in literacy development through family-school partnerships. *Curriculum Journal*, *15*(1), 5–18.

Wearmouth, J., Glynn, T., & Berryman, M. (2005). *Perspectives on student behaviour in schools: Exploring theory and developing practice*. London: Routledge.

Wearmouth, J., Soler, J., & Reid, G. (2002). *Meeting difficulties in literacy: Research, policy and practice*. London: Routledge.

Whitehead, D. (1993). *Factual writing think sheets*. Hamilton, New Zealand: Seaforth Education.

Williams, A. (2004). 'Right! get your book bags!' Siblings playing school in multi-ethnic London. In E. Gregory, S. Long, & D. Volk (Eds.), *Many pathways to literacy: Early learning with siblings, grandparents, peers and communities*. London: Routledge.

Wilkinson, I. A. G. (1998). Dealing with diversity: Achievement gaps in reading literacy among New Zealand students. *Reading Research Quarterly*, *33*(2), 144–167.

Wragg, E. C., Wragg, C. M., Haynes, G. S., & Chamberlain, R. P. (1998). *Improving literacy in the primary school*. London: Routledge.

Chapter 9
Enhancing Reading Comprehension Through Explicit Comprehending-Strategy Teaching

John Munro

Introduction

The capacity to understand written text and to learn from it is critical for living in the information-rich world of the 21st century. Many students have difficulty doing this. One approach to helping them is to give them more texts to comprehend. This usually does not work because it does not teach them how to comprehend each text; that is, how to act upon and think effectively about the information in the text. The focus of this chapter is on teaching students who have literacy learning difficulties how to comprehend text more effectively. The need for explicit instruction with this focus was noted in recent reports, both in the United States (NICHD, 2000) and in Australia (Rowe, 2005).

The theoretical framework used to examine reading comprehension strategy use

In order to comprehend a written text, readers need to identify its ideas and link them in the ways intended by the writer. Readers use various linking actions to do this. They link individual meanings in the text, link the meanings with what they know (that is predict and infer), link ideas in the text with images (that is visualise), identify and infer the 'main' ideas in the text and link individual ideas with these (that is summarise) as they read.

In this chapter, the linking actions are called 'comprehending actions' or 'strategies'. They are proposed to model the 'moment-by-moment', dynamic activity of able readers. While reading, readers paraphrase and visualise sentences, use their knowledge of written grammar, predict and speculate about what other ideas are consistent with the text, consolidate, summarise and identify the questions it answers. These actions are the 'conceptual tools' readers use to link the ideas in a written text.

J. Munro (✉)
The University of Melbourne, Melbourne, VIC, Australia
e-mail: jkmunro@unimelb.edu.au

C. Wyatt-Smith et al. (eds.), *Multiple Perspectives on Difficulties in Learning Literacy and Numeracy*, DOI 10.1007/978-1-4020-8864-3_9,

Paraphrasing involves readers retelling a sentence in their own words. They modify the sentence while retaining its meaning or 'proposition'. They replace either its content words or its syntactic make-up with similar forms (Best, Rowe, Ozuru, & McNamara, 2005). They move to paraphrasing two or more sentences at a time, and use the main ideas and key points within a text to modify their retelling. This strategy helps them to link the text with their existing knowledge. It is seen by some investigators as the most effective strategy to support reading comprehension (Katims & Harris, 1997) for both fiction and non-fiction (Fisk & Hurst, 2003) because it integrates ' ... all modes of communication—reading, writing, listening, and speaking—which lead to a deeper understanding of the text' (p. 182).

Visualising assists readers to create a mental image of a sentence by transforming its content words into visual images within context. To do this, they make links with their existing experiential knowledge. This assists readers to retain the meanings read more effectively, and to infer and predict more easily (Nielsen Hibbing & Ranking-Erickson, 2003).

Predicting involves linking one's interpretation of a text with one's existing knowledge and then extending comprehension by generating plausible inferences (Harvey & Goudivis, 2000; Keer, 2004; Munro, 2003, 2004). To do this, readers integrate the meaning of one or more sentences with global text features, such as the topic and disposition, as well as the discourse in which the sentences are located. Harvey and Goudivis (2000) link it with visualising, in the sense that visualising involves predicting with images rather than words.

Reading underachievers are less likely to use these comprehending strategies

Students who have reading difficulties have been conceptualised as non-strategic in their text-processing activity (Torgesen, 1986). While they may decode accurately, they have been described as 'passive' in their comprehension activity (Nielsen Hibbing & Ranking-Erickson, 2003). Their average learning peers use a broader range of comprehending strategies and are more likely to monitor actively their reading comprehension activity.

A psychometric 'cut-off point' frequently used to identify those at risk of literacy learning difficulty is the 25th percentile in reading recognition (Swanson, 1999). Their learning profiles include having immature word knowledge and vocabulary and inefficient short-term retention in working memory, due in part to a lack of automaticity in relevant knowledge (Gardill & Jitendra, 1999; Gersten, Fuchs, Williams, & Baker, 2001; Jitendra, Edwards, Sacks, & Jacobson, 2003; Pearson & Hamm, 2005; Saunders, 2001; Swanson, Hoskyn, & Lee, 1999). They may not understand key words or sentences, or may not integrate the text information in a meaningful way (Parker, Hasbrouck, & Denton, 2002).

Some 'passive' readers lack particular comprehending strategies, while others select inappropriately from their repertoire and still others fail to engage in self-monitoring efficiently while reading (Botsas & Padeliadu, 2003; Dreyer & Nel,

2003; Klassen, 2002; Swanson & De La Paz, 1998; Swanson et al., 1999; Vaughn, Gersten, & Chard, 2000). They may, for example, be less likely to visualise a text they have read accurately and therefore have greater difficulty with comprehension and recall. In other words, some students need to learn comprehending strategies, while others need to learn how to use their comprehending strategies more effectively.

The use of comprehending strategies illustrates the distinction between comprehension and comprehending. Comprehension is what readers know, having read a text. Comprehending refers to the use of a coordinated set of actions by readers to make the links and to comprehend the text.

This distinction is important in order to understand the reading comprehension process, diagnose reading comprehension difficulties and implement effective teaching. Giving students more texts to comprehend is unlikely to improve their comprehension, because it does not teach them how to link the ideas. They improve their comprehension when they improve the actions they use to represent text.

Strategies for achieving each level of comprehension

In order to form the links between meanings, readers act strategically on the text representations they form at any time, to generate other representations. A reader who has, for example, represented a text only by decoding or vocalising it may comprehend it less effectively than a reader who has linked the meanings in multiple ways.

It is useful to distinguish between comprehending strategies and comprehension skills. Strategies are used in a conscious, deliberate, attention-demanding way, while skills are used in a more automatic way without necessarily deliberately deciding to use them (Alexander, Graham, & Harris, 1998). This distinction is important for teaching. Instruction needs to begin with students learning to use a strategy and to have the opportunity to make its use automatic.

Teaching the comprehending strategies

Teaching comprehending strategies explicitly, such as visualisation, paraphrasing and predicting, improves reading comprehension (for example Harvey & Goudivis, 2000; Jitendra et al., 2003; Katims & Harris, 1997; Keer, 2004; Munro, 2003, 2004; Stahl, 2004; Swanson & De La Paz, 1998; Swanson, 1999).

It facilitates reading comprehension for students who have reading difficulties (Katims & Harris, 1997; Lee & von Collin, 2003; Munro, 2004; Pressley, Roehrig, Bogner, Raphael, & Dolezal, 2002; Reutzel, Smith, & Fawson, 2005; Swanson et al., 1999). The particular strategy most likely to assist at any time depends on the student's literacy learning profile, the teaching format and organisational factors, such as the duration of the teaching and the teaching style (Swanson, 1999).

Paraphrasing and visualising has improved reading comprehension for students with reading disabilities (Katims & Harris, 1997; Saunders, 2001; Stahl, 2004). It assists them to link the word meanings and then to visualise a sentence (Saunders,

2001). They benefit from guided scaffolding to 'make a picture' in their head of a sentence or paragraph that they hear.

Strategy teaching also enhances self-confidence as a reader. Beliefs about one's capabilities are critical to performance (Pajares, 2003). Many students with learning difficulties have low self-efficacy beliefs and do not take risks when reading and learning to read (Botsas & Padeliadu, 2003). Teaching them 'how to' comprehend text can improve their self-efficacy beliefs (Quirk, 2004) and guide them to see themselves as independent readers (Vaughn et al., 2000).

Various teaching regimes that combine visual imagery and verbal elaboration, such as self-questioning, enhance reading comprehension. The 'RIDER' procedure (Clark, Deshler, Schumaker, Alley, & Warner, 1984) teaches students to read the first sentence or section of the text, make an image of it, describe the image, evaluate it for its completeness and repeat the steps for the following sentences. The IEPC strategy (Wood & Endres, 2005) teaches a four-step procedure as one reads: to imagine, elaborate, predict and confirm.

A third regime, the RAP procedure, is based on paraphrasing (Katims & Harris, 1997). It focuses on summarising a paragraph and involves three steps: reading a paragraph; questioning the main idea and details; and saying the main ideas and details in one's own words. It enhances text comprehension for both able and underachieving readers (Parker et al., 2002).

Strategy instruction usually teaches students both how to do a strategy (their procedural knowledge) and when to use it (their conditional knowledge). The instruction makes them aware that there are actions they can use to enhance their comprehension of text, and teaches them (1) how to use each action; (2) when to use each; and (3) to use their strategies in an integrated way. This helps the students to self-regulate their strategy use.

The focus in this chapter

The above review has implications for how students who have reading difficulties can be assisted to improve their use of comprehending strategies. Some may need to be guided to learn more appropriate strategies when reading text, and others to use them more efficiently. This chapter examines two aspects of comprehending-strategy teaching as follows:

1. Is comprehending-strategy instruction effective when it is taught in regular classroom contexts that include both able and at-risk readers? While the earlier research has generally examined strategy teaching, either individually or in small groups, Deshler and Schumaker (1993) raise this as a possibility. The present chapter examines the efficacy of this possibility.

   This question has implications for classroom practice and for education providers. If at-risk readers can be shown to benefit from strategy teaching provided to a class as a whole, the instruction may offer a viable and economic means of dealing with particular types of reading underachievement. Rather than needing to implement small-group intervention that involves student withdrawal and additional financial resourcing, low-reading achievement can be ameliorated within regular teaching provision.

2. To what extent can readers who learn to use a comprehending strategy in one context transfer it to other contexts? This question, too, has implications for classroom practice and educational provision.
   Strategy instruction aims for students to use the strategies spontaneously in a self-directing, autonomous way. Instructional programs have tended not to monitor systematically the application of a strategy from scaffolding to self-management. This chapter examines the transfer of a strategy, from its use under direction with single sentences to its use in a self-directing way with connected prose.

The methodological approach used to research comprehending-strategy instruction

Research examining the effectiveness of teaching reading comprehension strategies uses versions of the intervention—matched control design (Swanson, 1999), in which the comprehension outcomes attributed to teaching a particular strategy are compared with the outcomes attributed to alternative teaching. This design is sometimes linked with readers being asked to 'think aloud' as they work through comprehension tasks.

The present investigation uses this design. Its data have been collected over the past decade and come in part from a series of action research projects completed by teachers studying a graduate course in early literacy intervention.[1] These projects examine the influence of teaching explicitly these reading comprehension strategies to students with learning difficulties in small-group and in classroom contexts. They examine a range of comprehending strategies and the conditions under which teaching a particular strategy is most likely to be associated with improved reading comprehension outcomes for students in Grades 3–6. In addition, they compared the efficacy of some comprehending strategies and the relationship between explicit strategy teaching and variables such as self-efficacy as a reader. Aspects have been published (Munro, 2003, 2004), used in doctoral and Masters of Education theses (Chan, 2006; Huggins, 2006; Verezub, 2005) and used in action research projects by teachers studying a graduate course in early literacy intervention.

A naturalistic classroom study of teaching comprehending

This chapter reports the extent to which explicit teaching of a reading comprehending strategy (visualising, predicting or paraphrasing) improves reading

[1] The set of action research studies is available on <http://online.edfac.unimelb.edu.au/LiteracyResearch/pub/Projects/P_list.htm>

comprehension outcomes for both students who are at risk of literacy learning difficulties in Grades 3–6 and their able reading peers. Unlike most studies in this area, the students comprised regular class cohorts. The readers attended primary schools in metropolitan Melbourne, Australia. Some of the classes provided the context for teaching one of the strategies and others provided a control. Each teaching class and its control class were matched on socioeconomic status and literacy performance of the class, as indicated by statewide assessment data over the previous 2 years. Each had the random diversity of student qualities that one would encounter in any classroom in contemporary metropolitan Melbourne. As such, the study is a naturalistic examination of the effectiveness of comprehending-strategy teaching to regular classroom cohorts.

Reading comprehension was assessed using the Tests of Reading Comprehension (TORCH) (Mossenson, Hill, & Masters, 2003). Whether any student was in the able reading or at-risk group was determined by their reading comprehension score on this scale. Students earning a score below the 25th percentile were categorized as 'at risk'.

The investigation comprises two studies.

Study 1: the influence of teaching visualising, predicting or paraphrasing

The first investigation examined the efficacy of the three strategies for improving reading comprehension. The number of students in each category of reading ability, intervention and grade range are shown in Table 9.1.

The strategy task: Prior to the comprehending-strategy instruction, students' ability to apply each strategy to sentences was examined using the strategy task. This provided an indication of the extent to which the students used the strategy prior to the instruction. The students heard a story about a familiar context: two boys going for a bike ride. They then read 16 sentences relating to the narrative and, depending on the strategy context to which they were allocated, were instructed either to paraphrase, visualise or predict, as follows:

Table 9.1 The number of students in each category of reading ability, intervention and grade range

		Grades 3–4		Grades 5–6	
		At-risk readers	Able readers	At-risk readers	Able readers
Paraphrasing	Teaching	21	79	12	45
	Control	15	45	12	48
Visualising	Teaching	10	56	13	50
	Control	14	44	11	52
Predicting	Teaching	14	62	11	43
	Control	17	65	10	39

1. For the paraphrasing task, the students were asked to 'say the sentence in another way with other words. Change as many words as you can but make sure it still says the same thing'.
2. For the visualising task, the students were asked to 'make a picture of it in your mind' and then to describe their picture in written words. Possible sentence prompts suggested were 'I can see in my mind . . . '
3. For the predicting task, the students were asked to 'think in your mind what might happen next' and then to describe what they thought in words. Possible sentence prompts suggested were 'I think . . . might happen', or ' . . . might . . . next'.

The same set of sentences was used for the three task conditions. The sentences increased in complexity, from describing one event to three events. Each response was scored on a three-point scale as follows: the application of the strategy to the sentence was comprehensively displayed (score = 2); the application of the strategy to the sentence was partially displayed (score = 1); the application of the strategy to the sentence was not displayed (score = 0).

For all conditions, the strategy application was modelled by the students' teacher prior to the assessment. This included the teachers 'thinking aloud' as they applied the strategy and the students practising applying it and being given corrective feedback.

Students' application of the strategy was also assessed following the completion of the instruction. This permitted an estimate of the extent to which application of the strategy changed over the instruction period.

The strategy instruction program: The strategy-teaching program taught students to apply explicitly one of the strategies to expository text. It followed the same sequence for the three strategies.

The strategy program was scripted and implemented by the students' regular teacher for 10 sessions of between 40 and 45 minutes duration in each classroom. The students applied the strategy first to sentences and then to paragraphs, first to reading aloud and then to reading silently. In each session, the teacher introduced the strategy and the students practised it and then reviewed it. The following aspects were developed over the 10 sessions, in each of the three contexts:

- The teacher defined the strategy (for example paraphrasing; 'After you have read each sentence, you say it in your own words').
- The teacher modelled the strategy first for contexts in which the students listened to a text being read and were scaffolded to apply the strategy.
- The teacher transferred the strategy to a reading context. In a whole-class activity the students listened to a sentences in an expository text being read aloud, heard their teacher model the strategy and were scaffolded to apply it themselves to successive sentences in the text. The teacher provided corrective feedback.
- The students read aloud the expository texts and practised applying the strategy sentence by sentence. They were asked to describe how they were applying the

strategy; that is they described in words what they did to apply the strategy. They also commented on how it helped them to comprehend the text.
- The students said that they would apply the strategy and described it in words before they began to read.
- The teacher scaffolded the students to apply the strategy to a paragraph of expository text that comprised three sentences.
- The teacher scaffolded the students to apply the strategy to sentences in expository texts that they read silently. The students practised applying the strategy. Again, they received corrective feedback. They repeated this for paragraphs. They described how they applied the strategy and how it helped them to comprehend when they read silently.

These aspects are characteristics of effective strategy teaching (Alder, 2004; Gersten et al., 2001).

The texts used at each grade range had a readability level that matched that level. As well, each text had a picture that illustrated the events described.

Procedure: The activities were administered in the following sequence:

1. The pre-assessment tasks (Strategy Task and TORCH) were administered to the control and teaching groups.
2. The teaching groups completed the strategy teaching comprising the 10 teaching sessions, each of 40–45 minutes, in their classes. The instructor was either their class teacher or the individual needs teacher at the school. The students in the control groups had their regular literacy education activities.
3. The post-assessment tasks (Strategy Task and TORCH) were re-administered to both the control and teaching groups.

Results

Applying the strategies to sentences: Our first question related to how well the students used the reading comprehending strategies with sentences. The mean performance of students using each strategy, both prior to and after strategy teaching, both for able and at-risk readers in each grade range, are shown in Table 9.2.

The results generally showed that the strategy teaching was associated with improved comprehension at the sentence level for both grade ranges ($p < 0.01$). Prior to instruction, the teaching group did not differ from the control group. Exceptions were the able reading group in Grades 3–4 and the at-risk group in Grades 5–6 taught to visualise. Following the teaching, the group taught to use the strategy achieved higher performance than its matching control group. Again, the only exception was the group taught to visualise at the Grades 3–4 range.

Able readers in Grades 3–4 showed a higher mean strategy score than their at-risk peers for all three strategies ($p < 0.01$). Able readers in Grades 5–6 showed higher mean paraphrasing and visualising scores than their at-risk peers ($p < 0.01$).

Table 9.2 The mean pre-comprehending-strategy scores (mean, standard deviation) for each strategy for able and at-risk reading comprehenders at each grade span

		Able readers				At-risk readers			
		Pre-teach		Post-teach		Pre-teach		Post-teach	
		Mean	SD	Mean	SD	Mean	SD	Mean	SD
Grades 3–4									
Paraphrasing	Teaching	18.25	7.18	28.78	2.35	8.20	3.51	19.78	4.38
	Control	19.21	5.16	22.28	6.45	9.07	6.37	12.67	2.21
Visualising	Teaching	17.36	5.17	19.78	5.31	10.19	7.87	16.89	3.41
	Control	18.21	3.83	20.69	5.46	8.34	1.63	10.23	4.67
Predicting	Teaching	19.15	4.76	29.32	4.54	10.00	3.81	21.00	9.21
	Control	19.12	6.18	24.29	2.31	11.46	4.51	14.44	5.56
Grades 5–6									
Paraphrasing	Teaching	23.62	5.50	28.34	3.07	18.08	8.14	26.35	2.74
	Control	23.05	4.69	24.07	2.67	19.13	7.59	21.16	3.21
Visualising	Teaching	21.09	8.64	27.41	4.63	18.21	4.81	27.72	4.21
	Control	27.32	4.23	24.78	4.02	24.36	5.57	21.42	3.25
Predicting	Teaching	23.69	5.20	26.73	3.89	18.15	3.19	25.39	2.36
	Control	23.26	3.54	25.18	2.92	19.02	4.58	21.93	4.47

The strategy training elevated the sentence-level comprehension of the at-risk readers to that of the able readers for most of the teaching conditions. The post-reading, sentence-level comprehension score for the at-risk readers did not differ from that of the able readers ($p > 0.05$). Only the Grades 3–4 at-risk readers taught to predict did not achieve the sentence-level comprehension of their able reading peers also taught to predict.

How did the three comprehending strategies differ in ease of use? First, at each grade range, prior to the strategy instruction, the three strategies did not differ in ease of application ($p > 0.05$). Second, the able readers in each range applied them more efficiently than their at-risk peers ($p < 0.01$). Third, the Grades 5–6 students used them more successfully than their matching Grades 3–4 peers ($p < 0.01$).

The error patterns of the at-risk readers can assist in showing the nature of their comparative difficulty. For paraphrasing, the Grades 3–4 at-risk students achieved at a lower level than their able reading peers on all tasks; their performance was not influenced by either the semantic or grammatical complexity of the sentence to be paraphrased. It was the capacity to paraphrase, regardless of the sentence to be paraphrased that accounted for their lower performance.

A similar pattern was displayed for predicting. It was the capacity to predict or to 'think ahead' rather than the comparative complexity of some sentences that accounted for their difficulty.

For visualising, on the other hand, the at-risk readers showed a lower performance only on particular items that had words or phrases that were comparatively unfamiliar and difficult to visualise, such as 'enjoying themselves', 'supposed to

Table 9.3 The mean pre-prose and post-prose comprehension scores (mean, standard deviation) for each strategy for able and at-risk reading comprehenders at each grade span

		Able readers				At-risk readers			
		Pre-teach		Post-teach		Pre-teach		Post-teach	
		Mean	SD	Mean	SD	Mean	SD	Mean	SD
Grades 3–4									
Paraphrasing	Teaching	37.81	6.2	41.8	11.9	19.63	7.6	30.11	6.6
	Control	43.00	8.8	45.88	12.4	21.00	7.1	25.17	11.1
Visualising	Teaching	35.85	6.3	34.77	9.1	21.33	8.0	19.33	10.9
	Control	36.92	7.5	36.92	11.4	13.00	2.3	19.00	1.7
Predicting	Teaching	36.63	6.6	38.21	8.2	13.71	10.0	21.95	3.5
	Control	38.24	6.3	35.29	9.3	19.38	8.3	17.88	9.2
Grades 5–6									
Paraphrasing	Teaching	54.14	11.8	52.21	9.3	35.50	5.5	45.50	9.0
	Control	53.47	11.5	54.33	6.9	38.74	3.2	41.60	15.0
Visualising	Teaching	49.38	10.6	43.58	10.0	35.50	5.1	35.33	11.8
	Control	53.71	8.5	47.74	9.0	31.67	2.9	33.56	6.4
Predicting	Teaching	59.17	7.9	57.83	9.1	32.07	6.14	44.67	4.50
	Control	57.28	8.2	55.28	12.9	38.00	3.4	37.00	4.5

be' or 'gasped'. It is possible that the students were familiar with the process of visualising and may, for example, have used it during listening activities. The items that caused more difficulty may have had images that were difficult to encode in words.

At the Grades 5–6 range, the differences were restricted to particular items for all of the strategies. This is consistent with the students having developed an initial awareness of each strategy and needing to learn how to apply it while reading.

These data show that comprehension of sentences can be enhanced by teaching these comprehension strategies. They are more influential for students who have comprehension difficulties.

Applying the comprehending strategies to connected prose: Our third question related to how well the students transferred the reading comprehending strategies to prose that comprised several connected sentences. This was investigated by examining the extent to which prose-reading comprehension measured by the TORCH improved following the strategy training. The mean performance of students prior to and following strategy instruction for each strategy (mean pre-TORCH and post-TORCH scores) for the at-risk and able reader groups at each grade range, with the matching control group performance, are shown in Table 9.3.

The extent to which the comprehending instruction transferred to improve connected prose-reading comprehension depended on the grade range and reading ability of the students, and the strategy taught. For the at-risk readers in Grades 3–4, the intervention was associated with improved comprehension for paraphrasing ($p < 0.001$, $\eta^2 = 0.73$) and for predicting ($p < 0.001$, $\eta^2 = 0.68$), while for the at-risk readers in Grades 5–6, the intervention was associated with improved comprehension for paraphrasing ($p < 0.001$, $\eta^2 = 0.64$) and for predicting ($p < 0.001$, $\eta^2 = 0.63$).

In other words, teaching the at-risk students at both the 3–4 and 5–6 grade ranges to paraphrase or to predict was associated with improved comprehension over their matched control peers ($p < 0.01$). The gains made in paraphrasing or predicting given single sentences transferred to the connected prose-reading context. The effect sizes (Cohen d) for paraphrasing and predicting were 1.5 and 1.3 for the Grades 3–4 students and 1.4 and 2.4 for the Grades 5–6 students, respectively.

A similar transfer effect was not observed for the able readers at either the 3–4 or the 5–6 grade ranges. The gains made in comprehension at the sentence context were not associated gains in the connected-prose context ($p > 0.05$). Teaching the able reading students in this grade range these strategies did not improve their comprehension. The effect sizes (Cohen d) for paraphrasing and predicting were less than 0.5 for all grades.

Teaching the visualising strategy did not facilitate prose comprehension for any condition (Cohen d and η^2 effect sizes were less than 0.2). This lack of influence could be attributed to a number of causes. It was noted earlier that one possible cause could be a difficulty in encoding images in words. Breaking this process into two steps, in which students first draw pictures of what they are imagining and then link words with these, may facilitate the transfer. This possibility was examined in Study 2.

The display of comprehension in the prose-reading comprehension context requires the reader to build both microscopic and macroscopic representations of the text. These data assist in unpacking how this construction can be enhanced by teaching these comprehension strategies.

Study 2: an alternative approach to teaching visualising

The second study examined a modified procedure for teaching Grades 3–4 to visualise. The students were taught to draw their images and to describe them in words as they read a text. This method draws in part from data reported by Miatello (2008) and by Munro (2004). In other respects its methodology was similar to that used in Study 1.

The participants were 97 students in Grades 3–4, aged from 7 years 11 months to 10 years 4 months. The teaching group and its matched control group attended the same school. Each group was a class, with approximately equal numbers of males and females, and a similar number of average and at-risk readers. Most of the students decoded relevant text with appropriate accuracy and fluency. The students had not previously received visualisation instruction.

The mean prose-reading comprehension for each group of students prior to and following the modified visualisation instruction (pre- and post-TORCH scores) for the at-risk and able reading groups are shown in Table 9.4.

Table 9.4 The mean pre- and post-prose comprehension scores (mean, standard deviation) for the visualising strategy for able and at-risk reading comprehenders in Grades 3–4

	Teaching group				Control group			
	Pre-teach		Post-teach		Pre-teach		Post-teach	
	Mean	SD	Mean	SD	Mean	SD	Mean	SD
At-risk readers	24.72	5.3	38.59	6.1	25.79	4.6	31.05	7.4
Able readers	45.38	5.2	49.7	6.9	43.00	3.8	45.88	4.8

In this case, the intervention was associated with improvement in prose-reading comprehension for both groups of readers ($p < 0.001$), with an effect size (partial eta squared) of $\eta^2 = 0.69$ and 0.47 for the at-risk and able readers, respectively. The data are consistent with the expectation that teaching students to visualise and to draw their imagery transfers to prose-reading comprehension. The support provided by this was greater for the at-risk readers. As indicated by the effect size, approximately 70% in the variability in their post-Torch scores can be attributed to the teaching.

Why was the modified teaching more successful in enhancing prose comprehension? The modified procedure taught explicitly the various steps in visualising; forming an image for each sentence, drawing it and naming its key parts, and then linking this with the images formed so far. Breaking it into these steps may have allowed the readers to complete each step and to use their working memory resources more effectively.

What the results mean for facilitating literacy learning for students who have learning difficulties

The study shows that reading comprehension, particularly for those who have reading difficulties, can be improved by implementing an explicit comprehending-strategy teaching program. This instruction can also benefit the comprehension of able readers, though not necessarily to the same extent.

The results show the efficacy of explicit comprehending-strategy teaching. Reading comprehension can be improved by teaching students how to act on the text they read. The results are consistent with the concept of students who have reading difficulties as non-strategic learners.

The results show also that small-group or individual instruction may not be necessary to teach these comprehending strategies. Instead, a well-structured teaching program that teaches aspects of a strategy explicitly and in appropriate steps or increments can achieve the improvement. The teaching program used in this study included a focus on retention. Towards the end of each teaching session, the students reviewed and described what they had done, how it had helped them and what they had learnt. The extent to which this procedure contributed to the success of the teaching program cannot be established from the data.

The study shows the potential contribution of a model of text comprehension typified by that of Kintsch and Kintsch (2005) to understanding the literacy learning characteristics of students who have learning difficulties, and to teaching these students to improve their text comprehension. Some students may be able to build microscopic representations of sentences that can be used for literal comprehension, but may not be as able to build macroscopic representations that permit the reader to infer in various ways about the prose. Similarly, some students may need strategy teaching that assists them to represent sentences, while others need teaching that assists them to represent inferences from these sentences.

Essential next questions

The comprehending-strategy teaching regime most appropriate for any student with learning difficulties will obviously depend in part on the student's existing knowledge. This includes the student's approach to learning, the student's knowledge of characteristics of text properties (such as letter clusters, vocabulary knowledge, written sentence and discourse conventions). Teaching paraphrasing or visualising for a particular type of text, for example, may not be appropriate for students whose word-reading accuracy for that type of text is comparatively low. It should be noted, however, that the use of these strategies can facilitate word-reading accuracy in prose. One question that warrants further research is whether teaching students with learning disabilities to use particular comprehending strategies leads to enhanced word-reading accuracy.

A second question that warrants future research is how to match comprehending-strategy instruction with a student's existing knowledge profile at any time. A wide range of comprehending strategies can be taught. These include vocabulary enhancement strategies, word-analogy strategies, phonological and phonemic strategies, inferencing, summarising, predicting, clarifying, questioning, monitoring, synthesising, evaluating and connecting, text-structure strategies, strategies that use text features and visual information in a text and consolidation-review strategies. Examples of action research projects examining instruction in some of these with learning disabled students are available online (see foot note 1). It is likely that these strategies will lie on a developmental trajectory, such that some are most easily learnt after others. It is expected that future research may investigate such developmental trends.

A third question relates to the conditions under which students learn to use these strategies comparatively automatically in an integrated way. Strategy use is mediated by the use of 'inner language', or self-talk. How students with learning difficulties can be assisted to learn to use self-talk comparatively automatically merits future research.

A fourth question relates to useful processes for communicating to teachers and schools an awareness of the potential contribution of teaching comprehending strategies to literacy enhancement, particularly for students who have learning disabilities. There are various ways in which this can be done.

One is to specify in curriculum standards the comprehending strategies that students are expected to acquire as they progress through school. The English Developmental Continuum P–10 for the Victorian Essential Learning Standards (VELS) in Australia provides an attempt at this. The Continuum specifies, for reading, speaking and listening, and writing, the comprehending strategies that students are expected to acquire at half-yearly increments along an English knowledge developmental pathway. The English Developmental Continuum P–10 for reading can be accessed at <www.education.vic.gov.au/studentlearning/teachingresources/english/englishcontinuum/reading/default.htm> (accessed 3 February 2009).

Some educators argue against this 'curricularisation' of strategy teaching (Fisher & Frey, 2008, p. 16). The data they use to support this position would seem to be due to inappropriate strategy teaching rather than to specifying the strategies as student knowledge outcomes. They seem to argue for students to learn to integrate or consolidate their use of strategies. The need for this integration has already been recognised in the VELS English Developmental Continuum P–10. As students progress along their knowledge pathway they are expected to display explicitly a synthesised use of the strategies they have learnt. They are also required, from Grade 3 through to Grade 10, to talk about their 'plan for reading' before they begin to read a text. Their plan is expected to refer to the strategies they intend to use as they read.

A fifth question relates to the application of the comprehending strategies to the comprehension of written hypertext. Students are increasingly required to learn by reading information presented in hypertextual contexts. Computer-assisted instruction and online teaching is often used with students who have learning difficulties. An appropriate issue here relates to similarities and differences between the reading processes involved in understanding conventional, written factual text and hypertext. This issue was investigated for students in Grades 3–6 required to comprehend factual information presented in written conventional and hypertextual formats (Verezub, 2005).

This chapter shows the value of interpreting reading-comprehension outcomes in terms of the comprehending actions that lead to them. It shows the effectiveness of improving the outcomes of students who have learning difficulties, by teaching them explicitly the relevant comprehending strategies. It also shows that this can be achieved in regular classroom contexts.

References

Alder, C. R. (2004). Seven strategies to teach students text comprehension. *Reading Rockets, 3479*, 1–5.

Alexander, P. A., Graham, S., & Harris, K. R. (1998). A perspective on strategy research: Progress and prospects. *Educational Psychology Review, 10*(2), 129–145.

Best, R. M., Rowe, M., Ozuru, Y., & McNamara, D. S. (2005). Deep-level comprehension of science texts. *Topics in Language Disorders, 25*(1), 65–83.

Botsas, G., & Padeliadu, S. (2003). Goal orientation and reading comprehension strategy use among students with and without reading difficulties. *International Journal of Educational Research, 39*, 477–495.

Chan, M. C. (2006, October). Using text at multiple levels: Comparing three reading comprehension strategies based on Kintsch's Construction Integration model of text comprehension. Thesis in partial completion of the Masters of Educational Psychology, University of Melbourne.

Clark, F. L., Deshler, D. D., Schumaker, J. B., Alley, G., & Warner, M. M. (1984). Visual-imagery and self-questioning strategies to improve comprehension of written material. *Journal of Learning Disabilities*, *17*, 145–149.

Deshler, D. D., & Schumaker, J. B. (1993). Strategy mastery by at-risk students: Not a simple matter. *The Elementary School Journal*, *94*(2), 153–167.

Dreyer, C., & Nel, C. (2003). Teaching reading strategies and reading comprehension within a technology-enhanced learning environment. *System*, *31*, 349–365.

Fisher, D., & Frey, N. (2008). What does it take to create skilled readers? Facilitating the transfer and application of literacy strategies. *Voices From the Middle*, *15*(4), 16–22.

Fisk, C., & Hurst, B. (2003). Paraphrasing for comprehension. *The Reading Teacher*, *57*(2), 182–185.

Gardill, M. C., & Jitendra, A. K. (1999). Advanced story map instruction: Effects on the reading comprehension of students with learning disabilities. *Journal of Special Education*, *33*(1), 522–544.

Gersten, R., Fuchs, L. S., Williams, J. P., & Baker, S. (2001). Teaching reading comprehension strategies to students with learning disabilities: A review of research. *Review of Educational Research*, *71*(2), 279–320.

Harvey, S., & Goudivis, A. (2000). *Strategies that work: Teaching comprehension to enhance understanding*. Maine: Stenhouse Publishers.

Huggins, B. (2006, October). The effect of explicit strategy instruction on reading comprehension outcomes and self efficacy beliefs for students with learning difficulties. Thesis in partial completion of the Masters of Educational Psychology, University of Melbourne.

Jitendra, A. K., Edwards, L. L., Sacks, G., & Jacobson, L. A. (2003). What research says about vocabulary instruction for students with learning disabilities. *Exceptional Children*, *70*(3), 299–323.

Katims, D. S., & Harris, S. (1997). Improving the reading comprehension of middle school students in inclusive classrooms. *Journal of Adolescent & Adult Literacy*, *41*(2), 116–123.

Keer, H. V. (2004). Fostering reading comprehension in fifth grade by explicit instruction in reading strategies and peer tutoring. *British Journal of Educational Psychology*, *74*, 37–70.

Kintsch, W., & Kintsch, E. (2005). Comprehension. In S. G. Paris & S. A. Stahl (Eds.), *Children's reading comprehension and assessment*. Matwah, NJ: CIERA.

Klassen, R. (2002). A question of calibration: A review of the self-efficacy beliefs of students with learning disabilities. *Learning Disability Quarterly*, *25*(2), 88–102.

Lee, S. W., & von Collin, T. (2003). *The effect of instruction in the paraphrasing strategy on reading fluency and comprehension* (pp. 2–28). Lawrence, KS: US Department of Education, University of Kansas.

Miatello, M. (2008). Teaching visualising improves reading comprehension. Unpublished research project submitted in partial fulfilment of Graduate Certificate in Literacy Intervention, Melbourne Graduate School of Education, University of Melbourne.

Mossenson, L., Hill, P. & Masters, G.N. (2003). *Tests of Reading Comprehension* (2nd ed). Hawthorn, Victoria: ACER.

Munro, J. (2003). Fostering literacy learning across the curriculum. *International Journal of Learning*, *10*, 327–336.

Munro, J. (2004). The use of reading comprehension strategies at multiple levels of text processing. *International Journal of Learning*, *11*, 835–843.

NICHD. (2000). *Report of the National Reading Panel: Teaching Children to Read* (NIH Pub. No. 00-4769). National Institute of Child Health and Human Development, National Institutes of Health.

Nielsen Hibbing, A., & Rankin-Erickson, J. L. (2003). A picture is worth a thousand words: Using visual images to improve comprehension for middle school struggling readers. *The Reading Teacher*, *56*(8), 237–242.

Pajares, F. (2003). Self-efficacy beliefs, motivation, and achievement in writing: A review of the literature. *Reading & Writing Quarterly*, *19*, 139–158.

Parker, R., Hasbrouck, J. E., & Denton, C. (2002). How to tutor students with reading comprehension problems. *Preventing School Failure*, *47*(1), 45–47.

Pearson, D., & Hamm, D. (2005). The assessment of reading comprehension: A review of practices—past, present, and future. In S. G. Paris & S. A. Stahl (Eds.), *Children's reading comprehension and assessment.* Matwah, NJ: CIERA.

Pressley, M., Roehrig, A., Bogner, K., Raphael, L., & Dolezal, S. (2002). Balanced literacy instruction. *Focus on Exceptional Children*, *34*, 1–14.

Quirk, M. (2004). Do supplemental remedial reading programs address the motivational issues of struggling readers? An analysis of five popular programs. *Reading Research and Instruction*, *43*(3), 1–19.

Reutzel, D. R., Smith, J. A., & Fawson, P. C. (2005). An evaluation of two approaches for teaching reading comprehension strategies in the primary years using science information texts. *Early Childhood Research Quarterly*, *20*, 276–305.

Rowe, K. J. (2005). *Teaching reading: Report and recommendations.* Report of the committee for the national inquiry into the teaching of literacy. Canberra: Australian Government.

Saunders, M. (2001). Impediments to reading comprehension. *Perspectives*, *27*(2), 5–8.

Stahl, K. A. (2004). Proof, practice, and promise: Comprehension strategy instruction in the primary grades. *The Reading Teacher*, *57*(7), 598–609.

Swanson, H. L. (1999). Reading research for students with LD: A meta-analysis of intervention outcomes. *Journal of Learning Disabilities*, *32*(6), 504–532.

Swanson, H. L., & De La Paz, S. (1998). Teaching effective comprehension strategies to students with learning and reading disabilities. *Intervention of School and Clinic*, *33*(4), 209–218.

Swanson, H. L., Hoskyn, M., & Lee, C. (1999). *Interventions for students with learning disabilities.* New York: The Guilford Press.

Torgesen, J. K. (1986). Learning disabilities theory: Its current state and future prospects. *Journal of Learning Disabilities*, *19*(7), 399–407.

Vaughn, S., Gersten, R., & Chard, D. J. (2000). The underlying message in LD intervention research: Findings from research syntheses. *Exceptional Children*, *67*(1), 99–114.

Verezub, E. (2005). Comparative analysis of comprehension of conventional written text versus hypertext for Grade 3 and Grade 6 students. Thesis in partial completion of the Doctor of Philosophy, University of Melbourne.

Wood, K., & Endres, C. (2005). Motivating student interest with the Imagine, Elaborate, Predict, and Confirm (IEPC) strategy. *The Reading Teacher*, *58*(4), 346–357.

Chapter 10
The Writing Achievement, Metacognitive Knowledge of Writing and Motivation of Middle-School Students with Learning Difficulties

Christina E. van Kraayenoord, Karen B. Moni, Anne Jobling, John Elkins, David Koppenhaver, and Robyn Miller

The importance of writing

Writing is a key aspect of literacy in the 21st century, and is seen as developing concurrently and reciprocally with other aspects of literacy, including reading, speaking, listening, viewing and shaping. Writing is 'a visible representation of thoughts and ideas using symbols of the writer's language system for the purpose of communication or recording' (Potato, 1980, p. 88). It may comprise the use of pictures and other symbols, and is not limited to the use of pencil and paper. Indeed, with changes in technologies, written language has been augmented with the use of writing in multimedia forms of communication.

The acquisition and use of writing is important for academic, personal and social development (Alber-Morgan, Hessler, & Konrad, 2007). As students grow and develop, they learn to write for different purposes and audiences. This learning develops in meaningful contexts with others, and thus writing is learned socially through interaction both in and out of school (Cairney, 2003; Rowe, 2008). However, not all students develop competence in writing print and digital texts. Problems with the acquisition and use of writing may be a characteristic of students who have difficulties in academic learning at school (Wakely, Hooper, de Kruif, & Swartz, 2006). In particular, those who are described as having learning difficulties in writing may have problems with learning to write, or of writing effectively in different situations. These students may also have problems related to motivation and writing.

Students with learning difficulties and writing

An examination of the literature related to the writing of students with learning difficulties has revealed that these students' problems include: a lack of planning, difficulties generating ideas and organising text, problems with mechanics (for

---

C.E. van Kraayenoord (✉)
The University of Queensland, Brisbane, QLD, Australia
e-mail: c.vankraayenoord@uq.edu.au

C. Wyatt-Smith et al. (eds.), *Multiple Perspectives on Difficulties in Learning Literacy and Numeracy*, DOI 10.1007/978-1-4020-8864-3_10,

example spelling, capitalisation and punctuation), difficulties with various aspects of metacognitive knowledge (for example they are less knowledgeable about the process of writing) and in the use of strategies and self-regulation of writing (for example monitoring and revision) (Chan & Dally, 2000; Englert, Raphael, & Anderson, 1992; Graham, 2006; Graham, Harris, & Larsen, 2001; MacArthur, Graham, & Schwartz, 1993; Swanson, 1999; Wong & Wilson, 1984). Graham et al. (2001) have also indicated that these students often overemphasise the importance of skills such as mechanics, rather than focusing on meaning.

For students with learning difficulties in the middle years, writing achievement is often lower than that of their peers (Forgan & Vaughn, 2000). With respect to the motivation and writing of students with learning difficulties, the research has indicated that these students have low motivation and maladaptive beliefs about the causes of success and failure (Sexton, Harris, & Graham, 1998). With respect to the self-concept, students with learning difficulties often have lower academic self-concepts than their normally achieving peers, and this is manifested in areas such as reading and writing (Chapman, 1988; Haager & Vaughn, 1997; Hay, 1996). According to Graham and Harris (2003), little is known about the attitude towards writing of students with learning difficulties.

Theoretical models of literacy learning

Current models of literacy learning are based on social–cognitive (Rowe, 2008; Vygotsky, 1962), sociocultural/social and cultural (Barton & Hamilton, 2000; Bos & Fletcher, 1997), and sociocritical (Cope & Kalantzis, 2000; Fairclough, 1995) theories or approaches. These theories have been applied to the acquisition and development of writing. Key elements in the development of writing are: a supportive environment comprising more knowledgeable writers as models; involvement by others that is matched to the students' needs to exert control over their learning; recognition of writing approximations as success; the use of supportive dialogue that shapes the students' thinking as they write; the development of planning strategies for creating text; use of editing and revising strategies; and publishing and sharing writing with real audiences (Bereiter & Scardemalia, 1987; Dyson, 1995; Englert & Mariage, 2003; Flower & Hayes, 1981; Hayes & Flower, 1980; MacArthur et al., 1993).

Theoretical framing and the *WriteIdeas Model*

The sociocultural/social and cultural model of writing has been the dominant model in recent years, guiding the research into difficulties in writing and into design and implementation of instructional programs in writing. The *WriteIdeas Model* (van Kraayenoord, Moni, Jobling, Koppenhaver, & Elkins, 2003) and associated teaching practices, learning activities and tools are located within the social–cultural/social and cultural model of literacy (see Fig. 10.1). As described in van Kraayenoord,

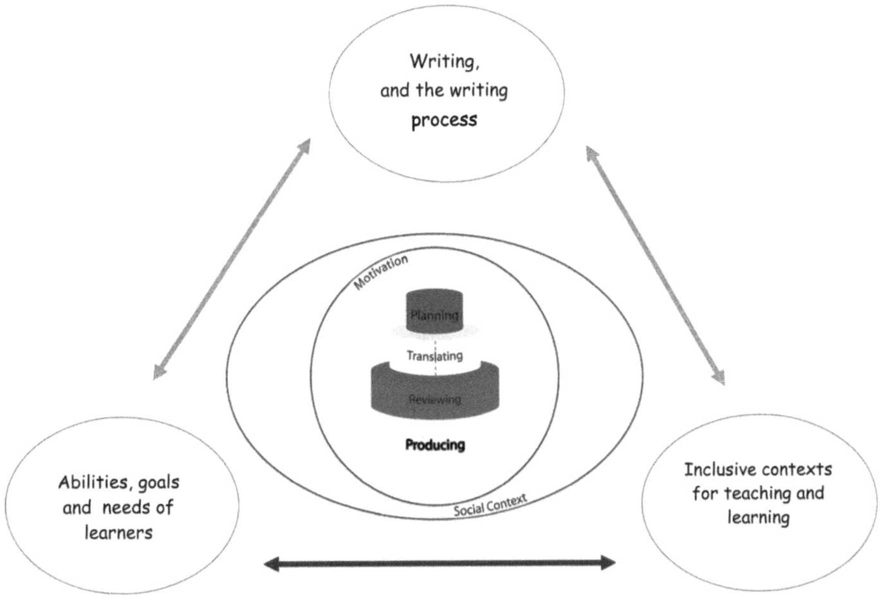

Fig. 10.1 The *WriteIdeas Model* and associated teaching practices, learning activities and tools. Source: Moni et al. (2007), van Kraayenoord et al. (2003, p. 21). Reproduced with permission from *Educational & Child Psychology*, © The British Psychological Society

Moni, Jobling, Koppenhaver, and Elkins (2004), the *WriteIdeas Model* acknowledges that writing production and sharing occurs in a social context. This means that, for the learning and use of writing to occur within the classroom context, there must be strong social support from others. In this model, the teacher creates a classroom environment that is supportive and rich with environmental print, and there are numerous opportunities for teaching about and through writing. The students are engaged in joint constructions of text with the teacher and as students together, and the students write and share collaboratively, in groups and pairs. The model also explicitly addresses the role of motivation. It fosters the idea that students must be engaged and motivated in order to write, and those who interact with the written text are also motivated by the writing they read, hear and see. In the *WriteIdeas Model*, motivation is promoted in various ways, including through an emphasis on high expectations, the belief or idea that all learners are writers and the notion that the learning activities should be related to the students' interests. The *WriteIdeas Model* connects the social and motivational understandings related to writing with what is known about the cognitive processes used when writing. The cognitive processes referred to in the model are planning, translating, reviewing and producing. As an embedded and recursive model, the elements are nested in one another and interact simultaneously with each other (van Kraayenoord et al., 2004). Thus, there is an interaction between the social, motivational, cognitive and metacognitve elements within the *WriteIdeas Model*.

The *WriteIdeas Model* is also concerned with pedagogy. Specifically, it refers to teaching practices, learning activities and tools. The teaching practices that are promoted in the professional learning workshop for teachers are evidence-based practices that are suited to inclusive classroom contexts. Several of the practices focus on the development of metacognitive knowledge and strategies. The teaching practices include explicit instruction, scaffolding, modelling and demonstration, thinking aloud, making adaptations through the application of differentiated instruction, and the use of praise and positive feedback. The learning activities are planned so that they are related to students' interests and needs. There is an emphasis on authentic learning activities that involve writing for real audiences and purposes. The teachers are encouraged to use activities that foster collaborative learning of, and about, writing in groups and peer partnerships, as well as the development of autonomous and independent writing. In the context of the *WriteIdeas Model*, tools refer to various technologies (for example computer hardware and software), templates, charts and environmental resources and prompts that support the students' writing in the classroom. These may be created by the teachers and/or the students, or may be commercially available products.

The WriteIdeas project

The *WriteIdeas* project involved adolescent students in the middle years of schooling with disabilities and learning difficulties (Grades 5, 6 and 7 in primary schools and Grades 8 and 9 in the English curriculum area in high schools) in inclusive settings. These students participated with their teachers, who were involved in professional learning about writing.

The project had three main aims. The first was to examine the abilities, skills and motivation of students with developmental disabilities and learning difficulties in middle-school classrooms in metropolitan (Brisbane), rural/remote (Longreach and Emerald) and regional (Cairns) areas in Queensland, Australia. Both Education Queensland (state/public) and Catholic Education schools were involved in the study. The second aim was the development, implementation and evaluation of professional learning for middle-school teachers, related to the teaching of writing and the integration of teaching practices, learning activities and tools. The third aim was documentation and evaluation of the teacher-developed, instructional intervention and support offered in units and lessons featuring writing as used in the classrooms of students with developmental disabilities, learning difficulties and their peers.

At each site, the project involved six phases of research across 12 months. The phases are detailed below.

- Phase 1: A member of the research team assessed students' writing achievement, metacognitive knowledge about writing, writing attitudes and self-perceptions, and surveyed the teachers' attitudes and practices related to teaching writing.
- Phase 2: The teachers participated in a 2-day professional development workshop and associated activities. The teachers discussed issues related to inclusive

education, contemporary models of literacy and current pedagogical practices used in Queensland schools, and learned about students with learning difficulties (and those with developmental disabilities) and their learning strengths and needs. These discussions and activities established the context for a focus on writing. Writing-related topics addressed in the workshop included: the development of writing, models of writing, the *WriteIdeas Model*, as well as associated teaching practices, learning activities and technological and other tools that support the development of writing (see Moni et al., 2007; van Kraayenoord et al., 2004).

- Phase 3: Teachers planned, developed and documented units of work and lessons that featured writing and comprised contemporary, student-centred teaching practices, learning activities, and technological and other tools for their students based on their learnings from the workshop. During this phase the teachers were supported by the researchers. The planned units and lessons were integrated into the regular curriculum and used in the next phase.
- Phase 4: The teachers implemented the units and lessons involving writing over a 10-week-period and monitored their teaching, the lessons and the ways in which their students responded to the activities, using a Teachers' Log (Muckert, Moni, & Jobling, 2003) and in a Weekly Account of Lessons (Muckert, van Kraayenoord, Moni, & Jobling, 2003). The Teachers' Logs and Accounts focused particularly on the students of interest. The researchers also conducted classroom observations (van Kraayenoord, Moni, Jobling, & Koppenhaver, 2004) and collected resources and artifacts during the observational visits.
- Phase 5: The students were assessed in a post-implementation phase.
- Phase 6: In a 6-month follow-up, both teachers and students were assessed again (see Phase 1).

The study

In this section of the chapter, we report on the writing performance, metacognitive knowledge about writing, attitude and self-perception about writing of students with learning difficulties following an intervention created by their teachers. The intervention was based on the *WriteIdeas Model* and associated teaching practices, learning activities and tools.

Participants

Eighteen students with learning difficulties were involved in the *WriteIdeas* project. The students comprised 15 males and 3 females. In both school sectors (that is state and Catholic education), students were identified as having a learning difficulty by 'appraisement', a process that is also used to determine the level of support these students should receive in schools (Education Queensland, 2006). The students came from three cohorts, each taking part in consecutive years of the project in Brisbane, the Longreach and Emerald areas and the Cairns area. Two students did not complete the post-implementation tasks due to absence. The students ranged in age from 9 years 9 months to 13 years 9 months (Mean = 12 years, SD = 14.3 months). The teachers of these students ($N = 12$) also participated in the study. Two of the teachers had nominated three students with learning difficulties in their classes, and two other teachers had two students with learning difficulties in their classes. These 12 teachers were involved in attending the workshop (Phase 2) in the respective years they participated, and in developing the units and lessons of work involving writing, with the support of the research team (Phase 3), and with the implementation of the units and lessons designed to develop and enhance writing over a 10-week-period (Phase 4).

Instruments

This study focused on assessing three domains of writing: writing achievement, metacognitive knowledge about writing and motivation, namely attitude towards writing and self-perceptions of writing.

Achievement in writing

Achievement in writing was determined by examining writing performance using the spontaneous writing subtests (Subtests 6, 7 and 8) of the *Test of Written Language—Third Edition* (*TOWL-III*) (Hammill & Larsen, 1996), and samples of student writing selected by their teachers (teacher-selected writing samples). The *TOWL-III* was used prior to the implementation of the classroom units and lessons, immediately following the 10-week implementation period and at follow-up. The *teacher-selected writing samples* were collected at pre-implementation and post-implementation only. No *teacher-selected writing samples* were obtained at follow-up.

1. *Test of Written Language—Third Edition*. The *TOWL-III* is a written language test comprising eight subtests, specifically five contrived subtests and three subtests that relate to a spontaneous writing task. In this study, only the spontaneous writing task of the *TOWL-III* was used. The *TOWL-III* is available in two versions, Form A and Form B. Form A was used in the pre-implementation phase, Form B in the post-implementation phase, and Form A again in the follow-up phase. Each student's written story was scored using the standardised scoring procedure described in the test manual. Subtest 6, *Contextual Conventions*, measures the use of arbitrary conventions of written language, such as spelling, punctuation and capitalisation. Subtest 7, *Contextual Language*, measures the use of language in the response, in particular sentence structure, grammar and vocabulary. Subtest 8, *Story Construction*, examines the use of prose, action, sequencing and theme in the written response. Inter-rater reliabilities of the coding on the Subtests 6, 7 and 8 of the *TOWL-III*

Writing qualities	1	2	3	4
Response to prompt (purpose)	• Attempt to respond to the prompt • Subject matter only partially relevant to the purpose	• Adequate response to the prompt • Contains subject matter relevant to the purpose	• Good response to the prompt • Subject matter is relevant to the purpose and sustained	• Good response to the prompt • Purpose introduced at beginning of composition and sustained throughout
Text development (relevant content building on theme, topic or idea)	• Theme not developed or sustained • Lack of connectivity between ideas	• Theme evident but not sustained • Some connectivity between ideas.	• Material included contributes to overall theme • May include an irrelevant description or limited explanation.	• Clear development of theme • No irrelevant descriptions or explanations
Organisation (relevant to genre)	• Not discernible or clear sequence of events • Not appropriate to genre	• Not completely discernible or clear sequence of events • Some attempt to address genre	• Sequence may not be entirely discernible or clear throughout composition • Good but may include too much emphasis on one part of the genre	• Good; discernible beginning, middle and end • Clear sequence of events and organised appropriately to genre
Vocabulary (word selection, communication of ideas with word choice, degree of word complexity)	• Inappropriate or incorrect word selection.	• Appropriate but basic word selection, may be repetitive.	• Good word selection with some complexity and variation	• Word selection is varied and interesting and enhances the quality of the text.
Details (quality and quantity of information, depth of information)	• Limited information	• Presented information without extension or elaboration	• Some elaborated and extended ideas	• Variety of elaborated and extended ideas
Sentence structures (grammatical components/SV agreement; & complexity of sentence structures - simple, compound, phrases, clauses)	• Incorrect and inappropriate throughout composition • Sentences may be incomplete	• Many incorrect and inappropriate • May include run-on sentences, or repeated structure	• Sentences mostly correct and appropriate • Limited variation	• Sentence structures accurate • Sentences varied and appropriate
Mechanics (punctuation, capitalisation and spelling)	• Many serious errors	• Serious errors	• Some errors	• Very few or no errors
Authenticity/Voice/Engagement (audience)	• The writer has not engaged with the task, the topic or the interests of the reader	• Not engaging, perfunctory, limited enthusiasm for the task, topic or the interests of the reader	• The writer is engaged in the task, topic and the writing engages the reader	• The writer demonstrates enthusiasm for the task, the topic and engaging the reader

Fig. 10.2 Writing analysis matrix. Source: Adapted from Schirmer and Bailey (2000, p. 54)

were calculated on the stories generated at the pre-implementation assessment. The inter-rater reliabilities for Subtests 6, 7 and 8 were 0.99, 0.99 and 0.93, respectively.

The writing analysis matrix (adapted from Schirmer & Bailey, 2000) was developed by the researchers as a holistic assessment of the students' stories obtained from the use of the *TOWL-III* (see Fig. 10.2). Scores could range from 4 to 32, with low scores indicating poor performance. The inter-rater reliabilities of the coding on the writing analysis matrix ranged from 0.83 to 0.96.

2. *Teacher-selected writing samples.* Teachers were asked to select one sample of the target student's written work undertaken independently as part of regular classroom lessons prior to the implementation phase, and to collect another sample under the same conditions following the implementation phase. These samples were scored holistically using the writing analysis matrix described above. The inter-rater reliabilities of the coding of teacher-selected writing samples using the writing analysis matrix ranged from 0.84 to 0.93. Teacher-selected writing samples were received from only 12 of the 18 students.

Metacognitive knowledge about writing

Four items about the 'self as a writer' from the *Student Writing Interview—Revised* (van Kraayenoord, Moni, & Jobling, 2004) were used in this study to obtain a measure of the students' metacognitive knowledge about writing. These items examined the students' metacognitive awareness about their achievement in writing and the reasons given for their answer, their awareness of the improvements they wished to make, and whether or not their teachers viewed them as writers. The items were Question 4a, 'How do you see yourself as a writer?', Question 4b 'Why?', Question 11, 'What would you like to improve in your writing?' and Question 15, 'Does your teacher think you are a writer?' The scores for the four questions were then added up, and this total score was called the student metacognition score. The inter-rater reliability of the coding of the responses on the full student writing interview ranged from 0.85 to 1.0.

Motivation comprising writing attitude and self-perceptions about writing

Three instruments were used to collect data about students' motivation, specifically about their writing attitudes and self-perceptions. These were: the *Writing Attitude Survey* (Kear, Coffman, McKenna, & Ambrosio, 2000), *the Perception of Writing Ability Scale* (Nicholls, 1978) and the *Writer Self-perception Scale* (Bottomley, Henk, & Melnick, 1997–1998).

The Writing Attitude Survey. This attitude survey comprises 28 items and examines students' feelings about writing. The items on the attitude survey use Garfield pictures to indicate feelings about writing ranging, from 1 (very upset) to 4 (very happy). Very slight changes were made to the language of the instrument so that it was useable in the Australian school context; for example for Question 3 'How would you feel writing a letter to a store asking about something you might buy there?', the word 'store' was changed to 'shop'. The students were required to provide a rating on a 4-point Likert-type scale. Internal consistency for the measure was calculated on the larger study's sample of 52 students, of whom the students with learning difficulties were a sub-group. The internal consistency was high ($\alpha = 0.882$).

The Perception of Writing Ability Scale. This scale assesses an individual student's perception of their writing ability within the class. The scale displays 30 smiley faces, with the face at the top representing the best writer in the class (30) and the face at the bottom representing the worst writer in the class (1). Students were asked to rank themselves as a writer in comparison to the other students in their class by putting a tick (checkmark) next to the face that represented them. This mark was translated into the score that the student received.

Writer Self-perception Scale. This scale assesses students' views about themselves as writers. The scale consists of 38 items. All items required a rating on a 5-point Likert-type scale, ranging from 1 (strongly disagree) to 5 (strongly agree). These items are distributed across the following five scales: general progress, specific progress, observational comparison, social feedback and physiological states. For details of scale composition, see Bottomley et al. (1997–1998). According to the administration manual for the scale, the rating provided in response to item 15 'I think I am a good writer' can be used as a stand-alone score. The internal consistency for the five subscales using the sample from the larger study ranged from $\alpha = 0.847$ to 0.891. The researchers added a further seven items that examined students' views about themselves as spellers. Five of the seven items were aligned with the five subscales (one item each) referred to above. Again, the students responded on a 5-point Likert-type scale, ranging from 1 (strongly disagree) to 5 (strongly agree). Thus, scores on the spelling scale ranged from 7 to 35. The internal consistency for the spelling scale using the sample from the larger study was $\alpha = 0.746$. Low scores indicated low self-perception.

Procedure

The school year in Australia runs from February through to December. The first assessments were collected in April of the first term of each year (Phase 1: pre-implementation), and again in October and November, at the end of the year of each student's involvement (Phase 5, post-implementation). Follow-up data were collected 6 months after the end of the school year (Phase 6).

All of the measures were administered in a one-to-one situation, with the student and a researcher or research assistant seated at a table in an unused classroom, staffroom or office, or on a quiet veranda at the school. The order in which the instruments were completed was as follows: the *Writing Attitude Survey*, the *Writer Self-Perception Scale*, the *Perception of Writing Ability Scale*, the *Student Writing Interview-Revised*, and the *TOWL-III*.

Data analyses

In order to study improvement over time, a series of repeated measures analyses of variance (ANOVAs) were conducted. We expected that students' performance would improve from pre-implementation to post-implementation, but decline from post-implementation to follow-up. Our reasoning was that the students would start a new academic year during this period and be taught by new teachers who had not received the *WriteIdeas* professional development. In addition, students with learning difficulties often require intensive and continuing support. It was unknown whether any of the students would receive such support. Thus, a curvilinear trend was anticipated. For the omnibus statistical tests, the quadratic *F*s are reported with an alpha level of 0.05. An alpha level of 0.01 for was used for second-tier (within-subjects) contrasts. Where significant differences were evident, effect sizes (eta-squared, η^2) were also calculated.

Results

Writing achievement

Table 10.1 presents the means, standard deviations and results of the statistical tests for the students' writing performance as assessed by the *TOWL-III*. Results for subtest 6: *contextual conventions* and subtest 7: *contextual language* at pre-implementation, post-implementation and follow-up showed significant effects, but not for subtest 8: *story construction*. Within-subject analyses revealed significant improvements between pre-implementation and post-implementation for *contextual conventions* and *contextual language*. In addition, as anticipated, there was a decline on both *contextual conventions* and *contextual language* between post-implementation and follow-up. This decline was only significant for *contextual conventions* (see Table 10.1). There was no significant difference in scores between pre-implementation and follow-up.

The writing analysis matrix was used to derive holistic scores from the spontaneously written stories that had been generated by the *TOWL-III* and teacher-selected writing samples for 12 students. A significant effect was seen in the students' scores on the *TOWL-III* at pre-implementation, post-implementation and follow-up. Further analyses revealed an improvement from pre-implementation to post-implementation, but not from post-implementation to follow-up. As expected, there was a decline between post-implementation and follow-up. There was no significant difference in scores between pre-implementation and follow-up. The analysis of the teacher-selected writing samples revealed significant improvement in students' writing performance between pre-implementation and post-implementation (see Table 10.2). No teacher-selected writing samples were collected at the follow-up phase.

Metacognitive knowledge of writing

Table 10.3 presents the means, standard deviations and results of the analysis of the student metacognition score. Of the 18 students, one student was absent at follow-up. The analyses revealed significant effects in the students' metacognitive

Table 10.1 Means, standard deviations and results of analyses for *TOWL-III* spontaneous writing task

	Pre-implement mean (SD)	Post-implement mean (SD)	Follow-up mean (SD)	Range	*df*	*F*	η^2	*p*
TOWL-III, Spontaneous writing task ($n = 16$)								
Subtest 6: contextual conventions	2.06 (2.20)	4.13 (2.39)	2.00 (1.63)	0–18	2.30	19.091	0.560	0.001**
Contrasts pre-post					1.15	12.933	0.463	0.003**
Contrasts post-F'up					1.15	18.766	0.556	0.001**
Contrasts pre-F'up					1.15	0.018		0.896
Subtest 7: contextual language	10.00 (3.58)	12.00 (4.18)	10.63 (3.22)	0–29	2.30	4.973	0.249	0.041*
Contrasts pre-post					1.15	5.647		0.031*
Contrasts post-F'up					1.15	2.366		0.145
Contrasts pre-F'up					1.15	0.772		0.393
Subtest 8: story construction	10.31 (3.36)	10.25 (3.57)	10.75 (3.53)	0–29	2.30	0.121		0.773

Note: Teacher-selected writing samples were not collected at follow-up.
Source: Moni et al. (2007). Reproduced with permission from *Educational & Child Psychology*, © The British Psychological Society.
* $p \leq 0.05$, ** $p \leq 0.01$

Table 10.2 Means and standard deviations for the metacognitive knowledge of writing, attitude and self-perceptions about writing

	Pre-implement mean (SD)	Post-implement mean (SD)	Follow-up mean (SD)	Range	*df*	*F*	η^2	*p*
TOWL-III,								
Spontaneous writing ($n = 16$)	17.25 (4.07)	20.00 (4.27)	19.44 (3.48)	4–32	2,30	7.286	0.327	0.016*
Contrasts pre-post					1,15	14.291	0.488	0.002**
Contrasts post-F'up					1,15	0.442		0.516
Contrasts pre-F'up					1,15	5.449		0.033*
Teacher-selected writing samples ($n = 14$)	18.83 (4.43)	24.58 (4.48)		4–32	1,13	14.660	0.530	0.002**

Note: η^2 reported for significant findings only
* $p \leq 0.05$, ** $p \leq 0.01$

Table 10.3 Means, standard deviations and results of analyses for the metacognitive knowledge of writing, attitude and self-perceptions about writing

	Pre-implement mean (SD)	Post-implement mean (SD)	Follow-up mean (SD)	Range	df	F	η^2	p
Student metacognition score ($n = 17$)	2.29 (.99)	2.88 (.99)	2.47 (1.28)	1–4	2, 32	6.217	0.293	0.025*
Contrasts pre-post					1,16	5.195		0.037*
Contrasts post-F'up					1,16	1.531		0.234
Contrasts pre-F'up					1,16	0.260		0.617
Writing attitude survey ($n = 17$)	76.82 (15.46)	71.53 (12.08)	77.65 (10.23)	28–112	2,32	6.49	0.289	0.022*
Contrasts pre-post					1,16	3.102		0.097
Contrasts post-F'up					1, 16	5.497		0.032*
Contrasts pre-F'up					1,16	0.252		0.622
The perception of writing ability scale ($n = 17$)	9.59 (8.46)	9.71 (6.58)	12.76 (6.69)	1–30	2,32	0.670		0.425

Note: η^2 reported for significant findings only
* $p \leq 0.05$

knowledge of writing at pre-implementation, post-implementation and follow-up. There was a trend suggesting improved metacognition from pre-implementation to post-implementation. However, second-tier contrasts of this pre-implementation to post-implementation trend using the more stringent alpha level of 0.01 indicated that the difference was not significant. No significant differences were found between pre-implementation and follow-up, or between post-implementation and follow-up.

Motivation

On the Writing Attitude Survey, there were a significant differences at pre-implementation, post-implementation and follow-up (see Table 10.3). There was a trend suggesting improved attitude from post-implementation to follow-up. However, second-tier contrasts of this post-implementation to follow-up trend using an alpha level of 0.01 indicated that the difference was not significant. No significant differences were found between pre-implementation and post-implementation, or between pre-implementation and follow-up. There were no differences in the students' self-perception on the perception of writing ability scale (see Table 10.3).

The Writer Self-perception Scale also examined students' views of their writing abilities. There were no significant differences between mean scores on item 15 'I think I am a good writer' at pre-implementation, post-implementation and follow-up on the writer self-perception scale. There were significant differences at pre- and post-implementation and follow-up on the observational comparison scale and the spelling scale. The analyses revealed that in these cases the differences were between post-implementation and follow-up only. There was no significant difference in scores between pre-implementation and follow-up (see Table 10.4).

Discussion

In examining the students' results on the *TOWL-III*, the findings indicated that the students' writing skills improved in two areas following the implementation of the teachers' instruction. First, they improved between pre-implementation and post-implementation in the use of contextual conventions, which are arbitrary conventions of written language. Similarly, the students improved in their use of contextual language between pre-implementation and post-implementation. These findings are consistent with other studies that have demonstrated that students' skills in writing can be improved through instruction (Baker, Gersten, & Graham, 2003; Graves, Montague, & Wong, 1990). Clearly, the units and lessons provided by the teachers' contributed to the students' development of skills in these areas.

The writing analysis matrix, adapted from the matrix of Schirmer and Bailey (2000), allowed the researchers to evaluate the students' writing performance against eight criteria or qualities of writing. Through the use of this matrix, the researchers were able to obtain detailed and comprehensive descriptions of how the students' writing performance varied in terms of quality and achievement. Such

Table 10.4 Means, standard deviations and results of analyses for writer self-perception scale

	Pre-implement mean (SD)	Post-implement mean (SD)	Follow-up mean (SD)	Range	*df*	*F*	η^2	*p*
Item 15	3.00 (1.28)	3.24 (1.09)	3.35 (.86)	1–5	2, 32	0.119		0.735
General progress	29.94 (7.88)	31.53 (4.39)	34.06 (3.72)	8–40	2, 32	0.191		0.668
Specific progress	26.06 (5.58)	26.65 (2.87)	28.18 (2.29)	7–35	2, 32	0.409		0.531
Social feedback	22.0 (5.48)	23.06 (5.61)	24.18 (5.15)	7–35	2, 32	0.002		0.966
Physiological states	19.88 (6.93)	18.65 (6.01)	20.06 (4.60)	6–30	2, 32	0.2.359		0.144
Observational comparison	22.88 (6.68)	21.24 (6.28)	23.94 (6.11)	9–45	2, 32	8.368	0.343	0.011*
Contrasts pre-post					1, 16	2.412		0.14
Contrasts post-F'up					1, 16	9.327		0.008**
Contrasts pre-F'up					1, 16	0.639		0.435
Spelling	21.53 (5.40)	20.76 (4.12)	27.76 (4.01)	7–35	2, 32	14.707	0.653	0.001**
Contrasts pre-post					1, 16	0.472		0.502
Contrasts post-F'up					1, 16	30.154		0.000**
Contrasts pre-F'up					1, 16	27.185		0.000**

Note: η^2 reported for significant findings only
* $p \leq 0.05$, ** $p \leq 0.01$

descriptions are useful to teachers as they can be used to inform instruction for individual students.

It is notable that the improvements in writing performance between pre-implementation and post-implementation were found in both the spontaneous stories produced by the students under standardised test conditions and in the samples of their classroom writing, although, interestingly, the students' improvement was more pronounced in the latter. This suggests that the classroom writing samples were effective in showing students' writing progress across time.

In this study, metacognitive knowledge about writing, expressed as student awareness about their achievement in writing, showed no improvement from pre-implementation to post-implementation. Thus, the instruction offered by the teachers did not enhance these students' metacognitive knowledge. The student metacognition score was derived from responses to four items of the student writing interview-revised. It is suggested that this measure of individuals' knowledge of their own cognition about writing may not adequately assess the student's knowledge (see Brown, 1987; Butler, 1995 and their discussion of the use of verbal reports).

The *WriteIdeas Model* and associated practices, activities and tools also focus on motivation in the development of writing through the concepts of engagement, high expectations and students' interests. In this study, there were no significant differences in the mean scores between pre-implementation and post-implementation on students' writing attitude. There were no significant differences on the perception of writing ability scale. While the findings in the current study do not support the results of investigations of intervention and its influence on the self-perception of students with learning difficulties, this may be due to the persistent and deeply held self-beliefs of this group of students, which the intervention was unable to influence sufficiently.

Differences between the mean scores between post-implementation and follow-up were found on the observational comparison scale of the writer self-perception scale and the spelling scale. These results indicated that the students' perceptions were higher at follow-up than post-implementation. However, in relation to the normative data provided by the authors of the scale, it is apparent that the students' self-perceptions were low at both post-implementation and follow-up. These findings indicating a low self-perception are consistent with research of students with learning difficulties, which has indicated that their self-concepts are typically lower than that of their peers (Haager & Vaughn, 1997). Enhancing these students' self-concepts in writing therefore remains a continuing challenge for teachers and researchers.

Limitations of the study and future research

The outcomes of this study should be interpreted within the context of the small number of students with learning difficulties who participated. Originally, the larger study was one that recruited teachers who were working with students with developmental disabilities in regular classrooms. However, at the first data collection point with the Brisbane, Queensland cohort (first year of the study), it became obvious that some teachers had selected students with learning difficulties to participate in the study. We retained these students and teachers in the study and sought documentation that indicated a designation that the student had a learning difficulty from the school system. From the following year of the study, the researchers opened up the sample population to include students with either learning difficulties or developmental disabilities. However, the number of students with learning difficulties selected by the teachers as target students across the 3 years of the study was small; that is 18 students, as reported here. Thus, any generalisation of the study's findings to other students with learning difficulties should be treated with caution.

Although the researchers believe the findings of improvement in writing performance for these students with learning difficulties may well be related to the teachers' intervention, only a controlled study would allow this hypothesis to be tested. Thus, there is a need for a more rigorously designed study with control groups.

The researchers contend that a longer duration of the workshop in Phase 2 and even more intensive and targeted researcher support for the teachers during Phases 3 and 4 may well enhance the professional development provided in any future study. Further, there may be a need for greater scrutiny of implementation fidelity among the teachers in future research.

Conclusion

In this study, the influence of teacher-planned and implemented units and lessons, based on the *WriteIdeas Model* and its associated teaching practices, learning activities and tools were examined. The writing achievement, metacognitive knowledge of writing and motivation of middle-school students with learning difficulties in inclusive classrooms were explored. The *WriteIdeas Model* connects social and motivational understandings related to writing with what is known about the cognitive processes that are used when writing (that is planning, translating, reviewing and producing), and what is known about the development of metacognitive knowledge and strategies associated with writing (van Kraayenoord et al., 2004). The study's findings have indicated that between pre-implementation and post-implementation the students with learning difficulties improved in various aspects of writing performance. The lack of improvement between pre-implementation and post-implementation on the metacognitive and motivational variables suggests a need for further research into metacognition and motivation of students with learning difficulties.

It is also argued that, in the future, projects that involve professional development and teacher participation as the vehicle for improvement in achievement in the writing of students with learning difficulties must be of extended duration (van Kraayenoord, Honan, & Moni, 2007a; van Kraayenoord, Honan, & Moni,

2007b), involve more intensive and direct interaction with the teachers, and seek new ways of attaining treatment fidelity and teacher compliance. Studies of the *WriteIdeas Model* involving control groups would also be valuable.

Essential next questions

Intervention

There has been a proliferation of new interventions for students with learning difficulties. Given the persistent difficulties of these students, interventions that involve middle and high school students are required. Examinations of the effectiveness of these interventions for students with learning difficulties in literacy need to be undertaken. The key questions that relate to investigations of the effectiveness of interventions for students with learning difficulties include:

1. What are the essential components of effective intervention?
2. In which contexts?
3. How should the intervention be delivered?

In order to answer these questions, researchers might initially focus on ‘best existence proofs’ (Pressley, 2002, p. 397), namely of those programs and locations where excellence and success are evident.

Teacher professional development

Effective professional development is important for the enhancement of teachers’ practices in relationship to the instruction of students with learning difficulties. This study involved the teachers in a professional learning program. Our experiences suggest that programs that focus on the teachers’ learning needs, that are collaborative and that provide active learning opportunities are warranted. However, there is much debate about effective models of professional learning. Research that examines different models of professional learning is required. Key questions in this research area are as follows:

1. What are the elements used to design successful and sustained professional learning programs?
2. What roles do the leaders of systems and school administrations play in these programs?
3. How can partnerships with universities be developed?

Researchers investigating these questions might begin with audits of a range of models of teacher professional development, investigating such issues as access, the learning environment related to local needs, pedagogy, modes of delivery, opportunities for sharing and feedback, timing and duration, systemic and school-based horizontal and vertical support, links with strategic priorities of the organisation, collaborations with external agencies and institutions, funding, and resources.

Acknowledgement Our special thanks go to Robyn Miller, one of the authors of this chapter and the *WriteIdeas* Project Manager, for undertaking the statistical analyses, and to Dr. Asad Khan and Dr. Michelle Haynes for statistical advice. This research was funded by the Australian Research Council, Grant Number DP0344749.

Glossary

Appraisement Appraisement is a process of data collection regarding the achievement and needs of students who may have learning difficulties. At the time of writing it was used in Queensland, Australia by schools to identify the educational support requirements of individual students. The process is undertaken to recommend the level and kind of support needed, beyond that provided by the regular educational program in the school, to help these students to access the curriculum more effectively, and to meet achievement expectations consistent with their age cohort.

Developmental disabilities Within the *WriteIdeas* project, students with developmental disabilities may be those with congenital disabilities, such as intellectual disabilities, autism, cerebral palsy, developmental apraxia and specific language disorders. Difficulties may also be the result of impairments associated with amytrophic lateral sclerosis (ALS), brain injury, spinal cord injury or stroke.

Disabilities The World Health Organization defines disabilities as a general term that comprises three aspects <http://www.who.int/topics/disabilities/en/index.html>. These are impairments, which refer to problems in body function or structure; activity limitations, which refer to difficulties experienced by individuals in completing tasks or actions in everyday life and participation restrictions, whereby individuals have problems in their involvement in life and social situations. Disabilities may arise or are evident early in life and may continue across the lifespan. They include conditions such as intellectual disabilities, vision disabilities, hearing disabilities and physical disabilities.

Learning difficulties This is a term used in Australia to describe students who experience problems in learning at school. These problems may be across the curriculum or in specific areas of learning, such as literacy or numeracy.

Normally achieving This refers to the measured performance of a student in an area of learning that is within the performance range of peers who are of the same age and in the same grade level.

References

Alber-Morgan, S. R., Hessler, T., & Konrad, M. (2007). Teaching writing for keeps. *Education and Treatment of Children, 30*(3), 107–128.

Baker, S., Gersten, R., & Graham, S. (2003). Teaching expressive writing to students with learning disabilities: Research-based applications and examples. *Journal of Learning Disabilities, 36*(2), 109–123.

Barton, D., & Hamilton, M. (2000). Literacy practices. In D. Barton, M. Hamilton, & R. Ivanič (Eds.), *Situated literacies: Reading and writing in context* (pp. 7–15). London: Routledge.

Bereiter, C., & Scardamalia, M. (1987). *The psychology of written composition*. Hillsdale, NJ: Lawrence Erlbaum.

Bos, C. S., & Fletcher, T. V. (1997). Sociocultural considerations in learning disabilities inclusion research: Knowledge gaps and future directions. *Learning Disabilities Research and Practice, 12*, 92–99.

Bottomley, D. M., Henk, W. A., & Melnick, S. A. (1997–1998). Assessing children's views about themselves as writers using the writer self-perception scale. *Reading Teacher, 51*(4), 286–296.

Brown, A. (1987). Metacognition, executive control, self-regulation and other more mysterious mechanisms. In F. E. Weinert & R. H. Kluwe (Eds.), *Metacognition, motivation, and understanding* (pp. 65–116). Hillsdale, NJ: Erlbaum.

Butler, D. L. (1995). Promoting strategic learning by postsecondary students with learning disabilities. *Journal of Learning Disabilities, 28*, 170–190.

Cairney, T. H. (2003). Literacy within family life. In N. Hall, J. Larson, & J. Marsh (Eds.), *Handbook of early childhood literacy* (pp. 85–98). London: Sage.

Chan, L. K. S., & Dally, K. (2000). Review of literature. In W. Louden, L. K. S. Chan, J. Elkins, D. Greaves, H. House, M. Milton, S. Nichols, J. Rivalland, M. Rohl, & C. van Kraayenoord (Eds.), *Mapping the territory – Primary students with learning difficulties: Literacy and numeracy* (Vol. 2, pp. 161–331). Canberra: Department of Education Training and Youth Affairs.

Chapman, J. W. (1988). Learning disabled children's self-concepts. *Review of Educational Research, 58*, 347–371.

Cope, B., & Kalantzis, M. (Eds.). (2000). *Multiliteracies: Literacy learning and the design of social futures*. London: Routledge.

Dyson, A. H. (1995). Writing children: Reinventing the development of childhood literacy. *Written Communication, 12*, 4–46.

EducationQueensland. (2006). *A chronology of special education in Queensland*. Retrieved 12 May 2009, from http://education.qld.gov.au/library/edhistory/state/chronology-spec/2000.html

Englert, C. S., & Mariage, T. V. (2003). The sociocultural model in special education interventions: Apprenticing students in higher-order thinking. In H. L. Swanson, K. R. Harris, & S. Graham (Eds.), *Handbook of learning disabilities* (pp. 450–467). New York: The Guilford Press.

Englert, C. S., Raphael, T. E., & Anderson, L. M. (1992). Socially mediated instruction: Improving students' knowledge and talk about writing. *The Elementary School Journal, 92*(4), 411–449.

Fairclough, N. (1995). *Critical discourse analysis: The critical study of language*. London: Longman.

Flower, L., & Hayes, J. R. (1981). A cognitive process theory of writing. *College Composition and Communication, 32*(4), 365–387.

Forgan, J. W., & Vaughn, S. (2000). Adolescents with and without LD make the transition to middle school. *Journal of Learning Disabilities, 33*, 34–44.

Graham, S. (2006). Strategy instruction and the teaching of writing: A meta-analysis. In C. A. MacArthur, S. Graham, & J. Fitzgerald (Eds.), *Handbook of writing research* (pp. 187–207). New York: Guilford Press.

Graham, S., & Harris, K. R. (2003). Students with learning disabilities and the process of writing: A meta-analysis of SRSD studies. In H. L. Swanson, K. R. Harris, & S. Graham (Eds.), *Handbook of learning disabilities* (pp. 323–344). New York: The Guilford Press.

Graham, S., Harris, K. R., & Larsen, L. (2001). Preventing and intervention of writing difficulties for students with learning disabilities. *Learning Disabilities Research & Practice, 16*(2), 74–84.

Graves, A., Montague, M., & Wong, Y. (1990). The effects of procedural facilitation on the story composition of students with learning disabilities. *Learning Disabilities Research, 5*, 88–93.

Haager, D., & Vaughn, S. (1997). Assessment of social competence in students with learning disabilities. In J. L. Lloyd, E. J. K. Kameenui, & D. Chard (Eds.), *Issues in educating students with disabilities* (pp. 129–152). Los Angeles, CA: Lawrence Erlbaum.

Hammill, D., & Larsen, S. (1996). *Test of written language* (3rd ed.). Austin, TX: Pro-Ed.

Hay, I. (1996). Understanding the interactions between self-concept and classroom practice for students with learning difficulties. *Specific Learning Disabilities Bulletin of New Zealand, 7*(3), 16–24.

Hayes, J. R., & Flower, L. S. (1980). Identifying the organisation of writing processes. In L. W. Gregg & E. R. Steinberg (Eds.), *Cognitive processes in writing* (pp. 3–30). Mahwah, NJ: Lawrence Erlbaum Associates.

Kear, D. J., Coffman, G. A., McKenna, M. C., & Ambrosio, A. L. (2000). Measuring attitude toward writing: A new tool for teachers. *Reading Teacher, 54*(1), 10–23.

MacArthur, C., Graham, S., & Schwartz, S. (1993). Knowledge of revision and revising behaviour among learning disabled students. *Learning Disability Quarterly, 14*, 61–73.

Moni, K. B., Jobling, A., van Kraayenoord, C. E., Elkins, J., Miller, R., & Koppenhaver, D. (2007). Teachers knowledge, attitudes and the implementation of practices around the teaching of writing in inclusive middle years' classrooms: No quick fix. *Educational & Child Psychology, 24*(3), 18–36.

Muckert, T., Moni, K. B., & Jobling, A. (2003). Teacher log. Unpublished assessment tool, School of Education, The University of Queensland, Brisbane, QLD.

Muckert, T., van Kraayenoord, C. E., Moni, K. B., & Jobling, A. (2003). Weekly account of lessons: Sequence of events and reflections. Unpublished assessment tool, School of Education, The University of Queensland, Brisbane, QLD.

Nicholls, J. G. (1978). The development of the concepts of effort and ability, perception of academic attainment, and the understanding that difficult tasks require more ability. *Child Development, 49*, 800–814.

Potato, J. A. (1980). Informal assessment of written expression. *Learning Disability Quarterly, 3*, 88–98.

Pressley, M. (2002). Conclusion: Improving comprehension instruction: A path for the future. In C. C. Block, L. B. Gambrell, & M. Pressley (Eds.), *Improving comprehension instruction: Rethinking research, theory and classroom practice* (pp. 385–399). Hoboken, NJ: Wiley.

Rowe, D. W. (2008). Social contracts for writing: Negotiating shared understandings about text in the preschool years. *Reading Research Quarterly, 43*(1), 66–95.

Schirmer, B. R., & Bailey, J. (2000). Writing assessment rubric: An instructional approach with struggling writers. *Teaching Exceptional Children, 33*(1), 54.

Sexton, M., Harris, K. R., & Graham, S. (1998). Self-regulated strategy development and the writing process: Effects on essay writing and attributions. *Exceptional Children, 30*, 1–12.

Swanson, H. L. (1999). Instructional components that predict treatment outcomes for students with learning disabilities: Support for a combined strategy and direct instruction model. *Learning Disabilities Research and Practice, 14*, 129–140.

Vygotsky, L. S. (1962). *Thought and language*. Cambridge, MA: MIT Press (Originally published in Russian in 1934).

Wakely, M. B., Hooper, S. R., de Kruif, R. E. L., & Swartz, C. (2006). Subtypes of written expression in elementary school children: A linguistic-based model. *Developmental Neuropsychology*, *29*(1), 125–129.

Wong, B. Y. L., & Wilson, M. (1984). Investigating awareness of and teaching passage organization in learning disabled children. *Journal of Learning Disabilities*, *17*(8), 447–482.

van Kraayenoord, C. E., Honan, E., & Moni, K. (2007a). *Improving and sustaining literacy through pedagogical change: The report*. Brisbane, QLD: School of Education, The University of Queensland.

van Kraayenoord, C. E., Honan, E., & Moni, K. (2007b). *Handbook of principles for improving and sustaining literacy through pedagogical change*. Brisbane, QLD: School of Education, The University of Queensland.

van Kraayenoord, C. E., Moni, K. B., & Jobling, A. (2004). *The student writing interview – Revised*. Unpublished assessment tool, School of Education, The University of Queensland, Brisbane, QLD.

van Kraayenoord, C. E., Moni, K. B., Jobling, A., & Koppenhaver, D. (2004). *The WriteIdeas classroom observation tool – Revised*. Unpublished assessment tool, School of Education, The University of Queensland, Brisbane, QLD.

van Kraayenoord, C. E., Moni, K. B., Jobling, A., Koppenhaver, D., & Elkins, J. (2003). *WriteIdeas: Enhancing writing – Teaching students in the middle phase of learning with developmental disabilities [and learning difficulties] in regular classrooms: The Manual*. Brisbane, QLD: School of Education, The University of Queensland.

van Kraayenoord, C. E., Moni, K. B., Jobling, A., Koppenhaver, D., & Elkins, J. (2004). Developing the writing of middle school students with developmental disabilities: The writeideas model of writing. *Literacy Learning: The Middle Years*, *12*(2), 36–48.

Chapter 11
The Role of Self-Monitoring in Initial Word-Recognition Learning

Robert M. Schwartz and Patricia A. Gallant

How to teach initial word recognition is one of the most contentious and hotly debated issues in literacy education. In *Learning to read*: *The great debate*, Jeanne Chall (1996) reviewed the history of this debate from 1910 through 1993. She contrasted meaning-emphasis and code-emphasis approaches to initial reading instruction, and suggested that the cyclic popularity of these different approaches seemed resistant to the mass of educational research demonstrating the benefits of an early code-emphasis approach.

Despite this research consensus, the literacy profession is so polarised regarding issues related to initial word-recognition instruction that the past 20–30 years have been characterised as 'the reading wars' (Moats, 2007; Pearson, 2004). The wars are fuelled by the deep concern of teachers, researchers and policymakers for children who struggle to learn to read. The varying viewpoints among these groups highlight different aspects of the complex teaching and learning tasks, and may blind literacy professionals to evidence from alternate perspectives (Chinn & Brewer, 1993).

The debate has been particularly damaging for children who struggle the most with initial literacy learning. For many of these children, the dichotomous debate has either limited their access to code-based information or condemned them to a steady diet of isolated skill instruction. Our goal in this chapter is to suggest that a more complex view of initial word-recognition learning and instruction renders the debate moot and leads to more productive approaches to instruction.

Grounded theory

As the developer of Reading Recovery, Marie Clay observed a multitude of struggling beginning readers, and discussed their learning and instruction with thousands of teachers working with these children. Clay (1991, 1998, 2001) wrote extensively

R.M. Schwartz (✉)
Oakland University, Rochester, MI 48309, USA
e-mail: rschwart@oakland.edu

C. Wyatt-Smith et al. (eds.), *Multiple Perspectives on Difficulties in Learning Literacy and Numeracy*, DOI 10.1007/978-1-4020-8864-3_11,

on this topic. Her theoretical position evolved from her early observational studies (1982), her continuing analysis of the research literature (Clay, 2001) and her analysis of intervention practices (Clay, 2005a, 2005b). This perspective led her to the belief 'that a view of complexity is the kind of understanding required to deliver results in an early intervention programme aiming to prevent subsequent literacy difficulties in as many children as possible' (2001, p. 138).

One aspect of this complexity emerged from Clay's (1982) year-long observations of 100 children, upon entry to school at age 5. She collected weekly records of children's oral reading of connected texts, and 'every response was categorized as true report, error, repetition, or self-correction' (p. 10). Clay found that 'high self-correction rates were associated with high reading progress and were inversely related to error rate' (p. 23). These observations provided her initial evidence for a grounded theory of strategic processing, related to early word recognition.

Errors and self-corrections are interesting reading behaviours. As Goodman (1969) argued, oral reading errors, or miscues, suggest the type of information a reader is attending to and the sources of information they ignore or fail to notice. Accurate responses are less helpful, since the reader could have used some or all of the redundant informational sources available in text to produce the response. Self-corrections are particularly informative behaviours, because they combine some searching activity to produce the initial attempt, some monitoring processes to evaluate that attempt, followed by additional searching that results in a correct response. From this perspective, searching and monitoring are theoretical constructs involved in all word-recognition decisions in continuous text. *The goal of word-recognition instruction is, therefore, to help each child develop fast and efficient searching and monitoring processes that support meaning construction and allow problem solving at the word level when necessary* (Clay, 2001; Schwartz, 1997, 2005).

How does this perspective change the nature of the great debate on initial word-recognition instruction? In the professional literature, Reading Recovery has, at various times, been positioned as either a code-emphasis or meaning-emphasis approach to early intervention. For example, Adams (1990) provided a very positive evaluation of the code-based instruction involved in the Reading Recovery lesson framework: 'The Reading Recovery program has been methodically designed to establish and secure the whole complex of lower-order skills on which reading so integrally depends' (p. 421). Moats (2007), on the other hand, lumps Reading Recovery with whole-language approaches, suggesting that, 'Reading Recovery is not effective unless modified with systematic, sequential instruction in decoding and phonemic awareness' (p. 23).

Clay's theory and Reading Recovery instruction avoids this debate by focusing upon change over time in word-recognition processing. Unlike other developmental models of word recognition (Chall, 1996; Ehri, 1990), the emphasis on observed changes in word recognition in continuous text allows for a broader use of information sources and strategic activity. Clay's (2001) theory is consistent with the major consensus finding in cognitive psychology (Stanovich, 1998; 2000) that proficient readers do not rely on context cues from the sentence meaning, or on syntax to produce word-recognition attempts. Proficient readers are fast and efficient at searching the visual information to generate word-recognition attempts. This efficient visual processing, however, represents the end point of a complex learning and developmental process (Clay, 2001).

How novice or struggling readers can be supported to reach this goal is still the central question for instruction. There is undoubtedly more than one path to fast and efficient visual processing (Clay, 1998). Many children have successfully learned to read with various whole-language and decoding approaches to initial instruction. Successful beginning readers learn not only what the program teaches, but also anything else they need to know to deal with the complexities of reading connected text (Clay, 2001). Struggling readers are far less likely to navigate the gaps left by a classroom program. One-to-one interventions by a trained teacher have been demonstrated to be highly effective for these children (Schwartz, Askew, & Gómez-Bellengé, 2007; Schwartz, Hobsbaum, Briggs, & Scull, in press; What Works Clearinghouse, 2008).

A developmental model of change over time in searching and monitoring processes aligns well with an emergent literacy approach to learning and instruction. An emergent literacy perspective includes the following basic principles: (1) acceptance of approximations while moving towards conventional forms of reading and writing; (2) instruction from this perspective focuses on identifying and building on the individual strengths of the child and (3) assessment of what the child can currently do is more important than identification of deficits, since there are many elements of a complex task that a novice will not yet be able to perform.

Each of these emergent literacy principles is central to word-recognition instruction in Reading Recovery (Clay, 2005a). An example helps to illustrate these points. Marty, a beginning first grader, was asked to read a simple patterned story as part of his assessment for Reading Recovery service. The story had one line per page, accompanied by a supportive illustration. After a brief book introduction, his teacher read the first page, 'The firefighter has a red hat', to establish the language pattern. The teacher continued to read the beginning of each sentence (shown in parentheses below), asking Marty to read the rest. Marty substituted 'black' for 'purple' in the sentence '(The pirate) has a purple hat.' The task became more difficult on the last two pages, where a woman was leaning a little too close to a monkey's cage. The woman has a yellow hat with different coloured fruits on it.

Text: The woman has a yellow hat. Now, the monkey has a yellow hat too.
Marty: (The woman) has a orange and red and green and purple and pink hat. (Now), the monkey has the green, a red, a pink, a purple and black hat too.

(After completing this page, Marty comments to his teacher)

Marty: I know it really says orange hat, but the colours!

We are pretty sure that what Marty meant was, 'You know I can't read the words. On each page when I needed a colour word I looked at the pictures and told you the colour I saw. This hat has too many colours, but I'm pretty sure there is only room in that sentence for one colour word!' Marty could not explain all that but he tried, and a knowledgeable teacher would easily infer from this interaction Marty's initial searching and monitoring strategies. He used information from meaning and language structure to search for word attempts. Some of this information came from the book introduction and the teacher's reading of the first sentence. Other meaning clues came from the pictures. That pirate should have had a black hat, but the author/illustrator saved black for the witch's hat on a later page. When Marty got to the last two pages, he tried to stay with his same searching strategy, but he realised it did not work. He was beginning to learn how to monitor his word-recognition attempts.

Even though Marty was one of the lowest performing children in his Grade 1 class, after 14 weeks of one-to-one instruction he could read at or above the level of his class, and had a set of searching and monitoring strategies that put him back on the path to developing fast and efficient word-recognition strategies. His errors at the point when the intervention was discontinued almost always started with the same letter as the word in the text, often had other visual similarities and fit the preceding meaning and sentence syntax. His self-correction rate was one self-correction for every three errors (1:3). He continued to make progress in his class and, by the end of his Grade 1 year, he could read text well within the average band for Grade 2 and showed a 1:2 self-correction rate. He moved from approximation towards convention, and appeared to be back on track for further literacy learning. This *recovery* of a normal developmental trajectory is the goal and meaning of Reading Recovery (Clay, 2001).

Information sources

Critics of Reading Recovery may look at the above discussion as exactly the type of dependence on multiple cueing systems that characterises whole-language approaches (Moats, 2007). The schism between reading educators and cognitive psychologists on this issue is illustrated by Adams' (1998) discussion of the three-cueing system. In this article, Adams was 'stunned' to realise that educators were using questions related to meaning, language structure and visual information to support monitoring and searching in relation to word recognition, rather than comprehension.

We can certainly agree with Adams' perspective that proficient readers should use multiple information sources to construct and monitor their text comprehension. We can even agree that the diagram that began Adams' inquiry was better suited to represent meaning construction. Still, we find it surprising that Adams, having written a seminal book on beginning reading (Adams, 1990), had not encountered any discussion of the use of meaning and language-structure prompts to support initial word recognition. We suspect that this schism results from the isolation, on the one hand, of education practice and theory grounded in that practice (Biemiller, 1970; Clay, 1982; Goodman, 1969; Weber, 1970) and, on the other hand, hundreds of cognitive studies of word recognition based on adults and children reading word lists. While many interesting questions about word recognition can be addressed by research that engages adults or children in reading word or nonsense-word lists, these studies do not reflect the complexity of word-recognition faced by children reading meaningful text.

Stanovich's (1980) work on context facilitation indicates that proficient readers do not use context to support their word recognition. Novice and struggling beginning readers, however, do benefit from context facilitation, as they internalise and learn to deploy knowledge of the phonemic and orthographic systems that support rapid visual processing of words. In fact, a compensatory processing system using meaning and syntactic information is exactly what Stanovich's (1980) model suggests we should expect from less skilled readers. Clay (2001) noticed that children

> were using 'predict and check' in many cases as a substitute for letter-sound decoding, in situations where their print knowledge was inadequate. These intermediate skills enable a reader to use prediction to narrow the field of possibilities and to reduce the decoding load (p. 254).

Developmental model

A developmental model of word recognition in connected text negates the distinction between meaning-based and code-based approaches to word recognition. A more complex and comprehensive understanding results from an analysis of change over time in how children use a variety of information sources for word recognition. Clay's theory and recent work in developmental psychology provide insights into this more complex view.

Siegler (2006) provides a view of strategic learning and development across a wide range of ages and experimental tasks. The tasks explored in this research are often conceptually less complex and easier to control than the study of children's word recognition in continuous text. Still, the microgenetic method and the perspective it provides on strategic development is instructive because of its similarity to the close observational methods used by Clay (1982, 2001) to study early literacy learning.

As Siegler (2006) explains, the *genetic* part of the microgenetic method relates to *genesis*, not *genes*. In this sense, the microgenetic method involves a close look at the origins of strategic behaviour. This method examines change during a period in which the particular competency is rapidly changing. During this period, the density of observation needs to be high, relative to the rate of change. Observations are analysed intensively, with the goal of inferring the mental processes involved with the change (Siegler, 2006). This method is similar to the progress monitoring in Clay (1982), and that which Reading Recovery teachers use as they take daily running records of children's word-recognition behaviours across the 12–20 weeks of their Reading Recovery program.

Siegler (2006) characterises the perspective on learning and development resulting from this research as *overlapping wave theory*. Its central findings are that early

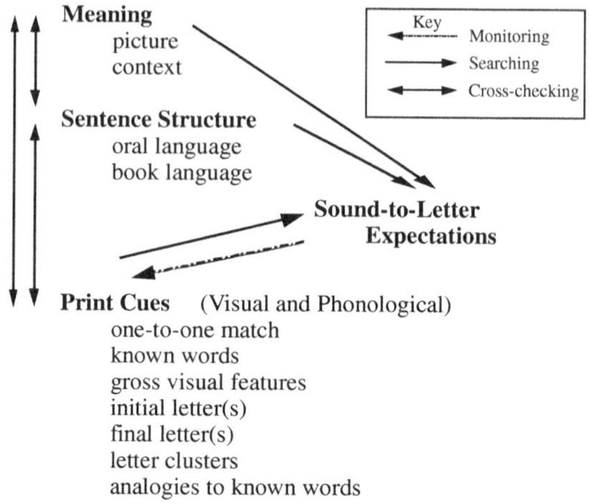

Fig. 11.1 Information sources Source: Reproduced from Schwartz (2005, p. 439), with permission

strategic approaches to cognitive tasks coexist and are gradually replaced by more advanced strategic processing. The onset and uptake of new strategic approaches can be slow or more rapid, thus leading to waves with gradual or steeper crests. Several waves may eventually give way to the mature form of processing, which approaches near 100% use. Siegler (2006) characterises the research on this overlapping wave theory as incorporating five dimensions of change: variability, path, rate, breadth and source.

Schwartz (1997, 2005) describes some of the changes in monitoring and searching behaviours observed in Reading Recovery. The complexity of these changes reflects the availability and use of different information sources in reading continuous text, including meaning, sentence structure, print clues and sound-to-letter expectations. Schwartz (1997) provided a graphic representation of these information sources (see Fig. 11.1). Changes in strategic activity in word recognition can be understood in terms of the five dimensions of Siegler's (2006) wave theory.

Variability of change

Variability of change suggests that a student might have several different ways to combine and employ these information sources for searching and monitoring at any given time. Faced with an unfamiliar word in a book, a child might: (1) choose to reread from the beginning of the sentence or even the beginning of the page, (2) attempt to sound the first letter or two, (3) look at the illustration for a meaning clue or (4) look to the teacher for help.

Any of these, or a variety of other, searching procedures may be employed at different spots within the same book. All of these approaches should eventually give way to a more mature form of searching that involves analysis of familiar visual components within the unrecognised word. This may require much more knowledge and rapid access to orthographic and phonological patterns than a novice or struggling reader is able to bring to bear. In Kaye's (2005) study of proficient students in Grade 2, students used a variety of overt solving behaviours, including more than 50 different multi-step solving behaviours. Rather than sounding phoneme-by-phoneme, they focused on larger units and used multiple sources of information.

A simplistic view of searching strategies posits that any familiar word can be recognised by proper application of sounding-out strategies. Moats (1999) made this argument in describing the new knowledge base for reading instruction coming from research psychologists. She suggests that the 'ability to sound out words is, in fact, a major underpinning that allows rapid recognition of words *by sight*' (p. 16). Rayner, Foorman, Perfetti, Pertsky, and Seidenberg (2001) present a more nuanced view. They conclude that phonics instruction is 'critically important because it does help the beginning reader understand the alphabetic principle and apply it to reading and writing' (p. 68). Like Chall (1996) they review research indicating that with groups of students, more letter–sound instruction is associated with higher outcomes. Still, they move somewhat away from a strict reliance on direct instructional methods for teaching letter–sound relationships. Rayner et al. (2001) indicate that the over 500 spelling-sound connections in English require some form of self-teaching as an important mechanism by which children move beyond a basic level of literacy learning.

There is no inherent conflict among these views. A strong instructional emphasis on one information source may influence a child's initial approach to searching or monitoring, but careful observation of word-recognition learning using a microgenetic approach will still demonstrate variability and change over time in strategic processing.

A simplistic view that suggests that practising sounding-out words letter-by-letter leads directly to the fast and efficient visual processing that proficient readers display is certainly an inadequate description of change over time in word-recognition strategies. Clay's (1982) observations of change showed that high progress children never used a letter-by-letter sounding approach through an entire word unless they were specifically instructed to use that method. Similarly, Kaye (2005) found that proficient readers in Grade 2 never attempted to solve a word letter-by-letter or phoneme-by-phoneme. Self-monitoring by sound-to-letter expectations provides a more efficient way to build the needed knowledge of phonemic and orthographic patterns (Clay, 2001; Schwartz, 1997).

Path of change

The path of change can be influenced by the instructional approach, which should be influenced by the knowledge and experience a particular child brings to instruction (Clay, 1998). For native speakers, meaning and syntax provide a reliable information source with which to monitor their word-recognition decisions. For English language learners (ELL), these information sources can be less reliable. For example, in the book *The Farm Concert* (Cowley, 1987), each page has a picture of an animal and a single sentence; for example 'The cow went moo'. When a Hispanic girl read this book, she read 'went' correctly, but then tried a variety of other words starting with the same letter, 'where', 'when', 'why' or 'was'. The meaning of 'went' required in this sentence was not part of her language, so she continued to search for a word that would make sense. If the text had read, 'The cow said moo', her expectations based on meaning and language structure would have confirmed her word-recognition decisions. When she read another book with the sentence, 'The farmer went back home', she had no problem with the word 'went'. English language learners, depending on their level of English proficiency, may rely more on visual cues for searching and monitoring. This creates a different path than that taken by many native speakers. Fortunately, encountering common words in multiple contexts supports their further language learning.

English language proficiency constitutes a large difference between native speakers and ELL in the knowledge they bring to the text and the word-recognition task. Much smaller differences in any of the information sources shown in Fig. 11.1 can influence a child's path towards proficient performance. A comprehensive literacy program should provide opportunities to refine and extend these knowledge sources to support strategic processing.

Rate of change

In Siegler's (2006) discussion of wave theory, rate of change is influenced by the initial discovery of new strategic approaches and the rate of uptake of that approach relative to other ways of solving the problem. For example, in early Reading Recovery lessons, Marty used meaning and syntax to generate many substitutions, like 'water' for 'river'. When he completed the sentences with that substitution, his teacher prompted for self-monitoring saying, 'That made sense, but check to see if it looks right.' Since the teacher had not indicated which word in the sentence did not look right, Marty reread the entire sentence, trying to monitor by visual information. When he noticed a discrepancy in his sound-to-letter expectation for 'water', he then reread again, this time combining the initial letter sound for 'r' with expectation about meaning and syntax, leading to a self-correction. A child who generalises from this processing activity to modify their future monitoring and searching strategies will demonstrate rapid uptake of the new processing and show accelerated learning. A child who requires many similar prompts, demonstrations or explanations to establish this process will make slower progress.

Learning difficulties appear when children persist with these intermediate strategic approaches rather than continuing flexibly to adjust their processing strategies as they gain more knowledge (Clay, 2001; Schwartz & Stanovich, 1981). For children who struggle with literacy learning, each of these partially effective approaches offers a possible oversimplification of a complex processing task (Spiro, Collins, Thota, & Feltovich, 2003). Either 'sounding out' or 'contextual guessing' can interfere with the development of fast and efficient word recognition. Unfortunately, these processing difficulties often result from children's responses to our instructional programs (Allington, 1994; Barr, 1974–1975; Poole, 2008).

Breadth of change

Breadth of change is also related to transfer and generalisability. New strategic approaches are likely to be taken up more quickly if they apply to a broad range of problems. For example, taking words apart in reading (Clay, 2005b) is a strategic approach to searching based on orthographic and phonological knowledge of units larger than a single letter. The teacher may prompt or the child may self-question, 'Do I know a word like that?' when encountering an unfamiliar word in print (Clay, 2005b). This processing approach is broadly applicable if the child has a large knowledge base of known words. When the child is able to quickly access words he knows with similar beginning or ending patterns, this processing approach may quickly supplant more letter-by-letter sounding approaches. As children begin to use this type of visual information for word recognition, their use of rereading to search for meaning and language-structure clues will decline. Rereading may still be used as a monitoring strategy to confirm the visual analysis.

Breadth of application seems to be part of the justification for the use of decodable texts in classroom programs that strongly emphasise direct instruction in letter-to-sound relationships (Rayner et al., 2001). Decodable texts try to ensure that students see the application of this instruction for word recognition in continuous text.

Instructional programs, like Reading Recovery, which attempt to build on a child's oral language, world knowledge and initially limited literacy knowledge are supported by the availability of a large number of texts arranged in a gradient of difficulty (Anderson, 2002; Clay, 2001). Thousands of book titles, numerically levelled from 1 to 20, based on field trials with children (Anderson, 2002) allow teachers a choice of highly supportive to more complex texts. As children learn to use more information sources for searching and monitoring they are able to successfully apply these strategies in more complex texts (Schwartz, 1997, 2005). This processing approach does not require the type of phonological control used in decodable texts.

Source of change

In Siegler's (2006) experimental tasks, sources of change often involved the type of feedback provided and the child's ability to explain their strategic approach. Since the goal of word-recognition instruction is to develop a system for rapid, automatic visual processing, Clay (2001) is much more cautious about the role of self-explanation in promoting change. Prompts as an instructional tool or as self-talk to guide processing are seen as temporary support to engage the child in effective processing, but the processing itself is the source of change that builds the child's system for word recognition and reading more generally. Within the word-recognition system, self-monitoring is a primary source of change. When the child notices that their current approach to searching and monitoring is not working, they will try to extend the set of information sources used for these monitoring or searching decisions (Clay, 2001; Schwartz, 1997, 2005).

Self-monitoring is a form of self-instruction that has the potential to build the intricate and complex knowledge of the phonological and orthographic system necessary to support fast and efficient processing. Children who have difficulty learning to read need to be encouraged to monitor their word-recognition decisions based on their increasing knowledge. They gradually learn ways of looking at print in continuous text that coordinate their phonological and orthographic knowledge with the visual information, while also dealing with the meaning of the story. This is what Clay (2001) described as a working system for literacy processing. Monitoring word-recognition attempts is a part of this system and operates both on correct and incorrect processing decisions. A large set of easy books that children can read provides a rich context for self-instruction on phonemic awareness, letter–sound relationships and orthographic patterns.

Word-recognition instruction

If self-instruction is seen as a primary way in which young and struggling readers learn to deal with the complexity of the word-recognition process, then what is the teacher's instructional role? Given the percentage of children who appear to struggle with early literacy learning, this is an important question. The relationship between teaching and learning is indirect, especially when the learning domain is complex. Clay (2005b) suggests that

> it is most helpful to think of the learner (who is successfully solving reading problems) as building a neural network for working on written text *and that network learns to extend itself.* Teachers can help this happen but what they call 'instruction' does not extend the neural network! It is the successful strategic activity called up by the learner that creates the self-extending system (p. 103).

Teachers are asked to deal with complexity on a number of different levels. Struggling readers bring different sets of strengths to the learning situation and encounter different challenges or difficulties as they change over time in their literacy learning. Prompting is a form of instruction that supports learning by engaging the child's processing system for word recognition. The prompt can focus on monitoring or searching decisions following an error. The decision to prompt for monitoring would depend on whether or not the child shows signs of having noticed the error and whether the teacher thinks there are information sources that the child could have noticed (Schwartz, 2005).

Table 11.1 Possible substitutions, support types, and strategy prompts

Text: 'Would you like a fat worm?' 'No I wouldn't'. (Cowley, 1987, p. 10)

Substitution for fat	Support type	Monitoring prompt[a]	Searching prompt[b]
Purple	Independence	Were you right?	What can you try?
Flat	Meaning support	Does that make sense?	Try that again and think what would make sense
Four	Structure support	Does that sound right?	Try that again and think what would sound right
Purple	Print support	Does that look right?	Try that again and get your mouth ready for the first sound

[a]Child makes the substitution and finishes the sentence. Showing no sign of noticing the error. Give the prompt without indicating where in the sentence the error might be. Use these prompts occasionally, following accurate reading so the prompt itself doesn't signal an error.
[b]Child notices that his or her initial attempt does not work or stops at the unknown word and makes no further attempt.
Source: Schwartz (2005, p. 441), reproduced with permission.

Table 11.1, from Schwartz (2005), shows a variety of prompts that could be used for searching and monitoring if the child makes a substitution for 'fat' in the sentence 'Would you like a fat worm?' Monitoring prompts are appropriate if the child finishes the sentence (or the page) showing no recognition of his substitution. If the child hesitates, appeals for help, makes multiple attempts or in other ways indicates that he is aware that the initial attempt was problematic, then prompting to search particular information sources may help the child refine his current searching strategies (Clay, 2005b).

Monitoring prompts have the added advantage of serving as a mechanism for building independence and change over time in searching strategies. Consider the substitution of 'purple' for 'fat' in Table 11.1. For a child who is just beginning to learn to look at print while reading, this is a common word-recognition attempt. The picture shows a bright purple worm. The pattern of this book facilitates emergent processing. On each page an animal suggests something it thinks the boy might like on his sandwich. Children at this level may not be able to decode the letter–sound relationships in 'grasshopper' or 'mouse', but they can use meaning cues from the pictures as a searching strategy for these final words in the language pattern. The page used in this example provides an additional challenge since it is the only page to include an adjective prior to the noun. The child could attempt to read this as 'Would you like a worm?' Because he already expects one-to-one correspondence between words in oral language and words in print, he is likely to notice this discrepancy. This monitoring may lead him to reread the sentence and search for additional information.

The picture suggests 'purple' as a possible description of the worm. This resolves the correspondence problem and, therefore, may meet the child's criteria for self-monitoring. The teacher's decision to prompt for additional monitoring is not based on the accuracy of his reading but rather a judgment that the child has the knowledge necessary to notice the gross visual discrepancy between 'purple' and 'fat' if attending more carefully to the visual cues. Since the goal is to prompt self-monitoring, it is important that the prompt follow the completion of the sentence and not indicate which word is the source of the difficulty. Prompting with 'That makes sense, but check to see if it looks right' is intended to reduce the sources of information the child needs to examine and thus let them focus attention on the visual component for monitoring.

For a child at this point in literacy development, phonics is a highly attention-intensive process. It requires the child sequentially to retrieve and blend two or more individual sounds while holding in memory some representation of the meaning and syntax of the prior portions of the sentence and story. Monitoring in response to this prompt may require less cognitive capacity. The child need only to judge each word against the type of sound-to-letter expectations they would use in writing. The teacher knows that if this child was trying to write the word 'purple' the attempt would likely have at least one 'p'. This expectation is sufficient to support the requested monitoring decisions.

The discovery that sound-to-letter expectations provide a way to monitor word-recognition decisions is an important milestone. If the child shows quick uptake of this strategy, it will lead to successful monitoring of many similar meaningful substitution like 'water' for 'river' or 'rabbit' for 'bunny'. More importantly, the application of this monitoring strategy to correctly read words provides a massive source of reinforcement and learning of letter–sound and orthographic information. For children who have struggled with phonemic awareness as a purely auditory task, these attempts to match sound-to-letter expectations, along with other procedures, like the use of Elkonin boxes for writing, may provide the necessary learning.

This increased monitoring capacity also supports strategic advances in searching. The child has no reason to modify their current searching procedures if they feel they are successful. In the above example, when the child notices that purple does not meet his sound-to-letter expectations, he rereads and uses the initial letter cue to suggest 'funny' for 'fat'. Even though this searching strategy does not produce an accurate response, it is a strategic approach that has transfer value and moves the child towards more proficient visual processing.

This form of instruction is consistent with a complex theory of literacy learning and instruction that builds on the child's individual knowledge and strengths to support growth from strategic approximations towards conventional forms of word recognition, based on fast and efficient visual processing. Direct instruction in letters, sounds and high-frequency words helps to establish knowledge that a child can bring to word-recognition processing, but this knowledge is only part of a developing processing system. Struggling beginning readers need to develop ways of accessing, coordinating, utilising and monitoring these processes and information sources, while reading for meaning. Skilled prompting during oral reading provides a way for teachers to support the child's construction of an effective processing system (Clay, 2001; Schwartz, 2005).

Teaching difficulties

Prompting effectively while listening to novice or struggling beginning readers is itself a complex and highly professional skill. Clay (2001) cautions that

> Young children are usually confused if we try to put these mental activities into words. Teachers cannot teach the learner how to orchestrate this complexity but they can support the child who is constructing effective reading processes, pointing up relevant information with direct and indirect prompts. If the teacher does not understand literacy processing, prompting will be hit or miss. (p. 128)

Oversimplification (Spiro et al., 2003) is a hazard for literacy teachers as they learn about the complexities of early literacy instruction. Teachers may favour one type of prompting over others. Some teachers always prompt for visual analysis, while others show a marked preference for meaning-level prompts (Lyons, Pinnell, & DeFord, 1993).

Schwartz (2005) used the tetrahedron model shown in Fig. 11.2 to describe some of the complexity involved in prompting decisions. Since a child's processing changes rapidly over time in early literacy learning, the teacher needs a way to track their response history. Daily running records provide information from which the teacher can infer the child's current strategic approaches, identify the types of substitutions the child is likely to make and plan prompts to support and extend the child's current processing.

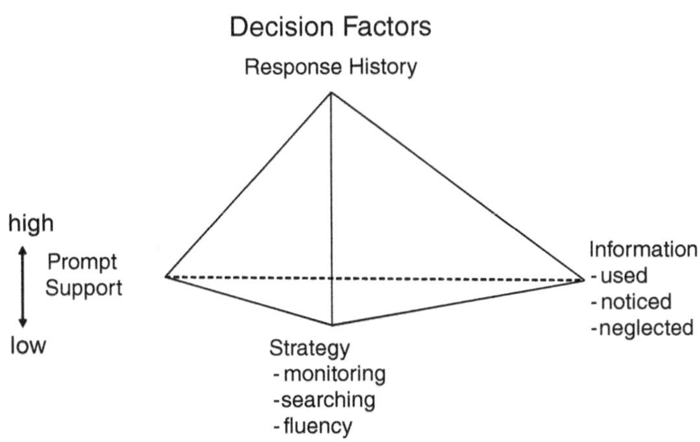

Fig. 11.2 Factors that affect teaching decisions during oral reading
Source: Schwartz (2005, p. 437), reproduced with permission

For example, towards the end of Marty's Reading Recovery program, he read a story about a giant and a ghost. Marty made several substitutions of 'giant' for 'ghost' and 'ghost' for 'giant' in his reading of the story. Since both the giant and the ghost were reacting to another character in the story, the substitutions did not affect the meaning and Marty showed no signs of noticing the substitutions. Based on these and other substitutions, Marty appeared to use meaning, syntax and initial visual cues for searching and to monitor substitutions by the final visual cue or ending sound (for example noticing the substitution of 'like' for 'love'). The 'ghost' for 'giant' substitutions met these criteria but reflected only a partial use of visual information for both monitoring and searching.

Errors of this type indicate that Marty has more to learn about looking at print in continuous text. A prompt to monitor for visual information would probably suffice to help Marty identify this error. His teacher can help Marty establish more mature forms of looking with this type of prompting, but he will also benefit from instruction and experience in breaking words based on onset, rime, syllable, prefixes and suffixes; writing words by analogy to known words and continuing development of a large reading and writing vocabulary.

Prompting for monitoring rather than searching strategies requires a principled and complex theory of literacy learning and instruction that incorporates both of these processes and change over time in the forms they may take (Gallant & Schwartz, in press; Schwartz, 2006). It is a tempting oversimplification for teachers to assume that early signs of monitoring and searching only need to be refined by additional letter–sound knowledge. Teachers often describe a child they are tutoring as being able to monitor and search. This seems to represent a partial understanding of monitoring and searching processes that needs to be further refined to include specification of the information sources used for searching and monitoring, and expectations about how these processes will change over time as the child's knowledge increases (Clay, 2001; Schwartz, 2005).

The need to support children's development of strategic processing in the context of reading connected text is too easily overlooked. Phillips and Smith (1997) and Smith (1999) found that teachers had great difficulty responding to children's undetected errors with prompts for monitoring. Instead, they drew the child's attention directly to the error and attempted to teach for searching strategies. These teaching decisions occurred in the context of one-to-one instruction.

The situation is more complex in small-group instruction. Poole (2008) examined the support provided for struggling readers in Grade 5 in mixed-ability reading groups. She argues that mixed-ability groups have become a common response to the widely disseminated negative research findings from studies of ability grouping. Unfortunately, these mixed-ability groups seem equally adept at maintaining the low status and skill level of struggling readers. The only form of word-recognition assistance these struggling readers received 'were limited to one-time corrective pronunciation of words they were unable to read' (Poole, 2008, p. 245). Teachers also ignored many oral reading errors made by the struggling readers. Poole (2008) recorded 34 unacknowledged miscues for one struggling reader across 2 days of oral reading turns. These types of responses to struggling students' oral reading will not help refine their monitoring and searching strategies.

Beyond the debate

It is tempting to declare the debate 'settled' if you can make an argument for your side winning! Adams and Bruck (1995) offer a resolution to the phonic versus whole-language debate by recognising positive contributions of the whole-language movement. These contributions include a variety of instructional procedures that go beyond word-recognition instruction and efforts to integrate instruction across the curriculum and to empower teachers with 'confidence, authority and self-esteem on which good teaching depend' (p. 18). Still, for Adams and Bruck, word recognition is all about 'viewing words as a sequence of letters and associating their spelling with sounds' (p. 15). Meaning and syntax have no role in their view of word-recognition teaching or learning.

The resolution we suggest differs from that proposed by Adams and Bruck (1995). Like them, we dismiss the more extreme forms of top-down word-recognition models (Goodman, 1969; Smith, 1971), in favour of a developmental model that progresses towards fast visual processing. We agree with Stanovich's (1998) view that it is risky to draw a strong connection between basic process research and practice. There are many different paths to the fast visual processing that proficient readers need to learn. Certainly, explicit instruction in letter–sound relationships is a necessary component of instruction for most struggling readers. Attention to how a child is able to deploy this knowledge to support searching and monitoring during reading of meaningful text shifts the focus of debate, allows for progress monitoring related to processing and provides cues to effective instructional approaches.

Similar to the debate in literacy education, developmental psychology floundered for years with fruitless debate over nature versus nurture. Developmental psychologists have moved beyond this debate with more complex theories that negate the either/or dichotomy (Lerner, 2006). Literacy educators need to make a similar transformation in our professional literature to a more complex view of early literacy learning and instruction. Clay's theory provides an excellent starting point for this transformation. Grounded in detailed observation, Clay (1982, 2001, 2005a) provides a view of waves of strategic activity (Schwartz, 1997, 2005; Siegler, 2006) that illustrate the interaction of monitoring, searching, phonological processes and changes over time in the child's ability to look at and use information from print.

Essential next questions

How would teaching from a complex theory of word-recognition change the nature of primary literacy instruction and learning? Reading Recovery has demonstrated that instruction based in a complex theory can support the learning of at-risk Grade 1 students. While this approach has been implemented in some small group and classroom settings, qualitative, quantitative and design research are needed to explore the factors that would influence implementation and effectiveness in a comprehensive literacy program. It would be particularly important to demonstrate this in school settings that appear to have a high proportion of at-risk students.

What types of professional development are effective in helping teachers to develop and implement instruction based in a complex theory of word recognition? Implementing instruction based on a complex theory requires considerable professional development. For example, reciprocal teaching (RT) has a deceptively simple dialogue format, but a complex set of theoretical principles. Twenty years of wide dissemination has often resulted in a series of mutations that completely abandoned the theoretical principles that made RT instruction effective (Palincsar, 2007). To support teacher learning, Palincsar, Spiro and colleagues, developed a hypermedia tool that allowed teachers to construct their understanding of the complexity of this method. By exploring video segments organised by themes, the teachers were able to develop more situation-sensitive understanding (Palincsar, 2007).

Unfortunately, the complex change over time in word-recognition processing described in this paper is almost invisible to teachers and researchers who do not individually teach struggling beginning readers. Would a hypermedia tool, which allows teachers and researchers to view change over time in children's word-recognition processes support professional development and help to re-focus the professional debate on early literacy instruction?

Like Marie Clay, we embrace complexity, realising that our current understanding is tentative and the way forward still uncertain. As literacy professionals, we would be wise to follow her lead, remembering that, as Clay always believed, the 'search for solutions has no end' (Clay, 2005b, p. 208). We do not claim that this model of initial word recognition is complete or that the instructional implications are fully developed. But it provides a way forward, a way of moving beyond unproductive debates of the past century and the 'reading wars'. Instruction based on close observation of children's reading behaviour, and interpreted in terms of strategic activity related to monitoring and searching, has the potential to support literacy learning for many of our most promising, but at-risk, children.

Glossary

Clues See information sources.

Cues See information sources.

Elkonin boxes A teaching procedure used to build phonemic awareness, sound-to-letter knowledge and orthographic awareness. . Clay (2005b) for a description of the procedure and how it changes as the child's knowledge increases.

Emergent literacy A theory of literacy learning that recognises the role of children's early approximations of reading and writing behaviours in the learning and development of conventional reading abilities. This approach is often contrasted with a readiness approach that specifies knowledge a child should have prior to the start of formal literacy instruction.

Information sources These sources of information include visual information coming from the text and all the information from prior experience stored in the child's brain. This stored information includes what the child has learned about orthographic and phonological relationships and recently constructed semantic and syntactic information from reading prior portions of the text. These information sources are also referred to as cues or clues in this chapter.

Microgenetic method A research method used in development psychology to study how learning occurs. Siegler (2006) describes the method, studies using this method and major findings about children's learning resulting from the application of this method.

Monitoring A set of strategic activities that a reader engages in to evaluate word-recognition attempts. The information sources a reader uses to evaluate these attempts gradually changes over the transition from emergent to beginning reading. As word-recognition processing becomes more automatic, attention shifts to monitoring comprehension decisions.

Searching A set of strategic activities that a reader engages in to make word-recognition attempts. The information sources a reader uses to generate attempts gradually changes over the transition from emergent to beginning reading. Searching includes processing activity to generate initial attempts as well as to make additional attempts when monitoring processes suggest that the initial attempt may not be correct.

References

Adams, M. J. (1990). *Beginning to read: Thinking and learning about print*. Cambridge, MA: MIT Press.

Adams, M. J. (1998). The three-cueing system. In J. Osborn, & F. Lehr (Eds.), *Literacy for all: Issues in teaching and learning* (pp. 73–99). New York: Guilford Press.

Adams, M. J., & Bruck, M. (1995). Resolving the 'great debate'. *American Educator, 19*(2), 7, 10–20.

Allington, R. (1994). What's special about special programs for children who find learning to read difficult? *Journal of Reading Behavior, 26*(1), 95–115.

Anderson, N. (2002). Reading Recovery and descubriendo la lectura book lists: Answers and important questions. *Journal of Reading Recovery, 1*(2), 10–12.

Barr, R. (1974–1975). The effect of instruction of pupil reading strategies. *Reading Research Quarterly, 10*, 555–582.

Biemiller, A. (1970). The development and the use of graphic and contextual information as children learn to read. *Reading Research Quarterly, 6*, 75–96.

Chall, J. S. (1996). *Learning to read: The great debate* (3rd ed.). Fort Worth, TX: Harcourt Brace.

Chinn, C. A., & Brewer, W. F. (1993). The role of anomalous data in knowledge acquisition: A theoretical framework and implications for science instruction. *Review of Educational Research, 63*(1), 1–49.

Clay, M. M. (1982). *Observing young readers: Selected papers*. Exeter, NH: Heinemann.
Clay, M. M. (1991). *Becoming literate: The construction of inner control*. Portsmouth, NH: Heinemann.
Clay, M. M. (1998). *By different paths to common outcomes*. York, ME: Stenhouse.
Clay, M. M. (2001). *Change over time in children's literacy development*. Portsmouth, NH: Heinemann.
Clay, M. M. (2005a). *Literacy lessons designed for individuals: Part one*. Portsmouth, NH: Heinemann.
Clay, M. M. (2005b). *Literacy lessons designed for individuals: Part two*. Portsmouth, NH: Heinemann.
Cowley, J. (1987). *What would you like?* Bethel, WA: Wright Group.
Ehri, L. C. (1990). Development of the ability to read words. In R. Barr, M. Kamil, P. Mosenthal, & P. D. Pearson (Eds.), *Handbook of reading research* (Vol. 2, pp. 383–417). New York: Longman.
Gallant, P., & Schwartz, R. (2010). Examining the nature of expertise in reading instruction. *Literacy Research and Instruction, 49*, 1–19.
Goodman, K. S. (1969). Analysis of oral reading miscues: Applied psycholinguistics. *Reading Research Quarterly, 5*, 9–30.
Kaye, E. (2005). Second graders' reading behaviors: A study of variety, complexity and change. *Literacy Teaching and Learning, 10*(2), 51–75.
Lerner, R. M. (Ed.). (2006). Theoretical models of human development (Vol. 1). In D. Kuhn & R. M. Learner (Series Eds.), *Handbook of child psychology* (6th ed.). Hoboken, NJ: Wiley.
Lyons, C. A., Pinnell, G. S., & DeFord, D. E. (1993). *Partners in learning: Teachers and children in Reading Recovery*. New York: Teachers College Press.
Moats, L. C. (1999). *Teaching reading is rocket science: What expert teachers of reading should know and be able to do*. Washington, DC: American Federation of Teachers.
Moats, L. C. (2007). *Whole language high jinks: How to tell when 'scientifically-based reading instruction' isn't*. Washington, DC: Thomas B. Fordham Institute.
Palincsar, A. S. (2007). Reciprocal teaching 1982 to 2006: The role of research, theory, and representation in the transformation of instructional research. In D. W. Rowe, R. T. Jimenez, D. L. Compton, Y. Kim, K. M. Leander, & V. J. Risko (Eds.), *56th yearbook of the National Reading Conference*. Oak Creek, WI: National Reading Conference.
Pearson, P. D. (2004). The reading wars. *Educational Policy, 18*(1), 216–252.
Phillips, G., & Smith, P. (1997). *Closing the gaps: Literacy for the hardest-to-teach*. Wellington, New Zealand: Council for Educational Research.
Poole, D. (2008). Interactional differentiation in the mixed-ability group: A situated view of two struggling readers. *Reading Research Quarterly, 43*, 228–250.
Rayner, K., Foorman, B. R., Perfetti, C. A., Pesetsky, D., & Seidenberg, M. S. (2001). How psychological science informs the teaching of reading. *Psychological Science in the Public Interest, 2*(2), 31–74.
Schwartz, R. M. (1997). Self-monitoring in beginning reading. *The Reading Teacher, 51*, 40–48.
Schwartz, R. M. (2005). Decisions, decisions: Responding to primary students during guided reading. *The Reading Teacher, 58*, 436–443.
Schwartz, R. M. (2006). Supporting teacher learning: Reading Recovery as a community of practice. *The Journal of Reading Recovery, 6*(1), 49–55.
Schwartz, R., Askew, B. J., & Gómez-Bellengé, F. X. (2007). What works? Reading Recovery: Analysis of the What Works Clearinghouse report. *Journal of Reading Recovery, 6*(3), 49–52.
Schwartz, R. M., Hobsbaum, A., Briggs, C., & Scull, J. (2009). Reading Recovery and evidence-based practice: A response to Reynolds and Wheldall (2007). *International Journal of Disability, Development and Education, 56*(1), 5–15.
Schwartz, R. M., & Stanovich, K. E. (1981). Flexibility in the use of graphic and contextual information in good and poor readers. *Journal of Reading Behavior, 13*, 264–269.

Siegler, R. S. (2006). Microgenetic analysis of learning. In D. Kuhn & R. M. Learner (Series Eds.), & D. Kuhn & R. S. Siegler (Vol. Eds.), *Handbook of child psychology: Vol. 2. Cognition, perception and language* (6th ed., pp. 464–510). Hoboken, NJ: Wiley.
Smith, F. (1971). *Understanding reading*. New York: Holt, Rinehart and Winston.
Smith, P. (1999). *The teacher, the child and the undetected error*. New Zealand: University of Auckland.
Spiro, R. J., Collins, B. P., Thota, J. J., & Feltovich, P. J. (2003). Cognitive flexibility theory: Hypermedia for complex learning, adaptive knowledge application, and experience acceleration. *Educational Technology*, *43*(5), 5–10.
Stanovich, K. E. (1980). Toward an interactive-compensatory model of individual differences in the development of reading fluency. *Reading Research Quarterly*, *16*, 32–71.
Stanovich, K. E. (1998). Twenty-five years of research on the reading process: The grand synthesis and what it means for our field. In T. Shanahan & F. V. Rodriquez-Brown (Eds.), *47th Yearbook of the National Reading Conference*. Chicago, IL: National Reading Conference.
Stanovich, K. E. (2000). *Progress in understanding reading: Scientific foundations and new frontiers*. New York: Guilford.
Weber, R. M. (1970). A linguistic analysis of first-grade reading errors. *Reading Research Quarterly*, *5*, 427–451.
What Works Clearinghouse. (2007). *Reading Recovery intervention report*. Retrieved January 30, 2009, from http://ies.ed.gov/ncee/wwc/reports/beginning_reading/reading_recovery

Chapter 12
Effective Instruction for Older, Low-Progress Readers: Meeting the Needs of Indigenous Students

Kevin Wheldall and Robyn Beaman

Introduction

Over recent years, we have witnessed mounting public concern in several Western countries regarding the quality of the literacy instruction children are receiving in school, their levels of performance in reading and related skills, and the efficacy of the instructional methods and/or philosophies underpinning the teaching they receive. This has led to national inquiries in these countries: the National Reading Panel (NRP) in the United States reported in 2000 (National Institute of Child Health and Human Development, 2000); the National Inquiry into the Teaching of Literacy (NITL) in Australia reported in 2005 (Department of Education & Science and Training, 2005) and the Rose Report in the United Kingdom was released in 2006 (Rose, 2006). In essence, all three of these reports found the current teaching of literacy wanting in several respects, not least a perceived reluctance by schools and teachers to employ instructional methods predicated upon the findings of sound scientific research into how children learn to read and the most effective forms of literacy instruction.

The purpose of this chapter is not to indulge in another battle in the continuing 'reading wars', but rather to explore these issues in the context of how we might best support socially disadvantaged and, particularly, Indigenous students to achieve at the literacy level of their peers from more favoured backgrounds. Whatever the case may be for literacy instruction for the population of school students more generally, it is surely beyond doubt that we, as a society, continue to fail these students who are most in need of intensive support from schools. While agreeing to disagree, perhaps, on how best to meet their needs, all educators must be aware and concerned that many of those students who most need our help seem to be receiving less than they deserve.

In a welcome move by the Australian Government in 2008, a commitment has been made to the forms of instruction advocated in the three reviews conducted in the United States, Australia and the United Kingdom as foundational to the teaching

---

K. Wheldall (✉)
Macquarie University, Sydney, NSW, Australia
e-mail: kevin.wheldall@mq.edu.au

C. Wyatt-Smith et al. (eds.), *Multiple Perspectives on Difficulties in Learning Literacy and Numeracy*, DOI 10.1007/978-1-4020-8864-3_12,

of literacy. As part of a new national curriculum for the teaching of English, the need for 'explicit and systematic teaching of sound–script correspondences is important, and not just for students who are in their first year or so of schooling, or for whom English is not a first language' (National Curriculum Board, 2008). The initial advice on the new curriculum released in October 2008, while acknowledging the longstanding debates about the teaching of the basics in literacy (also known as the 'reading wars'), has made it very clear that the Australian Government is determined to take notice of the 'extensive research literature on the topic', the importance of teaching 'the basics' in the early years of schooling and building on this knowledge throughout the school years—'grapheme-phoneme correspondence, phonological and phonemic awareness, alphabetic knowledge, grammar, spelling, and conventions of punctuation and so on' (National Curriculum Board, 2008, p. 7). While it remains to be seen how this will be operationalised across states/territories and educational systems, it is a most welcome initial development in Australian schooling.

A simple model of reading disability

Before embarking upon a discussion of reading difficulties and how best to teach literacy to socially disadvantaged students, it is important to define our terms. The first of these is literacy.

By 'literacy' we mean, quite simply, the ability to read and write; nothing more and nothing less. By 'reading' we mean the ability to decode text accurately and fluently with understanding of what has been read. This has been characterised as a 'narrow view' of literacy, coming from a cognitive–psychological approach (reviewed by de Lemos, 2002). Notwithstanding this, we are not persuaded by those who seek to subvert a clear and accurate definition into all-encompassing alternatives that serve merely to obfuscate the primary area of concern. It may well be the case that filmic and televisual 'texts' are in need of critical analysis, but this is not literacy except in the analogical sense, in the same way that we refer to 'computer literacy' to describe a familiarity with information technology. Similarly, we would not deny the importance of developing a sharp, critical faculty by which to appreciate, evaluate and challenge the messages inherent within texts, but this is not literacy per se. Rather, literacy, the ability to read and write, makes the exercise of this critical faculty possible. In order to criticise text you must first be able to read and understand it. The focus of this chapter is to discuss how best to help socially disadvantaged, and especially Indigenous, low-progress readers to become literate in this basic but essential sense.

The second key concept that needs to be defined for our current purposes is the nature of reading difficulties or reading disability. We favour the generic term 'low-progress readers' to refer to all students whose progress in learning to read is delayed *for whatever reason*. The term is neutral, non-pejorative and makes no assumptions regarding causality. It also aligns well with a *non-categorical approach to instruction* (discussed in a later section of this chapter), as we shall see. This is not meant to deny, however, that reading difficulties may be the result of several factors, operating both within the child and among low-progress readers. The argument is rather whether precise 'diagnosis' necessarily assists in the provision of effective instruction. Again, we shall return to this theme.

The first author (Wheldall) has offered a simple model of reading disability (see Pogorzelski & Wheldall, 2005) to help understand how delayed reading progress may result from two key factors. (This is not to deny that other factors may also be operating.) First, most reading scientists today subscribe to the phonological-deficit theory of reading difficulties (see Pogorzelski & Wheldall, 2005 for a thorough review and discussion). In essence, adherents of this theory (see, for example Stanovich, 2000) argue that reading disability is basically a problem with language, specifically the ability to discriminate (segment) and synthesise (blend) the constituent sounds within spoken words, known as phonemic awareness. This view is now too commonly known to require further elaboration here, except to emphasise that it currently enjoys widespread support among reading scientists (largely psychologists and linguists). If reading is critically dependent upon the ability to segment and blend the component sounds within words, then an inability or reduced ability to do so is likely to lead to difficulties in learning to read. The biological predisposition to be able to segment and blend words is what will make it more difficult for some children to learn to read and is probably the heritable component of reading disability (Bishop, 2001). Such biological causation, however, does not mean that learning to read becomes impossible (except in very rare cases), but it is likely to make learning to read far more difficult unless exemplary instruction is provided, based on scientifically determined, evidence-based best practice.

It is important to recognise and to accept, however, that not all reading difficulties will necessarily be predicated upon inherent underlying phonological processing problems. The language and literacy-learning environment is also critically important and plays a powerful role in determining how readily a child will learn to read. Children who have enjoyed a linguistically rich and stimulating environment and whose parents have read to them consistently from an early age will begin school with a huge advantage when it comes to learning to read. Many of the prerequisite skills of reading will already have been learned. Research has consistently demonstrated that the quality of the language and literacy-learning environment and its effects on children's language and literacy learning is highly correlated with social class (socioeconomic status). (See, for example the work of Hart & Risley, 1995, on the language experiences of young children in households varying in socioeconomic status. By the time children begin formal schooling, children from more favourable home backgrounds have experienced many more times the language directed specifically to them than children from socioeconomically deprived backgrounds, and this language is far more likely to be qualitatively more varied and less simply directive.)

Little wonder, then, that the 'Matthew effect' (Stanovich, 1986, 2000), whereby 'the rich get richer and the poor get poorer', applies equally to reading acquisition as to other areas of life. The implication of this is that even children who are biologically well-prepared by virtue of high intrinsic phonological processing ability, may

		Phonological Ability (PA)		
		High	**Average**	**Low**
Quality of literacy learning environment (QLLE)	**High**	High progress readers	Above average readers	Hidden and classic dyslexic readers
	Average	Above average readers	Average readers	Regular dyslexic readers
	Low	Below average readers	Below average readers	Doubly disadvantaged dyslexic readers

Fig. 12.1 A simple model of reading disability

still struggle initially in learning to read if their biological potential has not been realised by their literacy-learning environment.

Our simple two-factor model for reading disability (Pogorzelski & Wheldall, 2005), whereby phonological ability and quality of literacy-learning environment interact to yield differing levels of readers (see Fig. 12.1), aims to illustrate the likely consequences. These are spelled out by Pogorzelski and Wheldall (2005) in some detail, but for our present purposes we should note that among the generic group of low-progress readers from socially disadvantaged backgrounds, there will be those whose difficulties are largely environmental and those whose difficulties are due to both the environment and intrinsic, poor phonological processing ability. While it is more frequently the case that dyslexic students are diagnosed from among those whose verbal skills are (at least superficially) adequate or even advanced (and are typically middle class), there will also be students dismissed as being 'garden variety' poor readers (since their verbal skills are equally poor) who may also be truly dyslexic. Those poor readers from less-than-optimal language and literacy-learning backgrounds who are also evidencing poor levels of phonological skills may be those most difficult to help to learn to read. We shall return later to a discussion of how best we can identify students for particular forms of remedial intervention (see Essential Next Questions at the end of this chapter).

It may be seen, then, why the problems facing children from socially disadvantaged backgrounds are particularly severe, since they are ill-prepared when it comes to learning to read, in addition to any biologically determined phonological processing difficulties they may or may not have. This is the bad news facing students from poor, Indigenous and minority group backgrounds. The good news is that the reasons for their reading difficulties may not be relevant, and that their prognosis may be excellent when the reading interventions they are afforded are based on effective instruction. One such intervention, MULTILIT, has been researched by us since 1996 and has been shown to be highly effective with students from socially disadvantaged backgrounds.

What is MULTILIT?

We coined MULTILIT as an acronym for 'Making Up Lost Time In Literacy' in 1995, to refer to the continuing program of research and development into more effective instruction for low-progress readers conducted by the first author (Wheldall) and his colleagues at Macquarie University Special Education Centre. This enterprise, known originally as 'the MULTILIT Initiative', comprises scientific inquiry into how best to meet the instructional needs of students who are struggling to acquire basic reading and related skills, for whatever reason. It is now known as the MULTILIT Research Unit.

One of the products of this enterprise has been the development of the MULTILIT Reading Tutor Program (RTP), the second revised edition of which was released in 2007 (MULTILIT, 2007). MULTILIT RTP is an individualised program for one-to-one tutoring of low-progress readers. It comprises instruction in three major areas: explicit, synthetic, phonic, word-attack skills teaching; the systematic learning of 200 of the most frequently encountered sight words found in children's books; and one-to-one supportive tutoring in book reading.

The MULTILIT RTP has been repeatedly shown to deliver major gains in reading and spelling for low-progress readers, regardless of the reason for their delayed literacy acquisition—see Wheldall and Beaman (2000) for a detailed review of the early efficacy research. More recently, our focus has turned to the development of *small-group* versions of the program for both younger and older low-progress readers.

But MULTILIT is more than just the name of a literacy program for low-progress readers. It is an educational approach that has as its core belief the conviction that effective instruction is the key to growth in any area of the curriculum. In this sense, it is more accurately conceptualised as an approach, or a determination, to bring about rapid learning for low-progress students by whatever means scientific research has shown to be most effective. Consequently, it is a continually evolving approach to literacy instruction, changing as more scientific evidence becomes available from either within the MULTILIT research team per se or, more generally, from the international scientific reading research community. In essence, the MULTILIT approach may be characterised as being 'sensitive to data' and is predicated upon three core principles: a non-categorical approach to instruction; Positive Teaching and an integrated model of effective literacy instruction.

A non-categorical approach to instruction (Wheldall & Beaman, 2000; Wheldall & Carter, 1996; Wheldall, 1994) rejects the conventional but scientifically unsupported idea that fundamentally different approaches to instruction are necessary to teach different sub-groups within the community, whether this be students with different disabilities, cultural backgrounds, learning styles or preferences, or whatever. The non-categorical approach is predicated on the notion that effective instruction is effective instruction is effective instruction: that we all learn by the same basic routes and processes, regardless of our 'category', and that, consequently, it makes sense to concentrate on instructional methods that have been shown to be effective per se rather than to fixate on hypothesised needs for differing instructional approaches. The most effective instructional approaches for teaching basic literacy skills have been isolated as a result of scientific research and have been shown, in essence, to comprise instruction that is direct, explicit, structured and systematic, as confirmed by the report of the Australian National Inquiry into the Teaching of Literacy, entitled *Teaching reading* (Department of Education & Science and Training, 2005).

The second predicate, Positive Teaching (Merrett & Wheldall, 1990; Wheldall & Limbrick, 2005; Wheldall & Merrett, 1984, 1989; Wheldall, 1991) refers to the necessary (but not sufficient) condition for all students to learn: that they be paying attention and be actively engaged with the task and with the teacher. The student who continually exhibits troublesome behaviour is off-task and inhibits their own learning, the learning of others and the teacher's ability to teach effectively. The distracted student, who is continually (if more passively) off-task, is also inhibited from learning. MULTILIT embraces an approach known as Positive Teaching that employs applied behavioural reinforcement theory to 'catch students being good'. Highly contingent, positive reinforcement is deployed in the form of verbal praise (and very limited use of reprimands) and other reward strategies (as appropriate) to recognise and encourage improvements in both appropriate behaviour and academic outcomes, so as to help students to remain engaged with the instructional tasks and the teacher's requests and instructions. Similarly, Positive Teaching has a concern with classroom ecology, to design learning environments that facilitate appropriate student behaviour and improved learning. A concern with the most effective classroom seating arrangements is a good example of this.

The need for integrated literacy instruction is the third defining predicate of MULTILIT. Contrary to some misperceptions, it is not a wholly phonics-based approach to instruction. MULTILIT embraces all of the recommendations of the National Inquiry into the Teaching of Literacy (2005) regarding the need for intensive, systematic and explicit instruction in phonemic awareness, phonics, fluency, vocabulary and comprehension. The so-called 'simple view of reading' (Gough & Tunmer, 1986; Hoover & Gough, 1990; Tan, Wheldall, Madelaine, & Lee, 2007) makes clear that reading comprehension, being able to understand written language, is dependent upon both accurate decoding of the written word and understanding of spoken language; neither will lead to reading comprehension in the absence of the other.

There is little point being able to read, even fluently, if there is no understanding of what has been read. Instruction in phonemic awareness and phonics leads to effective decoding of text, reading fluency instruction promotes automaticity that aids comprehension, along with appropriate specific instruction in vocabulary and listening comprehension skills. In other words, all five 'pillars' of literacy instruction must be addressed if the reading (and spelling) instruction is to be effective. To this end, MULTILIT recommends the inclusion of other evidence-based instructional materials as well as those developed by the MULTILIT Research and Development Team. The use of SRA programs such as Spelling Mastery (Dixon, Engelmann, & Bauer, 1999) and Reasoning and Writing (Engelmann & Davis, 2001) are strongly recommended to supplement MULTILIT materials.

MULTILIT is, then, a data-driven approach to education, employing programs of proven efficacy and continually collecting data on the effectiveness of the programs deployed. As such, the basic model lends itself to all instruction more generally and not just to literacy instruction. The prerequisites for effective instruction elucidated and exemplified within MULTILIT form the foundation for all teaching of basic skills and subsequent more advanced teaching. Effective numeracy instruction logically follows a similar path, for example. By embracing an evidence-based approach, the most effective instructional methods may be identified and implemented so that low-progress students can be helped to make up lost time in all areas of the curriculum, not just literacy.

MULTILIT has been well-received by the education sector in Australia and was endorsed by the House of Representatives Standing Committee on Education and Training's Inquiry into the Education of Boys, in its report, entitled *Boys: Getting it right* (House of Representatives Standing Committee on Education and Training, 2002)

> 5.62 The knowledge and practical instructional techniques developed in MULTILIT by the researchers at Macquarie University should inform and enhance initial and remedial literacy instruction throughout Australia and form the core of remedial reading programs in primary and high schools. (p. 114)

By engaging with parents, teachers, schools and government, our aim is to bring about major educational change so that far fewer children are left to struggle in school with inadequate reading and related skills, the skills essential for academic success in all areas. As well as training parents and teachers to work with low-progress readers, we are also working on community projects in partnership with community organisations and government, to bring about fundamental change in the educational prospects of low-progress readers, particularly those from socially disadvantaged and Indigenous backgrounds.

The Schoolwise Program

Since 1996, we have been working with the Exodus Foundation, a charity dedicated to the welfare of homeless people (particularly 'streetkids'), to provide a tutorial centre for older low-progress readers who were potentially at risk for serious disaffection from school. The idea behind the program is to help students who are typically three or more years behind in reading and related skills to catch up with their peers, or at least to achieve functional literacy, before they reach high school. With increased literacy skills, these students are far more likely to be able to cope in high school and less likely to drop out from school.

Students attend the tutorial centre, mornings only, for two school terms, returning to their 'home' schools at lunchtime. During the morning sessions they receive a program of MULTILIT instruction. Originally, this had a heavy emphasis on one-to-one instruction, but this has been modified into largely small-group instruction over recent years.

At the commencement of each program, students are typically given a battery of standardised tests of reading and related skills ('the MULTILIT Battery'), administered by trained research assistants. The battery consists of measures of reading accuracy and comprehension (Neale Analysis of Reading; Neale, 1999), single-word recognition (Burt Word Reading Test; Gilmore, Croft, & Reid, 1981), spelling (South Australian Spelling Test; Westwood, 1999), oral reading fluency (Wheldall Assessment of Reading Passages; Wheldall, 1996) and more recently, non-word reading (Martin and Pratt Nonword Reading Test; Martin & Pratt, 2001). Students are tested again on the entire battery at the end of the typical two-term (20 weeks) program.

The success of the Schoolwise Program in teaching successive intakes of older low-progress readers to learn reading and related skills has been amply documented in the research report commissioned by the Australian Government Department of Education, Training and Youth Affairs (DETYA), *An evaluation of MULTILIT* (Wheldall & Beaman, 2000), which detailed the efficacy of MULTILIT over the years 1996–1998. In the 10 years following, successive intakes have typically continued to make very large average gains in reading accuracy, reading comprehension, single-word recognition, reading fluency and spelling, in less than 5 months of instruction. Most recently, in 2008, the 67 students who participated in the program (about 18 weeks of instruction between testings) made mean gains of 20 months in reading accuracy; 17 months in reading comprehension; 17 months in single-word reading; 22 months in spelling; 29 months in non-word reading (measuring phonological recoding); and, could now read 47% more words correctly per minute than when they commenced the program. These were students who had made little or no progress in recent years and who were typically three or more years behind in reading and related skills when they entered the program (Wheldall, 2009).

The progress of Aboriginal students in the Schoolwise Program

Research has shown that a large gap is evident between the reading and related skills performance of Aboriginal students, compared with that of their non-Indigenous peers, and that this gap increases over the primary years of schooling. Evidence from the *National report on schooling in Australia 2004* (MCEETYA, 2006) revealed that 31% of Indigenous students did not meet the minimum national benchmark for reading in Grade 5. This disturbing result is in the context of the report that 11% of the Grade 5 population overall failed to meet the minimum standard. In the 2004 New South Wales (NSW) Department of Education and Training (DET) *Report of the review of Aboriginal education* (NSW Aboriginal Education Consultative Group & DET, 2004), it was found that the average gap in the results of the Grade 3 Basic Skills Test between the reading ability of Aboriginal and non-Aboriginal students was 19 months, with the gap increasing to over 30 months by Grade 7. Clearly, students from Indigenous backgrounds are particularly at risk of poor literacy outcomes in Australia.

It follows logically, from the non-categorical perspective to effective literacy instruction that underpins and informs our overall approach, that there is no reason to suppose that Aboriginal students are in need of any different form of instruction from that offered to any other low-progress readers. The opportunity arose to test our assumption within the context of the second intake into the Schoolwise Program for 2004, since 14 of the 34 students who completed the program had families who were from Indigenous backgrounds. The results for this intake showed that the gains made by these Aboriginal students were just as high as those made by non-Aboriginal students, demonstrating that the MULTILIT Program delivered in Schoolwise is just as effective for Aboriginal students. A full account of this study is provided by Wheldall, Beaman, and Langstaff (2010).

The results of this study were highly influential in encouraging the development of a second tutorial centre, again funded by the Exodus Foundation in Redfern, Sydney, for Aboriginal students. These results also encouraged us to work with students in more remote Aboriginal communities.

The MULTILIT project on Cape York

In October 2004, we visited Coen on the Cape York Peninsula, at the invitation of Aboriginal leader Noel Pearson, Director of the Cape York Institute for Policy and Leadership based at Griffith University. The aim of the visit was to determine whether there was scope for a MULTILIT implementation in Cape York. Discussions with representatives of Cape York Partnerships and the local community members revealed considerable enthusiasm for such an implementation. The aim of this small-scale project was to demonstrate that students from Indigenous backgrounds can make large gains in reading and related skills when offered non-categorical, intensive, systematic and skills-based instruction daily for 3 hours over two school terms (as per the model in Schoolwise).

At the start of the pilot project in June 2005, all available primary-aged students in the school in Coen were assessed by trained research assistants, using the MULTILIT battery of tests to measure aspects of reading and spelling performance (described earlier). The data available for the 24 older students in Grades 4, 5, 6 and 7 showed that this group, whose average age was 10 years and 10 months, were typically over 3 years (40 months) behind their average age peers for reading accuracy, and nearly 4 years (46 months) behind for reading comprehension.

A MULTILIT Tutorial Centre was established on site in the school in mid-2005, with instruction being provided by MULTILIT instructors who had relocated from Sydney. The 10 low-progress readers from Grades 6 and 7, comprising the first intake into the MULTILIT Program in Coen, were, on average, aged 11 years and 6 months, and were about 4 years behind in both reading accuracy and comprehension at the commencement of the Program. Attendance in the program averaged 75%. After 17 weeks of instruction, these 10 students made average gains of: 13.6 months in reading accuracy; 4.3 months in reading comprehension; 12.0 months in single-word recognition; 24.3 months in phonological recoding; 15.9 months in spelling and could read 43% more words correctly per minute.

The second intake of eight low-progress readers, this time from Grades 5 and 6, had an average age of 10 years 6 months, and they were over 3 years behind their grade peers in terms of reading accuracy, and 4 years behind in reading comprehension, at program commencement. Attendance in the program for the second intake averaged 67%. After 18 weeks of instruction, the eight students made average gains of: 15.0 months in reading accuracy; 7.5 months in reading comprehension; 14.6 months for single-word recognition; 25.4 months in phonic decoding; 11.4 months in spelling and could now read 50% more words correctly per minute.

Various students joined and left the school during the course of the year, and all of the Grade 7 students left for high school at the end of 2005. Of the original 24 primary-aged students, 15 were present for the whole year and were assessed in mid-2005 and re-assessed in mid-2006. These 15 students made average gains over the year of: 21.4 months in reading accuracy; 10.7 months in reading comprehension; 19.0 months for single-word recognition; 25.9 months in phonic decoding; 22.8 months in spelling and could read 75% more words correctly per minute. Whereas this group of students was typically three and a quarter years behind their average age peers in reading accuracy in mid-2005, they were only two and a half years behind by mid-2006. Notwithstanding the fact that they still had a long way to go, these students had closed the gap substantially.

In 2007, further Australian Government support was secured to continue the MULTILIT trial activity in Coen. In this second pilot project, a main goal was to ensure the sustainability of effective instructional delivery, using MULTILIT approaches by embedding the approach within the classrooms as well as continuing to deliver the program to some students in the MULTILIT Tutorial Centre.

The participating teaching staff responded positively to the intervention, which included the embedding of a daily schoolwide peer-tutoring intervention (in Grades 3–7) of the MULTILIT Reinforced Reading program. In addition, the introduction of group teaching lessons in strategies such as MULTILIT Word Attack Skills, MULTILIT Sight Words and MULTILIT Reinforced Reading, as well as other direct instruction programs (for example Spelling Mastery) were implemented. The MULTILIT instructor provided remediation in the core skills of MULTILIT Word Attack Skills and Sight Words to selected students, with other program components being delivered by classroom teachers, assistants and aides. Over the course of the school year, all MULTILIT instruction was transferred to the school staff to deliver.

At the commencement of the year, all available students were assessed on the MULTILIT battery of tests of reading and related skills, as described. A few children who subsequently attended were not available for testing and a few others joined the school throughout the semester. Initial testing took place at the beginning of Term 1. At the end of Term 2 (Semester 1), all children in these Grade groups were re-assessed on the same test battery to determine progress made. Once again, not all children were available for testing. Some had left and gone to other schools, and some were not available during the testing period. This *mid-year testing* took place at the end of Term 2. Finally, towards the end of the school year, all students who were still enrolled at the school and were present during the testing period were assessed once more on the test battery. End-of-year testing took place in the final weeks of Term 4.

At the time of initial testing, 19 students in Grades 4–6 were available for assessment, but only 14 of these were subsequently available for mid-year testing (plus a further five students who had not been present at the beginning of the year). Similarly, only 16 students of those present at the mid-year testing were available for end-of-year testing. In total, only 13 students were available for testing on all three occasions. In order to maximise student numbers for the purposes of analysis, it was decided to analyse the two semesters separately and to treat the resulting statistics as best estimates of the performance of the group as a whole over the year, as we shall see.

At initial testing, the 14 students in the first semester group were aged, on average, 122 months (10 years 2 months), comprising six students from Grade 4, three students from Grade 5, one student from Grade 6 and four students from Grade 7. They were typically about 32 months behind what might be expected for their age for reading accuracy, as measured by the Neale, and 37 months behind in reading comprehension. Similarly, they were typically 21 months behind on the Burt Word Reading Test, 19 months behind on the South Australian Spelling Test and 24 months behind on the Martin and Pratt test of non-word reading (phonological recoding). In terms of general receptive verbal ability, as estimated by the Peabody Picture Vocabulary Test (PPVT; Dunn & Dunn, 1997), they were typically 34 months behind expectations for age and, in terms of reading fluency, could read 67 words correctly per minute on the Wheldall Assessment of Reading Passages (WARP).

After two terms of schooling (5 months, or about 17 weeks of MULTILIT-based reading instruction between testing sessions), students, on average, made the following gains: 9 months in reading accuracy; 5 months in reading comprehension; 10 months in single-word recognition; 20 months in non-word reading; 13 months in spelling and could read 39% more words correctly per minute. Thus, we can see that this group of students made good progress on all measures of decoding and spelling. The gain in phonological recoding is particularly impressive, these students having closed the gap between their age and reading age considerably, now being only 9 months behind, on average. These students gained, on average, 5 months in reading comprehension over 5 months, which is, of course, average progress but it should be emphasised that prior to this they had been making far less rapid progress than average, as witnessed by the fact that they were typically 37 months behind at the start of the year. This is almost certainly due to their poor verbal ability, which was considerably behind age expectations (as measured by the PPVT).

We turn now to the results for the second semester (Terms 3 and 4), but it should be emphasised again that the 16 students for whom we have complete data at mid-year and end-of-year testing do not comprise the same students as in Semester 1; only 13 students were present for all three assessment sessions and hence only 13 students are in common across the two groups. The second semester group was the same age at mid-year testing as the first semester group at this point, comprising seven students from Grade 4, three students from Grade 5, two students from Grade 6 and four students from Grade 7. Their average performance on the test battery was also very similar to that of the first semester group at mid-year testing (typically being only 1 month lower on average).

After two terms of schooling (5 months or about 15 weeks of MULTILIT-based reading instruction between mid-year and end-of-year testing sessions), semester two students, on average, made the following gains: 6 months in reading accuracy; 3 months in reading comprehension; 10 months in single-word recognition; 13 months in non-word reading; 9 months in spelling and could read 23% more words correctly per minute on the WARP. Once again, there were significant gains evident on all measures of decoding, although they were not as high as in the first semester.

If we combine the results for the two semesters (the groups being comparable as we have demonstrated), then the following gains over the year (10 months, or about 32 weeks of MULTILIT instruction) are apparent:

Reading accuracy	15 months
Reading comprehension	8 months
Single-word recognition	20 months
Spelling	22 months
Non-word reading	33 months
Reading fluency	69% increase

Before discussing these gains, it bears repeating that these students had been consistently falling further and further behind expectations for their age, prior to the introduction of MULTILIT, as witnessed by the magnitude of the gap between age and performance on all measures at the start of the year. Thus, the present gains must be appreciated in comparison with their previous rates of progress, which had typically been substantially well below average.

The most remarkable gain made by the group is the gain of 33 months in phonological recoding, or non-word reading. This test measures level of skill in decoding of phonically regular words. At initial testing these students were typically 2 years behind the average for their age peers. By the time of the end-of-year testing 10 months later, they were now only 2 months behind. In other words, they could now decode very nearly as well as the average for all Australian children. In terms of reading fluency, they could now read nearly 70% more words correctly per minute than they could at the beginning of the year. The gains for reading accuracy (15 months), single-word recognition (20 months) and spelling (22 months) over 10 months also bear witness to a further 'closing of the gap' between the decoding performance of this group of Aboriginal students from one school and what is typical for their non-Aboriginal peers. This is a remarkable achievement by the staff of the school, assisted by the MULTILIT team. The following table (Table 12.1) provides the mean gains between initial and final testings for the 13 students who were present all year.

Table 12.1 Mean gains on the relevant literacy variables (raw scores) for the 13 Coen students who were present all year in 2007

Variable	*N*	Mean	SD	*T*	*p* <
Neale accuracy	13	14.85	7.51	7.12	0.001
Neale comprehension	13	7.92	7.48	3.82	0.002
Burt	13	19.46	12.24	5.73	0.001
South Australian spelling	13	21.46	11.96	6.47	0.001
WARP (wcpm)	13	44.00	13.75	11.53	0.001
Martin and Pratt	13	34.31	23.58	5.24	0.001

Table 12.2 'Lag' between chronological age and literacy performance (months) for the Coen students, mid-2005 to end 2007

Variable	Lag mid-2005	Lag end 2007
(*N*)	(24)	(16)
(Age)	(130 months)	(132 months)
Neale accuracy	39 months	28 months
Neale comprehension	47 months	40 months
Burt	26 months	12 months
South Australian spelling	30 months	9 months
Martin and Pratt	28 months	2 months

Note: In terms of reading fluency as measured by the WARP, in mid-2005 the students could typically read 68 words correctly per minute, compared with 108 words at the end of 2007

In order to provide a broader context for these results, we may compare the 'lag' in performance with age now and when we first began work with the school in Coen at the beginning of Term 3, 2005 (based on the results for the 24 Grades 4–7 students present for testing at that point). Table 12.2 provides a summary of the relative lag between then and now.

Thus, it can be seen that a considerable improvement has been achieved in 'closing the gap' between typical performance for these remote area students and what might be expected for their age. This is particularly evident for phonological recoding, single-word recognition, spelling and reading fluency. This improvement in recoding skills is also flowing through into accuracy of text reading, as measured by reading accuracy on the Neale test. Where we still have a long way to travel, however, is in translating this increased decoding skill into improved reading comprehension, at least as measured by the Neale test of reading comprehension. We shall return to this theme in our concluding section—Essential next questions.

What we learned from the Coen trials

Valuable lessons were learned from each of the two Coen trials. First, while the establishment of the MULTILIT Tutorial Centre in the pilot project (the first trial in 2005–2006) delivered impressive results, there was little significant interaction between the staff in the school and the MULTILIT instructors. As a result, there was no opportunity for methods being used in the MULTILIT Tutorial Centre to be incorporated into classroom practice within the school. The students participating in the MULTILIT Program would typically come out of their regular classrooms for 3 hours each day and into the MULTILIT environment, often experiencing very different conditions for learning in each environment. This lack of consistency around such things as classroom behaviour management was problematic.

In addition, once students had participated in a two-term program in the pilot (first trial) program, while their literacy levels had improved substantially, they had generally not yet been consolidated. The regular classroom environment into which MULTILIT students returned was, at this stage, less able to support the growth or generalisation of these emerging literacy skills (even in a less intensive mode).

Another difficulty experienced in the first trial was the supply of appropriately trained MULTILIT staff to be deployed to Coen from our MULTILIT team in Sydney. The sustainability of the project was a concern, given the model adopted in the first trial in 2005–2006. For this reason, as well as for those articulated above, it was decided to attempt a training and development approach, embedding MULTILIT methods within the classrooms in the school as the second trial, with Education Queensland staff being taught how to deliver the instruction typically provided in a MULTILIT Tutorial Centre. This formed the basis of the 2007 phase two project (second trial) in Coen. The aim was to embed MULTILIT instructional practices within the regular classrooms across Grades 3–7, utilising teachers as well as classroom assistants and aides to provide small ability-based groups for tailored literacy instruction.

There were different but considerable challenges with this model, notwithstanding the fact that the students in the school were showing evidence of continued literacy growth. The issue of high staff turnover impacted significantly on the MULTILIT training and development initiative in the school in Coen. Another challenge we faced in this second trial was the absence of an intensive model of best MULTILIT practice that was evident in the first trial (in the MULTILIT Tutorial Centre). This meant that school staff did not have an opportunity to see first hand what exemplary MULTILIT practice looked like and how students respond to it. This has been particularly evident in terms of developing a Positive Teaching 'culture' in the school. A MULTILIT Tutorial Centre on site provides a model of what is required to achieve effective instruction and consequently increased student gains in reading achievement.

We have recently commenced work on a 'roll out' of MULTILIT across another three sites on Cape York Peninsula (Aurukun, Hope Vale and Mossman), plus Coen, as part of Noel Pearson's Cape York Welfare Reform Initiative, funded by the Australian Government. We are aiming to implement a hybrid model, drawing on the best from our experience to date with the two Coen trials: that is we are seeking to establish a MULTILIT Tutorial Centre on each of the four sites and to establish MULTILIT practices in the classrooms of the school over time, assisted by teachers from the schools who have been intensely trained and who will have had an opportunity to hone their skills working alongside a MULTILIT instructor for 6 months in the MULTILIT Tutorial Centre. This latest project commenced at the beginning of 2008, and will be completed over 4 years. It will provide the opportunity to test the effect of implementing MULTILIT more systematically across four different sites/schools serving Indigenous students.

Essential next questions

In this chapter we have focused on our research to date that has sought to meet the needs of socially disadvantaged and, particularly, Indigenous low-progress readers. We have shown how these students can be helped to make major gains in their reading and related skills as a result of scientifically based MULTILIT instruction. But there is still a way to go . . . . There are several key questions still to be answered, which we shall now address.

How can we improve reading comprehension? In the Coen studies we showed that we can get Aboriginal students almost up to grade level in phonological recoding, for example, but, while students do make reasonable gains in reading comprehension, they are not nearly as great as the gains in decoding. If we subscribe to the simple model of reading described earlier, this tells us that the remaining main problem is, then, one of listening comprehension rather than reading per se. Students will only ever be able to read and understand to the level that they can listen and understand. Consequently, a major focus for future research and development has to be in the area of enhancing receptive language development.

How can we facilitate receptive language development? We can certainly attempt to improve the language comprehension skills of older low-progress readers, and we do, indeed, include such instruction. But it makes far more sense to intervene early in promoting receptive language development, than it does to intervene early in teaching basic decoding skills before students become low-progress readers. If Aboriginal students had a greater level of skill in understanding English language, then they would at least be able to understand what they read once they had been taught to decode. To this end, we shall be researching a preschool program in Cape York for young children in the year prior to starting formal schooling, focusing on explicit phonemic awareness instruction, structured book reading to facilitate vocabulary and listening comprehension, and the 'incidental teaching' of expressive language (Dolly & Wheldall, 1987, 1988; Dolly, Glynn, & Wheldall, 1989).

How can we reduce the incidence of reading difficulties in Aboriginal communities? Clearly, providing remedial MULTILIT instruction for Aboriginal low-progress readers in schools where they are so prevalent is never going to address the question seriously. It is a temporary 'band-aid' solution for those students immediately in need, but it does not come close to solving the problem of low-progress readers being created more quickly than we can remediate or recover them. In our view, the Response to Intervention (RTI) model is the best way forward. A simple RTI model (there are several versions) is predicated on three tiers of intervention. The first tier of the RTI model consists of exemplary, scientific research-based, initial reading instruction provided from day 1 in Kindergarten, by teachers carefully trained and educated in the skills necessary for the initial teaching of reading. This program of initial reading instruction would have a strong phonics core, building on pre-existing phonemic awareness skills (either learned at home or specifically taught in preschool) and would be taught classwide to all students, intensively and systematically. A number of excellent programs of proven efficacy already exist that could be employed to achieve this, such as Jolly Phonics (Stuart, 1999). We can no longer afford to leave initial reading instruction to the vagaries of the whole-language approach if we are to ensure that all of our children learn to read.

After two to four terms of initial instruction, those making less than adequate progress would receive Tier 2 of the RTI model. Those students in, say, the bottom 25% of the population, making the least progress, would then receive small-group instruction, again based on what science has shown to be most effective in teaching reading but delivered even more intensively in small groups of four or five children. We have developed a version of the MULTILIT program for young children, known as MINILIT—Meeting Initial Needs In Literacy (Reynolds, Wheldall, & Madelaine, 2007a; Reynolds, Wheldall, & Madelaine, 2007b). The MINILIT program aims to provide a more cost effective (and effective) alternative to Reading Recovery, about which we have serious reservations (Reynolds & Wheldall, 2007; Reynolds, Wheldall, & Madelaine, 2009). Small-group instruction of this nature will typically bring most low-progress readers back 'up to speed' and making regular progress.

Those who are still not making adequate progress after Tier 2 intervention need Tier 3 of RTI. Tier 3 comprises individualised, one-to-one instruction by an expert reading teacher that is geared to their specific and idiosyncratic needs (again, based on scientific, evidence-based approaches to instruction). The MULTILIT Reading Tutor program for individual students, taught by a highly skilled reading teacher or special educator, would be appropriate for this. Relatively few students are likely to be left needing Tier 3 if Tiers 1 and 2 have been delivered properly. (We are probably talking about less than 5% of the school population.) These are the students whom we might choose to refer to as having a learning disability, or as being dyslexic. This three-tier RTI model is now becoming accepted as the most satisfactory way of 'diagnosing', and intervening with, learning disability, at least in the United States (Council for Exceptional Children, 2008).

In conclusion, we have reason to be optimistic about the likelihood of improving the life chances of socially disadvantaged and Indigenous students by ensuring that all of our children learn to read quickly, easily and well in their first few years of schooling. We have the instructional technology available to us to bring this about. Moreover, in Australia, the United States and in the United Kingdom, there is a growing political will to ensure that this happens.

References

Bishop, D. V. M. (2001). Genetic influences on language impairment and literacy problems in children: Same or different? *Journal of Child Psychology and Psychiatry*, *42*, 189–198.

Council for Exceptional Children. (2008). Responsiveness to intervention: A collection of articles from Teaching Exceptional Children. Washington, DC: Council for Exceptional Children.

de Lemos, M. (2002). *Closing the gap between research and practice: Foundations for the acquisition of literacy*. Camberwell, VIC: Australian Council for Educational Research.

Department of Education, Science and Training. (2005). *Teaching reading*. Canberra: Department of Education, Science and Training.

Dixon, R., Engelmann, S., & Bauer, M. (1999). *Spelling mastery Level B*. Chicago, IL: SRA McGraw-Hill.

Dolly, D., Glynn, T., & Wheldall, K. (1989). Increasing preschoolers' language use through incidental teaching. *New Zealand Journal of Educational Studies*, *24*, 81–92.

Dolly, D., & Wheldall, K. (1987). Talking to teacher – using incidental teaching to encourage child initiations in the nursery classroom: A case study. *Child Language Teaching and Therapy*, *3*, 277–292.

Dolly, D., & Wheldall, K. (1988). Developing functional language with young children from English-speaking and Punjabi-speaking home backgrounds: Incidental teaching and contingent access to materials. *Educational Psychology*, *8*, 101–116.

Dunn, L. M., & Dunn, L. M. (1997). *PPVT-III: Peabody picture vocabulary test*. Circle Pines, MN: American Guidance Service.

Engelmann, S., & Davis, K. L. S. (2001). *Reasoning and writing level A: A direct instruction program*. Columbus, OH: SRA.

Gilmore, A., Croft, C., & Reid, N. (1981). *Burt word reading test: New Zealand revision*. Wellington, NZ: New Zealand Council for Educational Research.

Gough, P. B., & Tunmer, W. E. (1986). Decoding, reading, and reading disability. *Remedial & Special Education*, *7*, 6–10.

Hart, B., & Risley, T. (1995). *Meaningful differences in the everyday experience of young American children*. Baltimore: Brookes Publications.

Hoover, W. A., & Gough, P. B. (1990). The simple view of reading. *Reading & Writing*, *2*, 127–160.

House of Representatives Standing Committee on Education and Training. (2002). *Boys: Getting it Right*. Report of the inquiry into the education of boys. Canberra: Commonwealth Government.

MCEETYA. (2006). National report on schooling in Australia 2004, Preliminary paper, National Benchmark results for Reading, Writing and Numeracy, Years 3, 5 and 7. Ministerial Council on Education, Employment, Training and Youth Affairs.

MULTILIT. (2007). *The MULTILIT reading tutor program (revised)*. Sydney, NSW: MULTILIT Pty Ltd.

Martin, F., & Pratt, C. (2001). *Martin and Pratt nonword reading test*. Camberwell, VIC: Australian Council for Educational Research.

Merrett, F., & Wheldall, K. (1990). *Positive teaching in the primary school*. London: Paul Chapman.

National Curriculum Board. (2008). *National English curriculum: Initial advice*. Canberra, ACT: National Curriculum Board.

National Institute of Child Health and Human Development, NIH, DHHS. (2000). *Report of the National Reading Panel: Teaching children to read* (00-4769). Washington, DC: US Government Printing Office.

Neale, M. D. (1999). *Neale analysis of reading ability*, 3rd ed. Hawthorn: Australian Council for Educational Research.

New South Wales Aboriginal Education Consultative Group Incorporated and New South Wales Department of Education and Training. (2004). *Report of the review of Aboriginal education: Yanigurra muya: Ganggurrinyma yaarri guurulaw yirringingurray: Freeing the spirit: Dreaming and equal future*. Darlinghurst: NSW Department of Education and Training.

Pogorzelski, S., & Wheldall, K. (2005). The importance of phonological processing skills for older low-progress readers. *Educational Psychology in Practice, 21*, 1–22.

Reynolds, M., & Wheldall, K. (2007). Reading recovery twenty years down the track: Looking forward, looking back. *International Journal of Disability, Development and Education, 54*, 199–223.

Reynolds, M., Wheldall, K., & Madelaine, A. (2007a). Developing a ramp to reading for at-risk year one students: A preliminary pilot study. *Special Education Perspectives, 16*, 39–69.

Reynolds, M., Wheldall, K., & Madelaine, A. (2007b, September 29). *Meeting Initial Needs in Literacy (MINILIT): Why we need it, how it works, and the results of pilot studies*. Paper presented to the annual conference of the Australian Association of Special Education, Sydney. *Australasian Journal of Special Education, 31*, 147–158.

Reynolds, M., Wheldall, K., & Madelaine, A. (2009). The devil is in the detail regarding the efficacy of reading recovery: A rejoinder to Schwartz, Hobsbaum, Briggs, and Scull. *International Journal of Disability, Development and Education, 55*, 17–35.

Rose, J. (2006). *Independent review of the teaching of early reading: Final report*. London: Department for Education and Skills.

Stanovich, K. E. (1986). Matthew effects in reading: Some consequences of individual differences in the acquisition of literacy. *Reading Research Quarterly, 21*, 360–391.

Stanovich, K. E. (2000). *Progress in understanding reading. Scientific foundations and new frontiers*. New York: The Guilford Press.

Stuart, M. (1999). Getting ready for reading: Early phoneme awareness and phonics teaching improves reading and spelling in inner-city second language learners. *British Journal of Educational Psychology, 69*, 587–605.

Tan, K., Wheldall, K., Madelaine, A., & Lee, L. W. (2007). A review of the simple view of reading: Decoding and linguistic comprehension skills of low progress readers. *Australian Journal of Learning Disabilities, 12*, 19–30.

Westwood, P. (1999). *Spelling: Approaches to teaching and assessment*. Camberwell: Australian Council for Educational Research Press.

Wheldall, K. (1991). Managing troublesome classroom behaviour: A positive teaching perspective. *International Journal of Disability, Development and Education, 38*, 99–116.

Wheldall, K. (1994). Why do contemporary special educators favour a non-categorical approach to teaching? *Special Education Perspectives, 3*, 45–47.

Wheldall, K. (1996). *The Wheldall assessment of reading passages: Experimental edition*. Sydney, NSW: Macquarie University Special Education Centre.

Wheldall, K. (2009). Effective instruction for socially disadvantaged low-progress readers: The Schoolwise program. *Australian Journal of Learning Difficulties, 14*, 151–170.

Wheldall, K., & Beaman, R. (2000). *An evaluation of MULTILIT: Making up lost time in literacy*. Canberra, ACT: Department of Education, Training and Youth Affairs.

Wheldall, K., Beaman, R., & Langstaff, E. (2010). 'Mind the gap': Effective literacy instruction for indigenous low-progress readers. *Australasian Journal of Special Education, 34*, 1–16.

Wheldall, K., & Carter, M. (1996). Reconstructing behaviour analysis in education: A revised behavioural interactionist perspective for special education. *Educational Psychology, 16*(2), 121–141.

Wheldall, K., & Limbrick, L. (2005). I can hardly believe it has turned out like this: A lighthouse school for boys education. *Education Today, 55*, 4–13.

Wheldall, K., & Merrett, F. (1984). *Positive teaching: The behavioural approach*. London: Allen & Unwin.

Wheldall, K., & Merrett, F. (1989). *Positive teaching in the secondary school*. London: Paul Chapman.

Chapter 13
Actualising Potential in the Classroom: Moving from Practising To Be Numerate Towards Engaging in the Literate Practice of Mathematics

Raymond Brown

Introduction

International and national curriculum documents, such as *The Salamanca Statement and Framework for Action on Special Needs Education* (United Nations Educational, Scientific and Cultural Organisation (UNESCO), 1994) and the *Teaching and Learning Mathematics: The Report of the Expert Panel on Mathematics in Grades 4–6 in Ontario* (Ontario Ministry of Education, 2004), acknowledge that access to mathematical knowledge is the right of all students. Documents such as the *Principles and Standards for School Mathematics* (National Council of Teachers of Mathematics, 2000) and *A National Statement on Mathematics for Australian Schools* (Australian Education Council, 1991), resonate this acknowledgement by highlighting that mathematical knowledge develops within classrooms that assist all students to make informed, numerate decisions in a balance of situations ranging from the everyday to the purely mathematical.

It is, therefore, not surprising that the term 'numeracy' has been defined by curriculum authorities as being 'the demonstration of practices and dispositions that accurately, efficiently and appropriately meet the demands of typical everyday situations that involve number, patterns and algebra, measurement, chance and data, and space' (Queensland Studies Authority, 2004, p. 5). According to this definition, being numerate means one is able to demonstrate practices and dispositions that efficiently and appropriately meet the demands of home, work and civic life (Australian Association of Mathematics Teachers, 1997). In other words, to be numerate means being able to display the characteristics of a mathematically literate person (Ontario Ministry of Education, 2004).

R. Brown (✉)
Griffith University, Brisbane, QLD, Australia
e-mail: ray.brown@griffith.edu.au

C. Wyatt-Smith et al. (eds.), *Multiple Perspectives on Difficulties in Learning Literacy and Numeracy*, DOI 10.1007/978-1-4020-8864-3_13,

Accessing the dispositions and practices of a mathematically literate person is a goal that all students have a right to pursue (Boaler, 2007; Cobb & Hodge, 2002; van Kraayenoord & Elkins, 2004). The universality of this right is reflected in international measures of mathematical knowledge, such as the Third International Mathematics and Science Study (see Mullis, Martin, Gonzalez, & Chrostowski, 2004), and in local documents and measures, such as those associated with the Numeracy Testing program (Queensland Studies Authority, 2007) conducted by the Queensland Studies Authority in Australia. However, within many classrooms there is reluctance to include students with learning difficulties in a balance of learning experiences that may promote the development of mathematical literacy (Woodward & Montague, 2002). This reluctance is not born out of a desire to deny students with learning difficulties access to a range of challenging tasks that span the different strands of a general mathematics curriculum, but out of a concern for all students to master basic mathematics facts and algorithms (Woodward & Montague, 2002). What is needed to address this reluctance is an approach to teaching and learning that recognises the social dimension of the mathematics classroom (Wright & Chapparo, 2008).

This chapter explores the social dimension of classroom practice and its importance in supporting the development of mathematical literacy by all students, including students considered by school authorities as displaying learning difficulties. First, a sociocultural (Vygotsky, 1987) framework for investigating and understanding the development of mathematical literacy in the everyday classroom is presented. Second, perspectives on the notion of 'learning difficulty' are identified that are in accord with a sociocultural framing of teaching and learning. Third, empirical data in the form of classroom learning episodes (see Brown, 2001) are analysed in accordance with strategies identified in the literature as being conducive to the development of mathematical literacy by students with learning difficulties. The chapter concludes by recognising the role that sociocultural approaches to teaching and learning can play in assisting students with learning difficulties to engage in the literate practice of mathematics.

Transforming numeracy into literacy through engaging in mathematical practice

For Vygotsky (1987), the dynamic, practical relationship between speaking and thinking embodies the learning process (Renshaw, 1996). Mediational means, such as language and other symbolic systems including mathematics, are transformed into internal tools of thinking and problem solving during the process of effective teaching. In summary, learner development is driven forward by effective teaching (Vygotsky, 1987). As such, a sociocultural approach to learning and teaching offers a theoretically distinct framework for interpreting and understanding the teaching–learning process in the everyday classroom. In comparison to theoretical approaches based on constructivist notions of development (cf. Piaget, 1952), sociocultural approaches view natural development as working in tandem with cultural development to form what Wertsch (1985, p. 42) refers to as a 'qualitatively unique explanatory framework' to account for learning and development. From this viewpoint, issues affecting the sociocultural organisation of mental functioning on the social plane (such as issues related to inclusion and exclusion, identity and participation, diversity and homogeneity) are seen as essential aspects of functioning on the personal plane.

Integral to a Vygotskian approach to learning and teaching is the notion of 'zone of proximal development'. The zone of proximal development (ZPD) can be interpreted from different but interrelated viewpoints. The ZPD is usually characterised as the distance between a student's actual ability, exhibited when working alone, and the student's potential ability, shown when working with a more-experienced other. However, authors such as Valsiner (1997) and Gutiérrez (2008) provide a more dynamic characterisation of the ZPD that emphasises the collective nature of learning where issues on the social plane, such as independence–dependence, leading–following, autonomy–compliance, are negotiated by participants, and where learning outcomes, both social and personal, are not exclusively predetermined by an institution or expert. For Valsiner (1997), the ZPD is an institutional space comprised of potentialities that are actualised through individuals negotiating learning relationships with others within and through the learning environment. For Gutiérrez (2008), the ZPD is an agentive space in which individuals and the social act together to develop sets of relations that promote learning. From these more dynamic characterisations of the ZPD, learning can be viewed as being more than simply a display of results, from pre-test to post-test. From these perspectives, the notion of learning expands to encompass the degree of engagement and influence that individuals and groups achieve within an activity setting. As such, patterns of engagement and influence that are enacted in learning environments are not seen as being incidental to learning, but as being crucial to it and as revealing the social, cultural and historical dimensions of learning. Hence, participation in a mathematics activity may be seen as being an extension of past participation in mathematics curricula—an extension that is directed towards accomplishing personal, social and cultural goals that have not yet been accomplished (Rogoff, 1995). Participation in mathematics, therefore, is always about knowing and doing something, and always incorporates values and dispositions. In other words, participation in mathematics is a literate practice (Gee, 1990); that is a practice that draws upon the ways of knowing, doing and valuing of a particular social group—for example mathematicians, shopkeepers, family members, peer group, etc.

The recognition that a student's present participation in the mathematics classroom is a literate practice mutually constituted through the interaction of past experience, continuing involvement and yet-to-be accomplished goals requires that a student's participation in mathematics be conceptualised as being both *product* and *process*—entities that do not merely alternate, but actually generate each other. 'Product' represents the momentary embodiment of the evolving relationship between learner and context (for example what is said, what is represented, the role that is assumed). Process represents sense-making through negotiating actions and through privileging certain ways of knowing and doing over others (for example the saying, the representing, the enacting of a role). The development of mathematical literacy, therefore, may be said to encompass more than mastering a basic set of foundational skills. From this point of view, being mathematically literate means mastering repertoires of practice that situate the learning of concepts and skills within ways of knowing, ways of doing and ways of valuing that may assist a person to participate in a multi-literate society (Brown & Hirst, 2005). According to the Ontario Ministry of Education (2004), these repertoires of practice need to focus on (a) making sense of mathematics, (b) developing depth and flexibility in thinking, (c) making connections between concepts, (d) developing a sense of numbers through problem solving, (e) persevering in order to understand and (f) communicating thinking and reasoning. This conceptualisation of mathematical literacy has implications for understanding the notion 'learning difficulty'.

Understanding the notion of learning difficulty

The Queensland Studies Authority (2007, p. 1) describes the term 'learning difficulties' as referring to 'barriers that limit student access to, and participation in, outcomes from the curriculum', and describes students who have learning difficulties as being 'those whose access to the curriculum is limited because of short term or persistent problems in one or more of the areas of: literacy; numeracy; learning how to learn'. By framing the notion of 'difficulties' in terms of access and participation, this description of students with learning difficulties is in accord with a sociocultural interpretation of learning and development, because it focuses on student potential rather than student ability, on student competencies rather than on student deficits. As such, the notion of 'learning difficulty' can be understood as being constructed in each curriculum encounter, based on what has gone before and what beliefs and dispositions teachers and students bring to these encounters regarding academic and social competence.

In terms of social competence and classroom learning, Wright and Chapparo (2008) maintain that issues related to resilience and social integration are crucial to the learning of students with learning difficulties. Investigating teacher perceptions of the social competence of children in primary classrooms, these authors found that students identified by their schools as having learning difficulties performed significantly poorer than their peers on measures of classroom survival skills, friendship-making, avoiding aggression and dealing with stress. Following a similar line of inquiry, Walker and Nabuzoka (2007) found that intervention strategies designed to assist students with learning difficulties need to target social-relationship difficulties that students experience in the classroom and school environments. Utilising attainment scores in mathematics and English, and sociometric and peer behavioural attribute scores for 236 (7–12 years old) children from a semi-rural school in England, Walker and Nabuzoka (2007) found a relationship between social adjustment and academic achievement. On measures of social adjustment, students with learning difficulties scored higher on negative behaviours than both high-achieving and low-achieving students. Also, low-achieving students and those with learning difficulties were perceived by their peers to be less co-operative and to have fewer leadership skills than high-achieving students. The authors concluded from these findings that intervention strategies in teaching mathematics and English need to target social-relationship difficulties that low-achieving students and students with learning difficulties experience in learning environments.

In terms of the development of mathematical literacy in school environments, Boaler (2005) and Zevenbergen (2000) maintain that students who experience difficulty in learning mathematics exhibit social and cultural disadvantages, in that teachers do not establish classroom environments that afford student engagement with the language of mathematics. By casting the analytical spotlight on the classroom environments in which students participate in mathematics and the nature of that participation, researchers such as Paul Cobb and Magdalen Lampert have shown that productive mathematical learning takes place in classrooms in which students are engaged in communal inquiry (Cobb et al., 1991; Lampert, 1990) and in which students' beliefs, emotional acts, obligations and expectations are viewed as being intimately related within the social context of doing mathematics (Cobb, Yackel, & Wood, 1989; Lampert, 1990).

As such, developing mathematical literacy in classrooms requires teachers to establish learning environments in which students are provided with multiple opportunities to (a) make sense of mathematics, (b) develop a depth and flexibility in their thinking, (c) make connections between concepts and to see patterns in the mathematics, (d) develop a sense of numbers and efficiency in their work so that their thinking builds as they progress towards a solution, (e) persevere in order to understand and solve problems, (f) communicate their mathematical thinking and (g) understand the mathematical reasoning of others (Ontario Ministry of Education, 2004, pp. 2–3). In other words, students with learning difficulties need to be learning how to learn as they engage with the mathematics of the classroom.

Insights into the nature of this engagement are provided by researchers such as Graham, Bellert, and Pegg (2007) and in curriculum documents such as the Queensland Studies Authority position paper on learning difficulties (Queensland Studies Authority, 2007). In reviewing research-validated strategies for assisting middle-school students (Grades 5–8) who have difficulties with mathematics to learn in the everyday classroom, Graham et al. (2007) advocate deliberate practice, focused feedback, the use of a 'step-by-step' approach to solving tasks, teacher and peer modelling, and explicitly teaching problem-solving strategies to students. According to the Queensland Studies Authority (2007), engaging students with learning difficulties in the mathematical practice of the classroom occurs over time and incorporates such things as (a) raising student awareness that persistence in learning can bring many benefits, (b) providing students with representations of 'steps' that may be used to complete tasks, (c) using 'talk aloud' protocols to work through a problem task and (d) providing teacher and student feedback.

However, teachers are often reluctant to allow students to engage in more than just practice and feedback when learning mathematics in the everyday classroom. Smith (1996) maintained that this may be because approaches to teaching mathematics that extend beyond transmission often fail to help teachers reconceptualise their sense of agency; that is knowing what to say, when to say it and how to say it, and thus fail to convince teachers to change their classroom practice. As such, what is needed for teachers to engage all students in strategies that may address difficulties with learning mathematics is a synthesis of productive student-centred pedagogies, derived from research-based classroom interventions, that can be clearly communicated to, and implemented by, classroom teachers (Woodward & Montague, 2002). The following section of this chapter draws on empirical data from learning episodes (see Brown, 2001) in order to synthesise the central elements of a student-centred pedagogy that may assist teachers to engage students with learning difficulties in the literate practice of mathematics.

Promoting mathematical literacy within the social context of an everyday classroom

The learning episodes referred to took place in a Grade 7 (11–12 years of age) classroom, located in a middle-sized (approximately 400 students) elementary school. The school drew its population of students mainly from suburbs of low socioeconomic status, located close to the centre of a large capital city in Australia. The episodes all took place during mathematics lessons taught by the same teacher (the author of this chapter). The class comprised 15 female and 11 male students, who were permitted to choose their own groups and what space those groups occupied within the classroom when doing mathematics.

In order to provide all students with access to the curriculum, I employed a student-centred approach to teaching and learning mathematics that utilised a pre-set model of collaboration. Pre-set models of collaboration often offer a 'step-wise' approach to attaining a collaborative solution to a problem task, and have been associated with promoting the attainment of intellectual quality, positive interdependence, accountability and face-to-face interaction in a variety of learning environments (Johnson, Johnson, Holubec, & Roy, 1984). The pre-set model of collaboration used in the classroom from which the episodes were taken was Collective Argumentation (Brown & Renshaw, 2000).

Collective Argumentation is designed to assist the teacher and students to move beyond the initiation–response–evaluation (I–R–E) (Mehan, 1979) format of classroom interaction. The I–R–E format of classroom interaction is usually evident when the teacher initiates an interaction with the student (for example asks a question), the student responds (for example answers the question) and the teacher evaluates the student's response. The predominate use of this format of classroom interaction has been used by researchers, such as Lemke (1990), to characterise 'didactic' classroom cultures. Collective Argumentation assists teachers to move beyond the predominate use of the I–R–E format by extending the range of teaching and learning opportunities made available to the teacher and students. This is brought about by engaging the teacher and students in the use of a 'key-word' format that requires students to *represent* a task or problem alone, *compare* their representations within a small group of peers, *explain* and *justify* the various representations to each other in the small group, reach agreement (*agree*) about a possible solution within the group, and to present (*validate*) the group's ideas and representations to the class to test their acceptance by their peers and the teacher. Each of the 'key' words—*represent*, *compare*, *explain*, *justify*, *agree* and *validate*—is derived from principles of collaboration (see Brown & Renshaw, 2000) that require a group of students to generalise their thinking about a task by constructing a personal representation, to objectify their thinking by comparing and explaining, to gain consistency in their thinking by justifying ideas, to gain consensus by agreeing on one representation to investigate further, and to recontextualise their thinking from the group to the class by presenting the agreed-upon representation to their peers and the teacher for discussion and reflection.

An important element of the Collective Argumentation model that supports the teacher and students to extend teaching and learning opportunities beyond the I–R–E format of classroom interaction is the negotiation of a class charter of norms. The norms negotiated reflect the social commitments of a learning community (see Bereiter, 1994; Lampert, 1990) and promote student engagement, courage, humility, honesty, restraint, persistence and generosity. Together with the key-word format, the negotiated norms guide activity and participation in mathematics lessons—lessons in which students are supported in displaying (a) the courage to represent their ideas to others, (b) the humility to accept that their ideas may not, at first, be adequate, (c) the honesty to give evidence-based feedback and reports, (d) the restraint necessary to maintain social cohesion, (e) the persistence to pursue ideas when an immediate solution is not evident and (f) the generosity to affirm the achievements of others. The negotiated norms and the 'key-word' format are represented on class posters and displayed in prominent positions in the classroom, so that the students and teacher can refer to them on a regular basis.

Researching the development of literate practice in the mathematics classroom

The learning episodes referred to in the remaining section of this chapter are taken from a study I designed to investigate how students may become literate in school mathematics in ways that challenge them to become socially involved in the practice of the classroom (Brown, 2001). The study extended over the length of one school year and employed a sociocultural methodology based on the 'teaching experiment' (Davydov, 1988).

The 'teaching experiment' methodology recognises social, historical, cultural and institutional aspects of the context of schooling as being crucial facets in the production of the sociocultural entity being studied (Davydov, 1995), namely the development of mathematical literacy by students. From this perspective, the activity of the students, the activity of the teacher/researcher and the activity of the classroom, which they co-construct, interrelate at a number of levels to create the 'life context' of learning episodes. As such, this methodology is concerned with the Vygotskian belief that studies into the transformation of the intrapsychological by the interpsychological should be involved in changing and not just interpreting activity, and that intervention efforts often provide the best setting for carrying out research (Wertsch, del Rio, & Alvarez, 1995). The Collective Argumentation model of collaboration described above provided the means through which the intervention was framed and implemented.

The study required the class of students and their teacher to be video/audio taped at the small-group and whole-class levels when doing mathematics, by a research assistant on a number of occasions throughout the school year. Anecdotal records relating to teacher–student and student–student interactions were made on a regular basis, and the teacher and students were asked to keep reflective journals. The videotaped group work and whole-class sessions were transcribed for analysis so as to enable an investigation of the literacy practice being displayed. As this chapter is concerned with exploring teacher engagement with students who have learning difficulties in the domain of mathematics, the transcripts were inspected to locate the teacher working with students described by school authorities as having learning difficulties in the area of mathematics. From those identified, episodes focusing on one student in particular were chosen for analysis. These episodes were chosen because they present a range of pedagogical interactions between the teacher, the student and the class representative of teaching and learning in this classroom.

The episodes chosen focus on a 13-year-old female student (Lynda) whose school records displayed a pattern of chronic absenteeism in school attendance. That is, Lynda was absent from school for over 50% of gazetted school days in any one school year without, what her teachers considered to be, a legitimate reason, and often with parental permission. Chronic absenteeism from school has been clearly linked to poor performance at every stage of schooling, from infancy to secondary (see, for example Scottish Council for Research in Education (SCRE), 1995; Reid, 2005). According to school-based records and interviews with school learning-support staff, Lynda displayed severe learning difficulties in the domains of social behaviour, numeracy and literacy. Transcripts from these episodes were then subjected to discourse analysis. Discourse analysis has been used by researchers in the field of mathematics to, among other things, situate teachers' instructional practices in institutional settings (Cobb, McClain, de Silva Lamberg, & Dean, 2003), and to study the development of students' critical awareness in the mathematics classroom (Wagner, 2007). Informed consent was provided by the teacher, Lynda and the parents of students in this classroom for the transcripts of lessons to be used for research purposes.

Engaging all students in the literacy practice of mathematics: negotiating space and level of engagement within the everyday classroom

Lynda, a student who had come to the school in Grade 6, chose, from day 1 of her Grade 7 studies, to sit in the far back corner of the classroom, with Mary and Cynthia. Mary was considered by school authorities to display average achievement in numeracy and literacy. However, Cynthia, a friend of Mary, was considered, like Lynda to be a below-average achiever in the domains of numeracy and literacy. Cynthia's presence in the group provided Mary with a work partner when Lynda was absent (a frequent occurrence).

Lynda arranged the space that this group occupied in the classroom so that it was clearly defined in relation to the rest of the class. Two walls defended the space at the back and side of the group, and to the other side was a table that restricted access by other students. Mary, Cynthia and Lynda placed their own desks in the opening to this alcove and thereby controlled access to themselves and to the back corner of the classroom. These three students appeared content to occupy their self-defined marginal space, and from this vantage point they resisted active participation in the mathematics of the classroom, with the result that students in the groups immediately in front and to the side of them turned their backs towards Mary, Lynda and Cynthia. This marginal arrangement persisted for some time. During this period, Mary and Cynthia engaged in group work when doing mathematics, with Lynda choosing to view their activities from the sideline. Within the group, Mary played the role of tutor to Lynda and Cynthia, and was the only member of the group to contribute to class discussions.

One role expected of everyone in Collective Argumentation is that of presenter, where the members of a group volunteer to represent and explain their thinking about a task to the rest of the class. Over a number of weeks, every other group in the class had presented their work, but even though Mary and Cynthia had accepted the teacher's direct invitation to present in collaboration with him, Lynda resisted, eliciting from the class direct statements that they wanted her to present.

Eventually, Lynda agreed to accompany Mary and Cynthia to the whiteboard during a presentation. Lynda stood by Mary and Cynthia as Mary assumed the roles of presenter and explainer for their group solution to the following two-part problem

> Light travels at three hundred thousand kilometres per second. How far does light travel in one minute? Use exponents to solve this problem. If light takes 8 minutes to travel from the Sun to the Earth, how far away is the Earth from the Sun? Use exponents to solve this problem.

We join the presentation (see Table 13.1) when Mary, Cynthia and Lynda are representing their solution to the first part of the problem (see Fig. 13.1) on the whiteboard.

The extract commences with the teacher highlighting two important features of this classroom's mathematics practice—explaining ideas and actively listening—and portraying the group to the class as being able to participate and deserving of serious consideration. After Mary (the spokesperson of the group) reads the problem, the teacher focuses the group's attention on thinking (Turn 5) and Mary proceeds to employ everyday language to give an inadequate account of how the group arrived at its answer (Turns 6–10). This account is improved by a re-phrasing by Annie, who introduces the term 'hundred thousand' (Turn 7) and the teacher who introduces the term 'number' (Turn 11).

Table 13.1 Mary, Cynthia and Lynda's presentation to the class

Turn	Speaker	Text
1	Teacher (to the class)	Lynda, Mary and Cynthia are representing (on the whiteboard). That's why they are not talking to us. Soon, they will turn around and explain their ideas, their representation to us. Now, remember, when they are explaining you are to be actively listening. Decide what you think about these ideas
2	Allan	I already have
3	Teacher (to the class)	Now, Mary and Cynthia and Lynda want to get up and do this more often, so don't go spoiling their time by coming up with 'smart' responses
4	Mary (reads the question)	Light travels at 300 000 kilometres per second. How far does light travel in 1 minute? Use exponents to solve this problem
5	Teacher (to the group)	Okay, can you tell us what your thinking is?
6	Mary	Well, we did 300 by 60
7	Annie	300 000
8	Mary	300 000, whatever!
9	Teacher	Yes, that's right
10	Mary	And we got this (points to the answer [18 000 000] on the board)
11	Teacher	Can you read that number Mary?
12	Mary	No
13	Teacher	Well, can your group help you?
(The group members talk among themselves)		
14	Teacher	So, 60 by 300 000 equals?
15	Group	Eighteen million
16	Teacher	Eighteen million. So, with what the girls have just told us, if light travels at 300 000 kilometres per second, how far does it travel in 1 minute? Tracey?
17	Tracey	They have told us that light travels at 18 million per minute
18	Teacher	Eighteen million kilometres per minute. Okay girls (the group), keep going
19	Mary	Um
20	Teacher (referring to the girls' representation)	That's just your working out, is it?
21	Mary	Yes
22	Teacher	How did you work it out?
23	Mary	Well, we had the correct numbers and we just multiplied
24	Ryan	How did you get it (referring to the 18 000 000)?
Mary gives no response		
25	Teacher	Well, they broke the sum up. They didn't want to multiply 60 by 300 000, so they multiplied it by 300 first
26	Mary	And then we got 18 000, so we added another three zeros

Table 13.1 (continued)

Turn	Speaker	Text
27	Teacher	Okay. Mary and her group knew that if you are going to multiply by 1 000 then you have to add three zeros. So, first of all, they multiplied by 300, but because it was 300 000, they added three zeros to their answer, which is a really good way of thinking about it. You are using your knowledge well. That's good, girls
28	Teacher	So, light travels 18 million kilometres in 1 minute. How far are we, therefore, away from the sun?
No response from the group		
29	Teacher (to the group)	Can I do it (the rest of the problem)?
30	Group	Yes

Fig. 13.1 Mary, Cynthia and Lynda's written representation of a task solution

300
× 60
18 000
18 000 000km

Unable to read their answer to the class, Mary is directed to access the resources within her own group (Turn 13), while the teacher uses conventional mathematical language to revoice the group's explanation (Turn 14). The group names its solution to the problem (Turn 15) and the teacher locates their response within the context of the original problem (Turn 16).

Employing the language of the teacher, Tracey then accredits the problem solution to the group (Turn 17), and the teacher asks the group to explain the procedure they used to arrive at their answer (Turns 18–22). Once again, Mary offers an inadequate explanation (Turn 23), and the group is re-focused on the task of providing an adequate mathematical explanation by Ryan (Turn 24). The teacher then proceeds to co-participate with the group (Turns 25–27), providing an explanation to the class that highlights the mathematical thinking of the group. At Turn 28, the teacher poses a question that challenges the group members to extend their thinking beyond the procedural context of their response and to apply their knowledge to a related problem. At this point, the group is unable to continue and, with the group's permission, the teacher takes centre stage in the presentation (Turns 29–30).

The above extract illustrates how the teacher framed the group's knowledge of the task within the literacy practice of the class. By strategically creating the 'fiction' that Mary, Cynthia and Lynda were keen to participate more fully in the collective work of the classroom, the teacher was able to orchestrate the group's speaking and thinking within a framework consistent with the conventional language of mathematics. In turn, this orchestration enabled a working consensus to be established between the group's need for social acceptance and the class's need to understand. This consensus, once established, facilitated the establishment of a mutual understanding between the group and the class—an understanding that permitted the presentation to proceed without interference from any unnecessary overt social or emotional conflict.

However, teacher accountability does not permit the mastery of mathematical concepts and skills by students to remain a 'fiction'. How the teacher fulfils the continuing role of mathematician within this episode demonstrates another way in which participation in the literate practice of mathematics was promoted within the classroom; namely, making explicit the mathematical language that scaffolds mature thinking within the context of completing a task.

Promoting thinking through language in use within the everyday classroom

We resume the presentation where the teacher takes centre stage and shares his thinking with the class (see Table 13.2).

The teacher commences this episode by making explicit what mathematical tool (exponents) he is going to use to work the task and by inviting others to join in his thinking by posing a rhetorical question (Turn 31). He then continues to think out loud: stating relationships, asking rhetorical questions, describing possible actions and providing justifications (Turn 33). At Turn 34, the class becomes caught up in his thinking and interrupts, providing a conclusion to one of his thoughts (Turn 34).

Table 13.2 Thinking with the group

Turn	Speaker	Text
31	Teacher	Well, I'm going to work this (the problem) out using exponents. So if I've got 300 000, now that's kilometres isn't it?
32	Group	Yes
33	Teacher	That's how far light travels in 1 second. How is that expressed in powers of 10? Yes, three times 10 to the power of five. I'm going to multiply that by 60 'cause that's how far light travels in 1 minute. So 60 is six by 10 to the power of . . .
34	Class	One
35	Teacher	One. Okay, and now I want to know how far light travels in 8 minutes, so I have to multiply that by eight, which is eight times 10 to the power of zero
(As the teacher has been thinking out loud he has been writing the following on the board: $3 \times 10^5 \times 6 \times 10^1 \times 8 \times 10^0$)		
36	Teacher (to the class)	Now, what do you know about multiplication?
37	Tracey	When using exponents just add the exponents

The class response is included in the teacher's thought, and his thinking aloud continues: restating the goal of the task, offering a possible procedure and translating his thoughts into the signs and symbols of mathematics (Turn 35). At Turn 36, the teacher signals that his time at centre stage has concluded by engaging the class through a direct questioning technique.

The employment of 'think-aloud' strategies is an element of teachers' professional craft knowledge; that is 'the knowledge that experienced teachers gather throughout their careers that enables them to make decisions about how best to approach professional tasks' (Cooper & McIntyre, 1995, p. 186). However, it is important to note that this teacher situates the 'think-aloud' strategy within the literacy practice of mathematics privileged in this classroom; that is within practices that may assist students to view mathematical thinking as being intertwined with the social nature of learning negotiated within the classroom (Ivey, 1996). This 'situatedness' allows the teacher to progress the discourse and to align students' actual speaking and thinking abilities with potential abilities in a fashion that is sensitive to developing mutual understanding, conceptual development and the appropriation by students of the role of 'learner' within the classroom.

Developing an openness to learning

During the above learning episode, Lynda was a part of the group process. Her participation fits Lave and Wenger's (1991) description of a Legitimate Peripheral Participant; that is one who participates in an activity simply by being acknowledged by others as belonging, even if she remains silent and inconspicuous. After this presentation, the group immediately in front of Mary, Cynthia and Lynda re-arranged their desks so that their backs would no longer be facing them. This suggests that in this classroom resistance and exclusion operate reciprocally. When Mary, Cynthia and Lynda chose a marginal position and Lynda initially refused to present, the groups nearest them acted in ways that reinforced their marginality and exclusion by sitting with their backs to them. However, Lynda's willingness to be involved in a presentation was reciprocated by the nearest groups' gesture towards inclusion and open communication. The teacher's role in this cannot be ignored, but it was indirect in the sense that he reminded the students regularly of the ground rules of Collective Argumentation; that is the assumption that everyone participates, nobody is excluded, and that members treat each other according to the social norms of Collective Argumentation.

As the year progressed, Lynda allowed other students to join her in the alcove, particularly one student, Terri, who, like Lynda, transferred to this class from a different school in Grade 6. During her time in the alcove, Terri displayed a similar solidarity with Lynda as Mary had displayed; that is not presenting work to the class when Lynda did not want to, playing the role of tutor to Lynda and seeking clarifications and explanations from the teacher when doing mathematics. However, Terri also displayed a solidarity with the class, frequently entering whole-class discussions to provide constructive criticism, asking inquiry type questions during group presentations and inviting the teacher to work jointly with her on difficult mathematics tasks. Lynda's acceptance of Terri into her group space was marked by an increased participation by Lynda in the group processes of Collective Argumentation; that is asking Terri questions, seeking and giving explanations and seeking to reach a consensus with Terri about solutions to tasks. Lynda also began to interact with other class members in a friendlier fashion; that is talking to others about everyday events and sharing items of property. Eventually, the table on the side of Lynda's group space was removed by students and the alcove became the back corner of the classroom. Into this opened space other students would often sit with Cynthia, Mary, Terri and Lynda. Access to the back corner of the classroom was no longer restricted, and the alcove became a shared space in which groups of students would gather to work on tasks.

Discussion

In terms of sociocultural theory, the teacher in the above classroom promoted learning mathematics to the students as 'an aspect of culturally, historically situated activity' (Lave, 1993, p. 30). The teacher was able to do this by orchestrating student discussions in a manner similar to that described by Forman, Stein, Brown, and Larreamendy-Joerns (1995), where the teacher's focus is on creating a space in the dialogue in which students can be aligned with each other and with the content of academic work, while simultaneously being socialised into particular ways of speaking and thinking. In other words, just as a director is used in a play to orchestrate dialogue and to co-participate with an actor until they are ready to act in the 'scene', if a student or group was not ready to engage in the literate practice of the classroom, the teacher would step in and orchestrate their interpretation of a task so that it reflected both the group's contribution and the mathematics of the classroom.

In particular, the teacher assisted Lynda in her journey towards participating in the literate practice of the classroom by (a) emphasising the importance of inclusion and participation in mathematics to Lynda and to the class, (b) respecting Lynda's decision to remain physically located within the alcove, but applying pressure on Lynda and other members of the class to improvise a new relationship, (c) assisting Lynda and the class in coming to know how to deal with the problem of Lynda's resistance and (d) identifying Lynda as being a member of this class through regularly highlighting the social norms of Collective Argumentation. In other words, the teacher approached the goal of catering for Lynda's individual learning and social needs by working in partnership with the students to deal with the 'real life' problem of Lynda's resistance to schooling. In doing so, the goal was re-represented for the class, from being a neo-behaviouristic shaping of individual behaviour born of the belief that a functional relationship exists between a stimulus (for example teacher reward/punishment) and a response (for example student compliance/non-compliance) to being one of building a learning environment in which certain ground rules were collectively accepted, but individuals were allowed the option of standing apart but not 'opting out'.

In terms of teaching students with learning difficulties, the teacher in the above classroom utilised many strategies promoted in the literature as being beneficial to student learning; for example the use of a 'step-by-step' approach to solving tasks, teacher and peer modelling, the use of 'talk aloud' protocols by teachers and students, and the use of teacher and student feedback. However, these strategies were not used in isolation from the literate practice of the classroom, the culture of student interaction evidenced in the classroom, nor in ignorance of the learning histories displayed by students. The strategies were used with the intention of providing all students with access to the literate practice of mathematics. As such, a sociocultural approach to teaching and learning offers teachers in everyday classrooms alternative ways of interacting with students who have learning difficulties, ways in which dilemmas related to issues of 'dominance', 'resistance' and 'personal learning difficulties' are addressed in terms of the development by students of (a) stable social roles, (b) multiple ways of participating, (c) reciprocal responsibilities in the learning process and (d) the development of perspectives aligned with the literate practice of mathematics.

This chapter has explored the social dimension of classroom practice and its importance in supporting the development of mathematical literacy by all students, including students considered by school authorities as displaying learning difficulties in mathematics. Conceptualising the development of mathematical literacy as being an aspect of the socially situated practice of an everyday classroom raises important questions/issues that have yet-to-be addressed in this important field of investigation.

Essential next questions

How can teachers integrate practice, feedback, modelling and explicit teaching with an approach to teaching that privileges reasoning and argumentation?

In terms of research, inclusive approaches to the development of mathematical literacy, such as Collective Argumentation, offer the potential for the implementation of more innovative pedagogies that may make participation in the literate practice of a classroom accessible to all students. However, such approaches need to balance the needs of students who have learning difficulties in mathematics with the needs of students who do not have learning difficulties in mathematics. As such, how to integrate deliberate practice, feedback, modelling and explicit teaching within forms of mathematical reasoning and argumentation that will provide students with access to future mathematics courses is a worthy course for future research to take.

The work of Paul Cobb provides insights into what direction such research may take. Advocating a 'design experiment' approach to conducting research (see Cobb, Confrey, diSessa, Lehrer, & Schauble, 2003), Cobb and his associates have provided, over time, research tools to help educators study the 'learning ecology' of school mathematics lessons. It is by taking such an approach to research that educators may provide the field of special-needs education with greater understandings of how best to integrate reasoning and argumentation into the everyday routine of school classrooms so that all students may be provided with the tools necessary to access the literacy practices of mathematics.

How can policy makers advocate the use by teachers of assessment tools that can assess and encourage student participation in the literacy practices of mathematics?

In terms of policy, this chapter has provided a view of 'learning difficulty' as being a socially constructed category that is based upon what has gone before and the beliefs and attitudes that are drawn upon to explain certain behaviours and to account for certain performances. In order to promote such a non-deficit view of learning, policy makers in the field of mathematics education need to advocate an approach to assessment that is inclusive of, and values, non-standardised ways of understanding students' ways of knowing and doing mathematics. As such, the advocacy of assessment tools that can assess and encourage student participation in the literacy practices of mathematics is imperative if administrators, teachers and parents are to take seriously the goal of moving students from practising to be numerate towards engaging in the literate practice of mathematics.

Miriam Amit and Michael Fried provide a view of what such advocacy may look like. In their research on the relationships between 'high-stakes' assessment and the development of mathematical literacy by all students, Amit and Fried (2002) provide educators with a framework in which to situate the tensions between society's ambition to privilege high achievement in mathematics and its aim for all students to attain a general mathematical literacy. In elaborating their framework, Amit and Fried (2002) emphasise the positive influences that assessment programs in mathematics can have on the promotion of student achievement in mathematics and on student development of mathematical literacy, if the construction of assessment tools is thoughtful and based on research in the field.

How might teacher development promote reflection on student and teacher participation in the literate practice of mathematics?

In terms of classroom practice, this chapter has provided a view of student agency as being embedded in contexts of participation; contexts that will allow students to stand apart, but not 'opt out'. In order to sustain the growth of such agency, teachers need to be provided with professional development opportunities that may assist them to gain insights into the efficacy of what they are doing or may do in their classrooms in order to provide all students with access to the literate practice of mathematics. As such, assisting teachers to reflect upon their own participation in the literate practice of mathematics is crucial in actualising student potential within the everyday classroom.

Providing insights into how teacher development may promote reflection by teachers on their participation in the literate practice of mathematics is the focus of my current research. In research, I am exploring the development of agency with mathematics by teachers and students as they participate in what Jaworski (2006) refers to as a research community of practice; that is a community of teachers, students and educators that uses inquiry into their own practice as a tool to engage with change and the development of mathematical literacy.

References

Amit, M., & Fried, M. N. (2002). High-stakes assessment as a tool for promoting mathematical literacy and the democratization of mathematics education. *Journal of Mathematical Behaviour, 21*, 499–514.

Australian Association of Mathematics Teachers. (1997). *Numeracy = everyone's business*. Adelaide: Author.

Australian Education Council. (1991). *A national statement on mathematics for Australian schools*. Melbourne, VIC: Curriculum Corporation.

Bereiter, C. (1994). Implications of postmodernism for science, or, science as progressive discourse. *Educational Psychologist, 29*(1), 3–12.

Boaler, J. (2005). The 'psychological prison' from which they never escaped: The role of ability grouping in reproducing social class inequalities. *FORUM, 47*(2&3), 135–144.

Boaler, J. (2007). Promoting 'relational equity' and high mathematics achievement through an innovative mixed-ability approach. *British Educational Research Journal, 34*(2), 167–194.

Brown, R. A. J. (2001). *A sociocultural study of the emergence of a classroom community of practice*. Unpublished PhD. thesis, University of Queensland, Brisbane.

Brown, R., & Hirst, E. (2005). Mathematical literacy: Promoting literate practice in a year 7 classroom. *Literacy Learning: The Middle Years, 13*(3), 29–38.

Brown, R. A. J., & Renshaw, P. D. (2000). Collective argumentation: A sociocultural approach to reframing classroom teaching and learning. In H. Cowie & G. van der Aalsvoort (Eds.), *Social interaction in learning and instruction: The meaning of discourse for the construction of knowledge* (pp. 52–66). Amsterdam: Pergamon Press.

Cobb, P., Confrey, J., diSessa, A., Lehrer, R., & Schauble, L. (2003). Design experiments in educational research. *Educational Researcher, 32*(1), 9–13.

Cobb, P., & Hodge, L. L. (2002). A relational perspective on issues of cultural diversity and equity as they play out in the mathematics classroom. *Mathematical Thinking and Learning, 4*(2&3), 249–284.

Cobb, P., McClain, K., de Silva Lamberg, T., & Dean, C. (2003). Situating teachers instructional practices in the institutional setting of the school and district. *Educational Researcher, 32*(6), 13–24.

Cobb, P., Wood, T., Yackel, E., Nicholls, J., Wheatley, G., Trigatti, B., et al. (1991). Assessment of a problem-centred second-grade mathematics project. *Journal for Research in Mathematics Education, 22*(1), 3–29.

Cobb, P., Yackel, E., & Wood, T. (1989). Young children's emotional acts while engaged in mathematical problem solving. In D. B. McLeod & V. M. Adams (Eds.), *Affect and mathematical problem solving: A new perspective* (pp. 117–148). New York: Springer.

Cooper, P., & McIntyre, D. (1995). The crafts of the classroom: Teachers and students accounts of the knowledge underpinning effective teaching and learning in classrooms. *Research Papers in Education, 10*(2), 181–216.

Davydov, V. V. (1988). Problems of developmental teaching (Part II). *Soviet Education, 30*(9), 3–83.

Davydov, V. V. (1995). The influence of L. S. Vygotsky on education theory, research, and practice (T. Kerr, Trans.). *Educational Researcher, 24*(3), 12–21.

Forman, E. A., Stein, M. K., Brown, C. A., & Larreamendy-Joerns, J. (1995, April). *The socialization of mathematical thinking: The role of institutional, interpersonal, and discursive contexts.* Paper presented at the annual meeting of the American Educational Research Association, San Francisco.

Gee, J. (1990). *Social linguistics and literacies.* Cambridge: The Falmer Press.

Graham, L., Bellert, A., & Pegg, J. (2007). Supporting students in the middle school years with learning difficulties in mathematics: Research into classroom practice. *Australasian Journal of Special Education, 31*(2), 171–182.

Gutiérrez, K. D. (2008). Developing a sociocultural literacy in the third space. *Reading Research Quarterly, 43*(2), 148–164.

Ivey, K. M. C. (1996). Mechanistic and organic: Dual cultures in a mathematics classroom. *Journal of Research and Development in Education, 29*(3), 141–151.

Jaworski, B. (2006). Theory and practice in mathematics teaching development: Critical inquiry as a mode of learning in teaching. *Journal of Mathematics Teacher Education, 9,* 187–211.

Johnson, D., Johnson, R., Holubec, E., & Roy, P. (1984). *Cooperation in the classroom.* Edina, MN: Interaction Books.

Lampert, M. (1990). When the problem is not the question and the solution is not the answer: Mathematical knowing and teaching. *American Educational Research Journal, 27*(1), 29–63.

Lave, J. (1993). The practice of learning. In S. Chaiklin & J. Lave (Eds.), *Understanding practice: Perspectives on activity and context* (pp. 3–33). Cambridge: Cambridge University Press.

Lave, J., & Wenger, E. (1991). *Situated learning: Legitimate peripheral participation.* Cambridge: Cambridge University Press.

Lemke, J. L. (1990). *Talking science: Language, learning, and values.* Norwood, NJ: Ablex Publishing.

Mehan, H. (1979). *Learning lessons: Social organization in the classroom.* Cambridge, MA: Harvard University Press.

Mullis, I. V. S., Martin, M. O., Gonzalez, E. J., & Chrostowski, S. J. (2004). *TIMSS 2003 international mathematics report.* Chestnut Hill, MA: TIMSS & PIRLS International Study Centre.

National Council of Teachers of Mathematics (NCTM). (2000). *Principles and standards for school mathematics.* Virginia: NCTM.

Ontario Ministry of Education. (2004). *Teaching and learning mathematics: The report of the expert panel on mathematics in grades 4 to 6 in Ontario.* Ontario: Ministry of Education. Retrieved August 21, 2008, from http://www.edu.gov.on.ca

Piaget, J. (1952). *The origins of intelligence in children* (Margaret Cook, Trans.). New York, NY: International Universities Press, Inc.

Queensland Studies Authority. (2004). *Mathematics: Years 1 to 10 syllabus.* Brisbane, QLD: State of Queensland.

Queensland Studies Authority. (2007). *Learning difficulties.* Brisbane, QLD: State of Queensland. Retrieved August 21, 2008, from http://www.qsa.qld.edu.au/syllabus/691.html

Reid, K. (2005). The cause, views and traits of school absenteeism and truancy: An analytical review. *Research in Education, 74,* 59–82.

Renshaw, P. D. (1996). A sociocultural view of the mathematics education of young children. In H. Mansfield, N. Pateman, & N. Bednarz (Eds.), *Mathematics for tomorrow's young children* (pp. 59–78). Amsterdam: Kluwer.

Rogoff, B. (1995). Observing sociocultural activity on three planes: Participatory appropriation, guided participation, and apprenticeship. In J. V. Wertsch, P. del Rio, & A. Alvarez (Eds.), *Sociocultural studies of mind* (pp. 139–164). New York, NY: Cambridge University Press.

Scottish Council for Research in Education (SCRE). (1995). *Understanding truancy: Links between attendance, truancy and performance*. Edinburgh: SCRE, Scottish Council for Research in Education.

Smith, J. P., III. (1996). Efficacy and teaching mathematics by telling: A challenge for reform. *Journal of Research in Mathematics Education, 27*(4), 387–402.

United Nations Educational, Scientific and Cultural Organisation (UNESCO). (1994). *The Salamanca statement and framework for action on special needs education* (ED-94/WS/18). Salamanca, Spain: Author.

Valsiner, J. (1997). *Culture and the development of children's action: A theory of human development* (2nd ed.). New York: John Wiley & Sons.

van Kraayenoord, C. E., & Elkins, J. (2004). Learning difficulties in numeracy in Australia. *Journal of Learning Disabilities, 37*(1), 32–41.

Vygotsky, L. S. (1987). *The collected works of L. S. Vygotsky, Volume 1: Problems of general psychology*. New York: Plenum Press.

Wagner, D. (2007). Students critical awareness of voice and agency in mathematics classroom discourse. *Mathematical Thinking and Learning, 9*(1), 31–50.

Walker, A., & Nabuzoka, D. (2007). Academic achievement and social functioning of children with and without learning difficulties. *Educational Psychology, 27*(5), 635–654.

Wertsch, J. V. (1985). *Vygotsky and the social formation of mind*. Cambridge, MA: Harvard University Press.

Wertsch, J. V., del Rio, P., & Alvarez, A. (1995). Sociocultural studies: History, action, and mediation. In J. V. Wertsch, P. del Rio, & A. Alvarez (Eds.), *Sociocultural studies of mind* (pp. 1–34). New York: Cambridge University Press.

Woodward, J., & Montague, M. (2002). Meeting the challenges of mathematics reform for students with LD. *The Journal of Special Education, 36*(2), 89–101.

Wright, M., & Chapparo, C. (2008). Social competence and learning difficulties: Teacher perceptions. *Australian Occupational Therapy Journal*, 55, 1–10.

Zevenbergen, R. (2000). 'Cracking the code' of mathematics classrooms: School success as a function of linguistic, social and cultural background. In J. Boaler (Ed.), *Multiple perspectives on mathematics teaching and learning*. Ablex: Westport, CT.

Chapter 14
Effective Instruction in Mathematics for Students with Learning Difficulties

Marjorie Montague

Achievement in mathematics is critical to success in school and, ultimately, to success in the adult world. As a consequence of the *No Child Left Behind Act of 2001* (NCLB, 2001), students in the United States must pass a 'high stakes' state assessment in mathematics to graduate from secondary school. However, poor performance remains evident at the state level (for example Florida Comprehensive Assessment Test: FCAT), the national level (for example National Assessment of Educational Progress, NAEP, 2003) and also at the international level (for example Trends in International Mathematics and Science Study, TIMSS, 2004), leaving many students at risk for failure and/or dropping out of school. Recognising the consistently dismal mathematics performance of students in the United States, in 1989 the National Council of Teachers of Mathematics (NCTM, 1989) published the *Curriculum and Evaluation NCTM Standards for School Mathematics* (NCTM, 1989) and revised those standards in 2000 as the *Principles and NCTM Standards for School Mathematics* (NCTM, 2000).

In 2006, the United States Department of Education convened a National Mathematics Advisory Panel (NMAP) to 'use the best available scientific research to advise on improvements in the mathematics education of the nation's children (NMAP, 2008, p. xv). *Foundations for Success: The Final Report of the National Mathematics Advisory Panel* was published in 2008, focusing on six principal recommendations: (1) the pre-K–8 curriculum should emphasise the most critical topics in the early grades, (2) what is known from research in how children learn should be incorporated into instructional methodology, (3) teachers must be knowledgeable and proficient in teaching mathematics, (4) high-quality research must inform instructional practices, (5) national and state assessments should be improved and include critical knowledge and skills leading to the development of algebraic reasoning and (6) research in mathematics teaching and learning should be rigorous and able to inform policy and practice. The government also recognised the need to close the gap between the mathematics achievement of non-minority and

M. Montague (✉)
University of Miami, Coral Gables, FL, USA
e-mail: MMontague@aol.com

C. Wyatt-Smith et al. (eds.), *Multiple Perspectives on Difficulties in Learning Literacy and Numeracy*, DOI 10.1007/978-1-4020-8864-3_14,

minority students and between students with and without disabilities, understanding that these low-achieving groups represent a significant challenge for teachers.

Low-performing students vary considerably in ability, achievement and motivation; however, given the current legislation in the United States (NCLB) and the growing demands in a technological society, like all children, they must develop the numerous mathematical concepts, skills and applications needed not only to perform well on mathematics assessments in school, but also to apply these skills successfully in the real world. As NMAP noted, teachers often do not have the necessary content, or pedagogical knowledge or understanding in how to teach mathematics, especially when students have diverse learning and behavioural characteristics. Given the rapid movement towards full inclusion of students with a wide range of learning needs in our classrooms and schools, it is incumbent on teachers to be cognisant of, and proficient in, proven instructional practices and procedures for teaching all students, including those with special learning and behavioural needs. Curricular materials and teaching methodologies based on NCTM standards have been found to be too complex, unstructured and confusing for most students with diverse learning needs (for example Baxter, Woodward, & Olson, 2001; Woodward, 2006). Fortunately, though, researchers have identified, and are continuing to identify, effective instructional practices that address the numerous attributes of students who are performing poorly in mathematics (see Montague & van Garderen, 2008).

The purpose of this chapter is to review several effective practices for teaching mathematics to students who are performing poorly. The cognitive and behavioural characteristics of children and adolescents that impede successful performance in mathematics are described first. Second, a theoretical context for instruction to improve mathematics learning for children and adolescents is presented. With respect to this theoretical context, several studies using effective instructional approaches for teaching students with mathematics difficulties are discussed. Third, instructional practices and procedures that promote learning for children and adolescents who perform poorly in mathematics are highlighted. Finally, good instruction is based on the ability to monitor the progress of children over time, to ensure they are learning and performing to their potential. The concept of 'best practice' is discussed as it relates to assessing and improving mathematics performance of students who have difficulty learning mathematics.

Student characteristics and mathematics achievement

A variety of cognitive and/or behavioural characteristics can interfere with successful performance in mathematics. These include problems in attention, memory and language, as well as limited self-regulation strategies and difficulties with problem representation (Montague, 2006). Geary (2003) proposed three sub-types of mathematical disabilities: deficient semantic memory, procedural deficits and visuospatial deficits. Children with semantic memory problems have difficulty retrieving mathematical facts or answers to simple arithmetic problems, make more errors when they do and vary considerably on reaction time for correct retrieval. These problems seem relatively stable over time, and children develop idiosyncratic characteristics that differentiate them from younger, academically typical children. In contrast, children with procedural deficits frequently use developmentally immature procedures similar to those used by younger, academically typical children. Procedural deficits may be associated with developmental delay and thus may improve over time. Characteristically, these children have working-memory problems and difficulties in counting. Visuospatial problems indicate a difficulty in spatially representing numerical and other mathematical relationships. Problems with number alignment and rotation, measurement, place value and geometry are typical of visuospatial deficits.

Students with mathematical difficulties characteristically have self-regulation problems that interfere with effective and efficient math performance, particularly math problem solving. Self-regulation problems translate into problems with organisation, knowing where or how to begin a task and an inability to locate and correct errors and self-evaluate. Problem representation involves translating and transforming numerical and linguistic information in mathematical word problems (Montague & Applegate, 1993). If students have difficulty in representing mathematical problems, they automatically have difficulty solving problems. Problem-representation difficulties may have to do with difficulties in reading and understanding the problem and attending to the information, identifying important information and then representing that information visually.

Behavioural characteristics also can interfere with math performance. Children and adolescents with attention deficit hyperactivity disorder (ADHD), for example, may have the necessary math skills but may not use them when they need to. Rather than having a performance problem, these students often have a production problem. In other words, they do not produce what is expected due to inattention, impulsivity and/or hyperactivity. Inattentive children are generally careless, off-task, disorganised, distractible and forgetful. They often appear not to listen, frequently do not finish what they start and consistently misplace or lose things, including their homework. Impulsive children lack self-control and often act without thinking. Hyperactive children seem to talk excessively and move constantly. Affective characteristics that can have a negative impact on math performance include low motivation, self-efficacy and confidence.

A theoretical context for teaching and learning mathematics

The theoretical context for improving mathematics teaching and learning is based on both behavioural theory (for example Adams & Carnine, 2003) and cognitive theory; that is developmental theory (for example Case, 1985; Flavell, Miller, & Miller, 1993), information-processing theory (for example Sternberg, 1985) and sociocultural theory (for example Vygotsky, 1989). Behavioural theory is grounded in principles and procedures that govern behaviour; for example reinforcement and punishment. Behaviourists believe that learning is basically external to the individual, and that when tasks and environments are structured and include appropriate rewards and consequences, individuals will acquire and maintain skills and behaviours. In contrast, cognitive theory considers the internal nature of the individual. Developmental theory postulates cognitive growth as a function of age. Thus, development depends on systematic, age-related stages of development that proceed in a predictable manner. Developmentalists basically believe that learning is cumulative and dependent on progressive maturation of cognitive abilities; that is mastery of lower-level skills is necessary before higher-level skills can be learned. Information-processing theory is predicated on the idea that particular cognitive abilities, for example perception, attention and memory, are critical in processing sensory input for learning to take place. The information-processing perspective focuses on the learner's cognitive and modality strengths and weaknesses, such as visual memory and auditory discrimination. Sociocultural theory views learning as interactive and the learner as an active participant. Often referred to as constructivists, advocates of sociocultural theory believe that children construct meaning as they are encouraged by their instructors to make sense of new information and ideas as a result of prior knowledge and experiences.

These theories provide the foundation for two instructional approaches that were found to be the most effective interventions for students with learning disabilities (LD) (Swanson, 1999): direct instruction and cognitive-strategy instruction. Direct instruction is based on behavioural theory, while cognitive-strategy instruction incorporates both behavioural and cognitive theory. Both instructional approaches include similar research-based practices and procedures, such as cueing, modelling, verbal rehearsal and feedback. They are both highly structured and organised, embedding appropriate cues and prompts that eventuate in mastery of mathematical concepts, skills and applications. Direct instruction is more didactic than cognitive-strategy instruction, and is associated more with basic skills instruction (Kroesbergen & van Luit, 2003). It utilises scripted lessons that are teacher-directed and fast-paced. In contrast, cognitive-strategy instruction is more interactive and uses explicit instruction that focuses on teaching students the processes involved in the application of skills; that is solving mathematical problems (for example Montague, 2003). One of the instructional procedures basic to cognitive-strategy instruction is cognitive, or process, modelling, whereby the teacher models how proficient problem solvers think through the problem, represent the problem, plan the solution and then work it out. The goal is to have students learn how to think and behave like successful problem solvers.

Direct-instruction research

Research using direct instruction to improve mathematics has focused primarily on drill and practice for improving math fact recall and computational skills (Miller & Mercer, 1997). Typically, drill and practice, a crucial aspect of academic remediation for students with LD (Fuchs & Fuchs, 2001), has been used to improve recall and automaticity of facts because it provides challenge, appropriate time on task and numerous response opportunities. One typical example of a drill-rehearsal model is Burns' (2005) incremental rehearsal, which was used in one study to teach multiplication facts to three Grade 3 students with LD. Like other drill-and-rehearsal models, this model uses a gradually increasing ratio of known to unknown items, until students achieve at least 90% accuracy. Ten new facts are introduced each time. A criterion is preset to determine whether the fact is known or unknown. One concern about drill and practice is that students with disabilities do not usually generalise what they have learned to similar tasks; for example other similar but unfamiliar math facts. Poor number sense, a characteristic associated with mathematical learning difficulties, has been suggested to contribute to students' inability to generalise learning in mathematics (Gersten & Chard, 1999). Students need a fundamental understanding of numbers before they can sufficiently learn the facts and procedures needed to add, subtract, multiply and divide (Woodward, 2006). Concrete manipulatives have been used to develop number sense (Funkhouser, 1995); for example, before working on basic fact knowledge, students with LD in kindergarten and Grade 1 were taught to recognise the number of objects in a set without counting them. They were given vertical displays of rectangles divided into equal squares with zero to five dots or jellybeans placed within the squares. Students then counted the dots or jellybeans and identified the number represented. Different combinations that could be made using the configuration were discussed. Eventually, students were introduced to the addition symbol and the basic facts of addition.

Other direct-instruction studies for teaching basic math skills and concepts to elementary school children with LD involve a progression from concrete materials to visual representations, to abstract or symbolic representations. These studies focused on teaching place value (Peterson, Mercer, & O'Shea, 1988), coin sums (Miller, Mercer, & Dillon, 1992), basic facts (Mercer & Miller, 1992) and multiplication procedures (Harris, Miller, & Mercer, 1995; Miller, Harris, Strawser, Jones, & Mercer, 1998). The concrete level involves the use of 3D manipulative devices to demonstrate specific mathematical concepts. The representation level involves using visual displays to represent mathematical concepts introduced at the concrete level. The goal is to promote conceptual understanding. Finally, instruction shifts to the abstract level, which involves solving problems using number symbols. At this level, students are required to memorise the facts and develop fluency (Hudson & Miller, 2006). The four-step lesson format includes an advance organiser (connect the lesson to a previous lesson, identify the lesson skill and provide a rationale for the skill being learned), demonstration of the skill, guided practice with feedback and independent practice.

Another direct-instruction study (Tournaki, 2003) compared a specific technique for addition to drill and practice. Students with LD were taught to use the minimum addend technique, whereby students first determined the larger addend and then used the counting-on strategy to count up from the higher-number the units specified by the smaller addend; for example in $4 + 5 =$, the students starts from the 5 and adds 4 more units. Students who were taught the technique improved significantly, compared to those who learned basic facts through drill and practice. More encouraging, however, was that students who learned the technique transferred what they had learned to three single-digits addition.

Cognitive-strategy instruction research

Although direct instruction has been the traditional approach for teaching basic math skills to students with LD, a few studies have been conducted using a more cognitive approach (Kroesbergen & van Luit, 2003). Instruction in self-regulation is an important component of cognitive-strategy instruction. Self-regulation was the foundation for a comprehensive program to improve Grade 3 students' basic multiplication and division skills (Van Luit & Naglieri, 1999); that is the MASTER Training Program (MASTER Program). One study investigating the effect of the MASTER Program included 42 elementary school students with LD. The MASTER Program includes an introductory phase, group practice phase and an individual practice phase. Following discussion between the teacher and students of all possible solution procedures, students are taught to identify various strategies and then to decide which are most effective for solving specific problems. Teachers model self-instruction. Students with LD improved significantly compared with students using only the general instructional program and also generalised the self-instruction procedure to more difficult problems.

In another study, the MASTER Program was used with 75 students, 27 with LD, aged 7–13 years (Kroesbergen & van Luit, 2002, 2003). Students received the MASTER Program, direct instruction or the regular curriculum. As part of the MASTER Program, self-instruction was promoted by means of a strategy decision sheet that contained various questions students asked themselves. Students taught using the MASTER Program outperformed those who received only the regular curriculum, and performed better than both the direct-instruction and regular curriculum groups on a task that required them to transfer what they learned. Interestingly, this study provided additional support for direct instruction as an effective approach for teaching basic skills to students with mathematical learning difficulties. The results showed that students without LD benefited more from the MASTER Program instruction than direct instruction in learning basic multiplication and division, while students with LD benefited more from direct instruction.

Cognitive-strategy instruction has been the primary intervention for teaching students with mathematical difficulties to solve math word problems (Montague & Dietz, 2009). *Solve It!* is one research-based example of a mathematical problem-solving instructional program (Montague, 2003). The program incorporates both cognitive processes and metacognitive strategies that are integral to problem representation and problem execution (Mayer, 1985). Problem representation can be defined as a combination of something written on paper, something existing in the form of physical objects and a carefully constructed arrangement of an idea in one's mind (Janvier, 1987; Mayer, 1985). Problem-representation processes and strategies are needed to comprehend and integrate problem information, maintain mental images of the problem in working memory and develop a viable solution path, often by finding alternative and unusual approaches to the problem (Silver, 1987). As such, problem representation involves translating and transforming linguistic and numerical information into verbal, graphic, symbolic and quantitative representations that show the schemata or relationships among the information in the problem (Mayer, 1985; Montague & Applegate, 1993; van Garderen & Montague, 2003) before generating appropriate mathematical equations or algorithms for problem solution. After the problem solver accepts the solution plan, problem-execution strategies are implemented. Problem execution requires the problem solver to work forward and backward without resorting to trial-and-error approaches to problem solving. *Solve It!* places particular emphasis on teaching students how to represent mathematical problems by paraphrasing problems, using visualisation strategies, such as diagram drawing or mental imaging, and hypothesising or setting up a plan. The ultimate goal of the program is to have students internalise the cognitive processes and metacognitive strategies so that they become automatic during problem solving. *Solve It!* incorporates the following cognitive processes and activities:

1. Reading the problem (reading, rereading, identifying relevant/irrelevant information)
2. Paraphrasing (translating the linguistic information by putting the problem into one's own words without changing the meaning of the 'story' or 'situation')
3. Visualising (transforming the linguistic and numerical information to form internal representations in memory through a drawing or image that shows the relationships among the components of a problem)
4. Hypothesising about problem solutions (establishing a goal, looking towards the outcome and setting up a plan to solve the problem by deciding on the operations that are needed, selecting and ordering the operations, and transforming the information into correct equations and algorithms)
5. Estimating the outcome or answer (validating the process as well as the product by predicting the outcome based on the question/goal and the information presented)
6. Computing the outcome or answer (recalling the correct procedures for the basic operations needed for solution—calculator skills are taught/reinforced here)
7. Checking (becoming aware of problem solving as a recursive activity and learning how to check both process and product by checking understanding and representation as well as accuracy of the process, procedures and computation).

Cognitive processes can be interpreted as the proactive, or 'to do', strategies, whereas metacognitive strategies are more reflective and require problem solvers to question what they are doing and what they have done. Problem solvers use metacognitive strategies to tell themselves what to do, ask themselves questions, recall what they know, detect and correct errors and monitor performance. These strategies help problem solvers gain access to strategic knowledge, guide their application and regulate use of strategies and overall performance as they solve problems. They can be used overtly (talking out loud or whispering to oneself) or covertly (thinking to oneself). Metacognitive strategies included in *Solve It!* include self-instruction (SAY what to do before and while performing actions), self-questioning (ASK questions of oneself while engaged in an activity to stay on task, regulate performance and verify accuracy) and self-monitoring (CHECK to make certain that

READ (for understanding)
Say: Read the problem. If I don't understand, read it again.
Ask: Have I read and understood the problem?
Check: For understanding as I solve the problem.
PARAPHRASE (your own words)
Say: Underline the important information. Put the problem in my own words.
Ask: Have I underlined the important information? What is the question? What am I looking for?
Check: That the information goes with the question.
VISUALIZE (a picture or a diagram)
Say: Make a drawing or a diagram. Show the relationships among the problem parts.
Ask: Does the picture fit the problem? Did I show the relationships?
Check: The picture against the problem information.
HYPOTHESIZE (a plan to solve the problem)
Say: Decide how many steps and operations are needed. Write the operation symbols (+, -, x, and /).
Ask: If I …, what will I get? If I …, then what do I need to do next? How many steps are needed?
Check: That the plan makes sense.
ESTIMATE (predict the answer)
Say: Round the numbers, do the problem in my head, and write the estimate.
Ask: Did I round up and down? Did I write the estimate?
Check: That I used the important information.
COMPUTE (do the arithmetic)
Say: Do the operations in the right order.
Ask: How does my answer compare with my estimate? Does my answer make sense? Are the decimals or money signs in the right places?
Check: That all the operations were done in the right order.
CHECK (make sure everything is right)
Say: Check the plan to make sure it is right. Check the computation.
Ask: Have I checked every step? Have I checked the computation? Is my answer right?
Check: That everything is right. If not, go back. Ask for help if I need it.

Fig. 14.1 *Solve It!* Math problem-solving cognitive routine. Source: Montague (2003). Copyright Exceptional Innovations; reproduced with permission. Permission to photocopy this figure is granted for personal use only

everything is done correctly throughout the problem-solving process). Figure 14.1 presents the *Solve It!* cognitive routine.

Another well-researched and documented cognitive approach to improving mathematical problem solving for children and adolescents is schema-based strategy instruction (see Fuchs & Fuchs, 2003; Xin & Jitendra, 2006). Schema-based strategy instruction has been demonstrated as effective with students with mathematical difficulties in Grades 2–8 (Jitendra & Hoff, 1996; Jitendra, DiPipi, & Perron-Jones, 2002; Jitendra et al., 1998; Jitendra, Hoff, & Beck, 1999). Instruction focuses on helping youngsters to identify the problem type or problem schema; that is 'a general description of a group of problems that share a common underlying structure' (Xin & Jitendra, 2006, p. 53) and to use that structure as a map to solve the problem.

First, students are taught to identify the problem type, then represent it before developing a plan to solve it (Xin & Jitendra, 2006). Several areas of conceptual knowledge are essential to problem solving using this model. These include schema knowledge, elaboration knowledge, strategic knowledge and execution knowledge. Students learn the various problem types for addition, subtraction, multiplication and division. For example, problems that use the phrase 'three times as many as' are 'compare' problems. After identifying the correct schema, the student generates a schematic representation that fits the schema by elaborating on the principal features of the schema. To illustrate, 'Janet has 2 cats. Her friend gave her 3 more. How many cats does Janet have now?' represents a change problem. The student maps the details of the problem or the information onto the representation: '2 cats' and '3 cats' and '? cats'. Planning the solution involves establishing a goal (what is the outcome?), selecting appropriate operation(s) and writing the algorithm or equation. Finally, the plan is carried out using execution knowledge to compute and check the solution (Xin & Jitendra, 2006).

Instructional practices and procedures

As discussed previously, direct instruction and cognitive-strategy instruction share several instructional procedures: structured, organised lessons; appropriate cues and prompts; guided and distributed practice; immediate and corrective feedback on learner performance; positive reinforcement; overlearning; and mastery (Montague, Warger, & Morgan, 2000). However, cognitive-strategy instruction is more explicit than direct instruction. It is less didactic and more flexible than direct instruction because it relies upon continuous interaction between teachers and students. It also allows teachers to adapt the teaching routine and tailor instruction to accommodate the strengths and weaknesses of students. Explicit instruction as exemplified in *Solve It!* (Montague, 2003) includes proven, effective instructional strategies such as the following:

1. Active student participation and interaction among students and between students and teachers (Peterson, 1988). A guided discussion technique ensures that all students participate. Role reversals and problem solving teams/groups engage all students in the process (Jenkins & O'Connor, 2003).
2. Verbal rehearsal, a mnemonic technique, helps students learn the processes and strategies until they can recall the problem-solving routine from memory (Scruggs & Mastropieri, 2003). Cue cards and wall charts are phased out as students become more proficient problem solvers (Hutchinson, 1993).
3. Cognitive modelling, or thinking aloud while performing an activity to demonstrate how successful problem solvers solve math problems, is used repeatedly during initial learning and then periodically during practice. Teachers and then students model problem solving, demonstrating both correct and incorrect behaviour (Wong, Harris, Graham, & Butler, 2003).

4. Distributed practice and performance feedback provide students with continuing evaluation and correction as they become more competent. Students learn how to evaluate their own performance and set learning goals from the outset (Adams & Carnine, 2003).
5. Reinforcement is essential for students as they make progress towards mastery. They need specific information about when and why they are successful. Labelled praise indicates precisely the behaviours that contribute towards effective problem solving. As students gain confidence, they can begin to reinforce one another and themselves (Eisenberger & Cameron, 1996).

Assessing and monitoring student progress

Evaluation of student learning is essential to program planning and policy decisions. State, national and international standardised assessments of educational progress in mathematics are conducted regularly to ascertain the various levels of performance of samples of students who are representative of the population. These types of evaluations are used to make decisions about individual students at the local level, policy at the local and state level and to make relative comparisons across countries at the international level. However, there is a different approach to assessment of student learning that has direct implications for instruction. This approach towards progress monitoring is referred to as curriculum-based measurement, or CBM, which has been developed in a research context beginning in the 1980s, particularly in reading and writing (Deno, 2003). Only lately has CBM been extended to monitoring the progress of students in mathematics (Foegen, Jiban, & Deno, 2007; Fuchs, 2004).

CBM is a progress-monitoring system that uses data from regularly administered measures of academic performance; for example addition and subtraction, single-digit computation, or proportions, ratios and probability problems, to formatively evaluate progress on specific concepts, skills or applications within a given domain. CBM entails setting learning goals for students, administering brief probes to determine whether goals are being met and then using the results of the CBM to make instructional decisions regarding specific students. Generally, graphs are used to plot progress across time for individual students who may or may not meet criteria for mastery. If a student meets the learning goal, that is reaches mastery, then new goals are set and progress towards those goals is measured. If a student does not meet the learning goal, then the goal or the instruction is adjusted and data collection is continued pertaining to the new or adjusted goal. Several researchers have developed CBM for mathematics instruction in basic skills (Fuchs, 2004), estimation (Foegen & Deno, 2001), middle-school math problem solving (Montague, 2008) and algebra (Foegen, 2008). However, as with standardised measures, CBM must be technically adequate; that is the measures must be reliable and valid. Additionally, these measures must be easily administered and scored, inexpensive, and effective and efficient.

Teachers, administrators and schools are increasingly being held accountable for the academic performance of students, regardless of their ability and achievement level. Assessment tools are needed to measure students' progress over time to determine whether students are making adequate progress towards learning goals and are meeting the standards and benchmarks needed to advance academically in the curriculum. This is especially important in mathematics, as this academic domain consists of multiple strands and topics. As a result, students must progress both horizontally across topics and vertically as they are challenged by more complex concepts, skills and problems to be solved. CBM is an assessment technique that should assist teachers in making important curricular and instructional decisions for students.

Determining and selecting effective instruction or 'best practice'

In the past decade, there has been a push for education to become like other fields, such as medicine, to rely on proven methods for educating children and, thus, to adopt programs and practices that are 'scientifically based' (NCLB, 2001) or 'research-based'. The most common descriptor of programs and practices that meet certain criteria usually determined by a committee of experts is 'evidence-based'. The United States Department of Education has several initiatives to identify 'evidence-based' programs and practices, the most notable being the What Works Clearinghouse (WWC). The WWC is managed by a research organisation under contract to synthesise research-based instructional programs and practices in a given education area, for example early reading, elementary math, middle-school math, early childhood education, and then to make recommendations about the use of these programs and practices in educational settings, based on the supporting 'scientific evidence'. The mission of the WWC and other synthesising efforts (Braden & Shernoff, 2008; Slavin, 2008) is to guide educators in selecting programs, practices and interventions that have a rigorous research base and are considered exemplary.

Unfortunately, the research base in mathematics and students with learning difficulties in mathematics is relatively sparse and often does not meet the criteria for high-quality research (for example Montague & Dietz, 2009). Many of the studies have not been replicated and, thus, are limited in advancing knowledge of effective practice. There is clearly a critical need for research to identify effective, evidence-based practice for teaching mathematics to students with learning difficulties. What we know is that students with special learning needs fall further behind in mathematics as they progress through school (NMAP, 2008). Geary (2003) estimated the prevalence of mathematical learning disabilities at between 5 and 8% of the school-age population, similar to the estimated prevalence of reading disabilities. Unlike reading, however, performance in mathematics actually may worsen as children progress through school, due to the nature of mathematics learning. Mathematics is based on the acquisition and application of multiple concepts and skills across numerous topics in mathematics; for example geometry, algebra and, most importantly, the ability to apply these concepts and skills to solve numerous types of mathematical problems.

The intervention studies in mathematics with students with LD have focused primarily on basic skills; that is math facts and algorithmic procedures, generally without regard for developing students' conceptual understanding and ability to apply what they have learned. The ultimate test of proficiency in mathematics is successful problem solving at school, at home and in the community. It is well-acknowledged that problem solving should be the focus of mathematics instruction in mathematics classrooms (NCTM, 2000). With the mandate to identify and utilise evidence-based practice in schools, for example NCLB (2001), it is imperative that future research provide direction for making instructional decisions at the individual, school and district levels.

Have past research studies provided any useful information? Do we know anything about what constitutes effective instruction in mathematics? As noted, the research is limited and often flawed, but it does provide some insight into effective practice. Slavin and Lake (2008), in their extensive review of research on programs in elementary mathematics, concluded that programs focusing on effective instructional practices and designed to change teaching behaviours and practices in the classroom are more promising than programs focusing simply on the curriculum or technology supports. Past research with students with LD tells us that effective instruction generally includes instructional principles and practices associated with direct instruction and cognitive-strategy instruction, and these principles and practices are embedded in most math interventions; for example, demonstration and modelling, verbal rehearsal, guided practice, corrective and positive feedback, independent practice, mastery and distributed practice. Several researchers are dedicated to conducting a series of studies using a particular approach or model; for example Jitendra's schema-based problem-solving model and Montague's *Solve It!*, which further substantiates the effectiveness of an intervention for different groups of students under different conditions. This perhaps is the most promising avenue if we are to identify what works, with whom it works and under what conditions. The research in mathematics interventions for students with learning difficulties, albeit limited, does indeed provide a base for moving to the next level; that is large-scale studies that provide evidence of effectiveness in authentic classroom settings.

The field of education is at a turning point; that is, it is going through a series of rapid changes that will presumably enlarge the knowledge base and improve the research methods to test that knowledge base. The resultant 'evidence base' could be used to make solid and informed decisions about what to teach children, how to teach children and under what conditions to teach them in order that all children achieve to potential. This is especially important in mathematics education, an area that has been relatively neglected but is emerging, along with science education, as critical to the academic and vocational success of students.

Essential next questions

When evaluating 'evidence-based' instructional programs and practices, it is important to describe the children who benefited from instruction, the settings and conditions under which the children received instruction, and the background and expertise of the teachers who delivered the instruction. So, the essential next questions for researchers have to do with who will benefit from the instructional program or practice, what conditions are important for optimal results and by whom instruction should be provided. These are essential questions because evidence-based instruction may have differential effects for children, for a variety of reasons. Educators who invest the effort, time and money in evidence-based instructional programs and practices must be assured and confident that they have made the most appropriate selections for the children they teach. To conclude this chapter, I provide suggestions regarding future research directions and examples from my intervention research in math problem solving that may help to answer these questions.

Who will benefit from the instructional program or practice?

Understanding the characteristics of the students who participated in the studies that led to the conclusion that a particular instructional program or practice is effective, based on empirical evidence, is critical to the selection of programs. For example, *Solve It!* was designed for, and implemented with, middle-school students with LD. The participants in the validation studies met certain criteria; that is average IQ, at least a Grade 3 reading level, low math problem solving performance and proficiency performing the four basic math operations with whole numbers and decimals. The participating students made substantial gains in problem solving and performed as well, as their normally achieving peers (Montague & Bos, 1986; Montague, 1992; Montague, Applegate, & Marquard, 1993). Additional studies, however, were needed to determine whether the intervention, perhaps with modifications, was beneficial for younger students, students with other types of disabilities or students with and without LD who have math difficulties. To illustrate, a 3-year, federally funded study currently underway is investigating the effects of *Solve It!* for students without disabilities who have math problem-solving difficulties. Preliminary findings suggest that low- and average-achieving students who have difficulty solving mathematical word problems may also benefit from the instructional program (Montague, 2008). Studies such as this lend support for the generalisability of instructional programs and are needed to establish the external validity of 'evidence-based' practices.

What conditions are important for optimal results?

A complete description of the setting and conditions under which the intervention was implemented in research studies is critical to understanding where, when and how the intervention was found to be effective. To continue with the previous example, in the early validation studies of *Solve It!*, the students with LD were removed from their general education classrooms and taught either individually (Montague & Bos, 1986; Montague, 1992) or in mid-sized groups of 8–14 students (Montague et al., 1993). Whether this instructional program is effective in other settings, such as the general education classroom, and how that setting may affect the performance of students with LD as well as other students are important concerns that need to be investigated. The current study (Montague, 2008) is being conducted in middle-school, general education classrooms. Results thus far suggest that the intervention can be effective if implemented with reasonable integrity in the context of general education classrooms.

By whom should instruction be provided?

The teacher is perhaps the most important variable in delivering 'evidence-based' instructional programs and practices. In mathematics instruction, quality teaching is particularly critical. That is, teachers must possess the pedagogical content knowledge and be able to implement instructional programs with the knowledge and expertise required to replicate evidence-based programs and practices. For example, research on algebra teaching has indicated that teachers' pedagogical content knowledge is questionable and needs to include knowledge of different representations, alternative ways of developing and strengthening conceptual knowledge, a basic repertoire of examples, knowledge of mathematics and algebra concepts and, very importantly, knowledge of students' ways of thinking about various aspects of algebra, particularly functions (Even, 1993) and functional relationships (Star, Herbal-Eisenmann, & Smith, 2000).

How much professional development is required before teachers can implement an instructional program or practice with fidelity needs to be established through research. Again, to use the *Solve It!* intervention studies as an example, teacher variability was a paramount concern. The general education, middle-school math teachers participating in the study were nominated by their administrators. The researchers had requested high-quality, certified math teachers who were open to new ideas and teaching practices. Despite the criteria and considerable professional development, teacher behaviour was highly variable, as determined by frequent classroom observations. Additionally, because the focus was on low-performing students with math problem-solving difficulties, particularly students with LD, the teachers needed to have some background and knowledge of the characteristics of these students. Results indicated that these students did not improve at the same rate or to the same degree as the average-achieving students who also had math problem-solving difficulties (Montague, 2008), which suggests that general education teachers may need additional training in specialised techniques that address the characteristics of students with LD, in particular.

In conclusion, these are some of the essential questions that must be asked before an instructional program or practice can be deemed effective or 'evidence-based'.

Claims about effective instruction must be warranted and validated by solid research that meets stringent evaluation criteria. In recent years, various groups in the United States have been engaged in establishing criteria upon which studies and bodies of research can be judged (see Braden & Shernoff, 2008; Odom et al., 2005; Slavin, 2008). Future research in mathematics instruction for students with mathematical difficulties not only must meet these criteria, but also must be able to ascertain and describe for whom the instructional programs and practices are effective, as well as the instructional conditions and settings that promote learning for these students.

Glossary

Academically typical children Children who are average achievers and typically score within the average range; that is plus or minus one standard deviation, on standardised academic measures.

Cognitive-strategy instruction Instruction that focuses on teaching youngsters a range of cognitive and metacognitive processes, strategies or mental activities to facilitate learning and improve performance utilising explicit instructional procedures; for example cognitive modelling.

Curriculum-based measurement (CBM) Often referred to as progress monitoring, CBM is an 'empirically developed approach to formative evaluation that relies on frequent assessment using brief measures that serve as indicators of general proficiency in a content area' (Foegen, 2008, p. 66).

Elaboration knowledge Knowledge of how to develop a schematic diagram or template that corresponds with the representation of a math problem by interpreting the information in the problem and elaborating on the main features of the schema (Xin & Jitendra, 2006).

Execution knowledge Consists of techniques that lead to solutions for math problems, including computation and checking the problem (Xin & Jitendra, 2006).

Learning disabilities 'A disorder in one or more of the basic psychological processes involved in understanding or in using language, spoken or written, that may manifest itself in an imperfect ability to listen, think, speak, read, write, spell or do mathematical calculations' (Individuals with Disabilities Education Act, 1999, p. 12422).

Mathematics A science that deals with the relationship and symbolism of numbers and magnitudes, and that includes quantitative operations and the solution of quantitative problems (Webster's Third New International Dictionary, 1986, p. 1393).

Procedural deficits sub-type A developmental delay that underlies frequent use of immature mathematical procedures and errors in their execution, poor conceptual understanding of procedures and difficulties in sequencing multi-step procedures (Geary, 2003, p. 205).

Schema knowledge Knowledge of the various math problem types in addition, subtraction, multiplication and division, and understanding their distinct core features (Xin & Jitendra, 2006).

Semantic memory deficits sub-type A heritable developmental deficit that manifests in an inability to retrieve mathematical facts, resulting in high error rates and unsystematic retrieval times (Geary, 2003, p. 205).

Strategic knowledge Developing a plan to solve a math problem by setting up goals and sub-goals, selecting the appropriate operation and writing the math sentence or equation (Xin & Jitendra, 2006).

Visuospatial sub-type Appears to be genetic, manifesting in 'difficulties in spatially representing numerical and other forms of mathematical information and relationships' and 'frequent misinterpretation or misunderstanding of spatially represented information' (Geary, 2003, p. 205).

References

Adams, G., & Carnine, D. (2003). Direct instruction. In H. L. Swanson, K. R. Harris, & S. Graham (Eds.), *Handbook of learning disabilities* (pp. 403–416). New York: Guilford.

Baxter, J., Woodward, J., & Olson, D. (2001). Effects of reform-based mathematics instruction in five third-grade classrooms. *Elementary School Journal, 101*, 529–548.

Braden, J. P., & Shernoff, E. S. (2008). Why the need for evidence-based interventions? In R. J. Morris & N. Mather (Eds.), *Evidence-based interventions for students with learning and behavioral challenges* (pp. 9–30). New York: Routledge.

Burns, M. K. (2005). Using incremental rehearsal to increase fluency of single-digit multiplication facts with children identified as learning disabled in mathematics computation. *Education and Treatment of Children, 28*, 237–249.

Case, R. (1985). *Intellectual development: Birth to adulthood.* San Diego, CA: Academic Press.

Deno, S. L. (2003). Developments in curriculum-based measurement: The emerging alternative. *Exceptional Children, 52*, 219–232.

Eisenberger, R., & Cameron, J. (1996). Detrimental effects of reward: Reality or myth? *American Psychologist, 51*, 1153–1166.

Even, R. (1993). Subject-matter knowledge for teaching and the case of functions. *Educational Studies in Mathematics, 21*, 521–544.

Flavell, J. H., Miller, P. H., & Miller, S. A. (1993). *Cognitive development.* Englewood, NJ: Prentice Hall.

Foegen, A. (2008). Algebra progress monitoring and interventions for students with learning disabilities. *Learning Disability Quarterly, 31*, 65–78.

Foegen, A., & Deno, S. L. (2001). Identifying growth indicators for low-achieving students in middle school mathematics. *Journal of Special Education, 35*, 4–16.

Foegen, A., Jiban, C. L., & Deno, S. L. (2007). Progress monitoring measure in mathematics: A review of the literature. *Journal of Special Education, 41*, 121–139.

Fuchs, L. S. (2004). The past, present, and future of curriculum-based measurement research. *School Psychology Review, 33*, 188–192.

Fuchs, L. S., & Fuchs, D. (2001). Principles for the prevention and intervention of mathematics difficulties. *Learning Disabilities Research and Practice, 16*, 85–95.

Fuchs, L. S., & Fuchs, D. (2003). Enhancing the mathematical problem solving of students with mathematics disabilities. In H. L. Swanson, K. R. Harris, & S. Graham (Eds.), *Handbook of learning disabilities* (pp. 306–322). New York: Guilford.

Funkhouser, C. (1995). Developing number sense and basic computational skills in students with special needs. *School Science and Mathematics, 95*, 236–239.

Geary, D. C. (2003). Math disabilities. In H. L. Swanson, K. R. Harris, & S. Graham (Eds.), *Handbook of learning disabilities* (pp. 199–212). New York: Guilford.

Gersten, R., & Chard, D. (1999). Number sense: Rethinking arithmetic instruction for students with mathematical disabilities. *The Journal of Special Education, 44*, 18–28.

Harris, C. A., Miller, S. P., & Mercer, C. D. (1995). Teaching initial multiplication skills to students with disabilities in general education classrooms. *Learning Disabilities Research and Practice, 10*, 180–195.

Hudson, P., & Miller, S. P. (2006). *Designing and implementing mathematics instruction for students with diverse learning needs*. Boston: Allyn and Bacon.

Hutchinson, N. L. (1993). Effects of cognitive strategy instruction on algebra problem solving of adolescents with learning disabilities. *Learning Disability Quarterly, 16*, 30–39.

Individuals with Disabilities Education Act (IDEA). (1999, March 12). Final regulations and comments from the United States department of education. *Federal Register, 64*(48), 12406–12672.

Janvier, C. (1987). *Problems of representation in the teaching and learning of mathematics*. Mahwah, NJ: Laurence Erlbaum.

Jenkins, J. R., & O'Connor, R. E. (2003). Cooperative learning for students with learning disabilities: Evidence from experiments, observations, and interviews. In H. L. Swanson, K. R. Harris, & S. Graham (Eds.), *Handbook of learning disabilities* (pp. 417–430). New York: Guilford.

Jitendra, A., DiPipi, C. M., & Perron-Jones, N. (2002). An exploratory study of schema-based word-problem-solving instruction for middle school students with learning disabilities: An emphasis on conceptual and procedural understanding. *The Journal of Special Education, 36*, 23–38.

Jitendra, A. K., Griffin, C. C., McGoey, K., Gardill, M. C., Bhat, P., & Riley, T. (1998). Effects of mathematical word problem solving by students at risk or with mild disabilities. *The Journal of Educational Research, 91*, 345–355.

Jitendra, A. K., & Hoff, K. (1996). The effects of schema-based instruction on the mathematical word-problem-solving performance of students with learning disabilities. *Journal of Learning Disabilities, 29*, 422–431.

Jitendra, A. K., Hoff, K., & Beck, M. M. (1999). Teaching middle school students with learning disabilities to solve word problems using a schema-based approach. *Remedial and Special Education, 20*, 50–64.

Kroesbergen, E. H., & van Lui, J. E. H. (2002). Teaching multiplication to low-math performers: Guided versus structured instruction. *Instructional Science, 30*, 361–378.

Kroesbergen, E. H., & van Luit, J. E. H. (2003). Mathematics interventions for children with special educational needs. *Remedial and Special Education, 24*, 97–114.

Mayer, R. E. (1985). Mathematical ability. In R. J. Sternberg (Ed.), *Human abilities: Information processing approach* (pp. 127–150). San Francisco: Freeman.

Mercer, C. D., & Miller, S. P. (1992). Teaching students with learning problems in math to acquire, understand, and apply basic math facts. *Remedial and Special Education, 13*, 19–35.

Miller, S. P., Harris, C., Strawser, S., Jones, W. P., & Mercer, C. (1998). Teaching multiplication to second graders in inclusive settings. *Focus of Learning Problems in Mathematics, 20*, 50–70.

Miller, S., & Mercer, C. (1997). Education aspects of mathematics disabilities. *Journal of Learning Disabilities, 30*, 47–56.

Miller, S. P., Mercer, C. D., & Dillon, A. (1992). Acquiring and retaining math skills. *Intervention, 28*, 105–110.

Montague, M. (1992). The effects of cognitive and metacognitive strategy instruction on mathematical problem solving of middle school students with learning disabilities. *Journal of Learning Disabilities, 25*, 230–248.

Montague, M. (2003). *Solve it! A mathematical problem solving instructional program*. Reston, VA: Exceptional Innovations.
Montague, M. (2006). Self-regulation strategies for better math performance in middle school. In M. Montague & A. Jitendra (Eds.), *Teaching mathematics to middle school students with learning difficulties* (pp. 89–107). New York: Guilford Press.
Montague, M. (2008). *Improving mathematics performance of at risk students and students with learning disabilities in urban middle schools*. Unpublished raw data.
Montague, M., & Applegate, B. (1993). Mathematical problem-solving characteristics of middle school students with learning disabilities. *Journal of Special Education, 27*, 175–201.
Montague, M., Applegate, B., & Marquard, K. (1993). Cognitive strategyninstruction and mathematical problem-solving performance of students with learning disabilities. *Learning Disabilities Research and Practice, 29*, 251–261.
Montague, M., & Bos, C. S. (1986). The effect of cognitive strategy training on verbal math problem solving performance of learning disabled adolescents. *Journal of Learning Disabilities, 19*, 26–33.
Montague, M., & Dietz, S. (2009). Evaluating the evidence-base for cognitive strategy instruction and mathematical problem. *Exceptional Children, 75*, 285–382.
Montague, M., & van Garderen, D. (2008). Effective mathematics instruction. In R. J. Morris & N. Mather (Eds.), *Evidence-based interventions for students with learning and behavioral challenges* (pp. 236–257). New York: Routledge.
Montague, M., Warger, C., & Morgan, H. (2000). Solve it! Strategy instruction to improve mathematical problem solving. *Learning Disabilities Research and Practice, 15*, 110–116.
National Assessment of Educational Progress. (2003). *NAEP 2002 mathematics report card for the nation and the states*. Princeton, NJ: Educational Testing Service.
National Council of Teachers of Mathematics (NCTM). (1989). *Curriculum and evaluation NCTM standards for school mathematics*. Reston, VA: The National Council of Teachers of Mathematics, Inc.
National Council of Teachers of Mathematics (NCTM). (2000). *Principles and NCTM standards for school mathematics*. Reston, VA: The National Council of Teachers of Mathematics, Inc.
National Mathematics Advisory Panel (NMAP). (2008). *Foundations for success: The final report of the National Mathematics Advisory Panel*. Washington, DC: United States Department of Education.
No Child Left Behind Act. (2001). Reauthorization of the Elementary and Secondary Education Act. Pub. L. 107–110, 2102(4).
Odom, S. L., Brantlinger, E., Gersten, R., Horner, R. H., Thompson, B., & Harris, K. R. (2005). Research in special education: Scientific methods and evidence-based practices [Special issue]. *Exceptional Children, 71*, 2.
Peterson, P. (1988). Teachers' and students' cognitional knowledge for classroom teaching and learning. *Educational Researcher, 17*, 5–14.
Peterson, S. K., Mercer, C. D., & O'Shea, L. (1988). Teaching learning disabled students place value using the concrete to abstract sequence. *Learning Disabilities Research and Practice, 4*, 52–56.
Scruggs, T. E., & Mastropieri, M. A. (2003). Science and social studies. In H. L. Swanson, K. R. Harris, & S. Graham (Eds.), *Handbook of learning disabilities* (pp. 364–381). New York: Guilford.
Silver, E. A. (1987). Foundations of cognitive theory research for mathematics problem solving. In A. Schoenfeld (Ed.), *Cognitive science and mathematics education* (pp. 33–60). Mahwah, NJ: Laurence Erlbaum.
Slavin, R. E. (2008). What works? Issues in synthesizing educational program evaluations. *Educational Researcher, 37*, 5–14.
Slavin, R. E., & Lake, C. (2008). Effective programs in elementary mathematics: A best-evidence synthesis. *Review of Educational Research, 78*, 427–515.

Star, J. R., Herbel-Eisenmann, B. A., & Smith, J. P. (2000). Algebraic concepts: What's really new in new curricula? *Mathematics Teaching in the Middle School, 5*, 446–451.

Sternberg, R. (1985). *Beyond IQ: A triarchic theory of human intelligence*. Cambridge, UK: Cambridge University Press.

Swanson, H. L. (1999). Instructional components that predict treatment outcomes for students with learning disabilities: Support for a combined strategy and direct instruction model. *Learning Disabilities Research & Practice, 16*, 109–119.

Trends in International Mathematics and Science Study (TIMSS). (2004). *Highlights from the Trends in International Mathematics and Science Study: TIMSS 2003*. Washington, DC: National Center for Education Statistics.

Tournaki, N. (2003). The differential effects of teaching addition through strategy instruction versus drill and practice to students with and without learning disabilities. *Journal of Learning Disabilities, 36*, 449–458.

van Garderen, D., & Montague, M. (2003). Visual-spatial representations and mathematical problem solving. *Learning Disabilities Research & Practice, 18*, 246–254.

Van Luit, J. E. H., & Naglieri, J. A. (1999). Effectiveness of the MASTER program for teaching special children multiplication and division. *Journal of Learning Disabilities, 32*, 98–107.

Vygotsky, L. S. (1989). *Thought and language*. Cambridge, MA: MIT Press.

Websters Third New International Dictionary. (1986). (p. 1393). Springfield, MA: Merriam-Webster, Inc.

Wong, B. Y. L., Harris, K. R., Graham, S., & Butler, D. (2003). Cognitive strategies instruction research in learning disabilities. In H. L. Swanson, K. R. Harris, & S. Graham (Eds.), *Handbook of learning disabilities* (pp. 383–402). New York: Guilford.

Woodward, J. (2006). Making reform-based mathematics work for academically low-achieving middle school students. In M. Montague & A. Jitendra (Eds.), *Teaching mathematics to middle school students with learning difficulties* (pp. 29–50). New York: Guilford Press.

Xin, P. V., & Jitendra, A. K. (2006). Teaching problem-solving skills to middle school students with learning difficulties: Schema-based strategy instruction. In M. Montague & A. Jitendra (Eds.), *Teaching mathematics to middle school students with learning difficulties* (pp. 51–71). New York: Guilford Press.

Chapter 15
Language, Culture and Learning Mathematics: A Bourdieuian Analysis of Indigenous Learning

Robyn Jorgensen (Zevenbergen)

Indigenous students in Australia perform poorly on testing measures (MCEETYA, 2009). This is of national concern and a priority for government, as evidenced in the 'Closing the Gap' initiative (FaHCSIA, 2009). Geographical location and poverty compound issues of indigeneity, so that Indigenous students in remote locations are most at risk of performing poorly on measures of literacy and numeracy. In this chapter, I seek to challenge the orthodoxy that poor performances among remote/Indigenous students are a consequence of constructs of ability or learning difficulties per se. Rather, I seek to illustrate how the mathematics curriculum delivered to Indigenous students represents a particular cultural form. This is particularly poignant as Australia moves to a national curriculum (National Curriculum Board, 2008). The difficulties in learning mathematics experienced by many Indigenous students can be thought of as a confrontation of language differences (and, by implication, culture). From this perspective, coming to learn mathematics is about 'cracking the code' through which mathematical concepts and processes are embedded and relayed, so that learning difficulties are viewed as structural difficulties rather than individual difficulties. By reconceptualising the 'learning difficulties' experienced by Indigenous learners in mathematics/numeracy, a more inclusive approach to educational reform can be envisaged and enacted.

For this chapter, the notion of 'learning difficulties' is understood to be a subversive process through which the failure of Indigenous students becomes reified through various practices such as curriculum, pedagogy and assessment. In so doing, this reification engenders a view of failure in mathematics as a characteristic of the learner rather than a process of structural exclusion. Using the theoretical constructs offered through the writings of Bourdieu, I explore the notion of practice, which in this case relates to the field of mathematics education, and how it is implicated in the exclusion of Indigenous learners, particularly those living in remote areas. The

R. Jorgensen (Zevenbergen) (✉)
Griffith Institute for Educational Research, Griffith University, Brisbane, QLD, Australia
e-mail: r.jorgensen@griffith.edu.au

C. Wyatt-Smith et al. (eds.), *Multiple Perspectives on Difficulties in Learning Literacy and Numeracy*, DOI 10.1007/978-1-4020-8864-3_15,

specific focus is on how the language practices of school mathematics become a barrier for Indigenous students.

Drawing on a number of projects in which I worked with Australian Indigenous communities,[1] where the focus has been on improving numeracy learning, the intersection of literacy and numeracy becomes important. It is this intersection that is the focus of this chapter. Depending on their geographical location, Indigenous students may come to school with a home language spoken in their community; a Kriol that allows them to interact with speakers of other Indigenous languages with whom they come in contact; and as they enter the school context, they encounter English as the medium of instruction. Many students are multilingual, with minimal support being offered for the English as a second language (ESL) status. For many remote Indigenous students, English is spoken only in the school and in contexts such as regional shopping centres or in interactions with agencies. In this way, coming to learn English and the cultural practices of formal schooling represent challenges for many Indigenous learners. Thus, the 'learning difficulties' faced by these students are not biological or inherently individual features but rather represent one of cultural and linguistic conflict. In this chapter, I use the notion of 'learning difficulties' to mean the problematic clash of language and culture as Indigenous students encounter Western mathematics education.

'Learning difficulties' as symbolic violence

Rather than viewing 'learning difficulties' as characteristic of the individual, I draw on Bourdieu's work to argue that the 'learning difficulties' experienced by Indigenous learners are acts of symbolic violence. I propose that the practices within which mathematics are taught/learned by students are structured in subtle and coercive ways that limit the success of particular groups of students. Taking such a perspective makes it possible to understand more fully the systematic exclusion of some groups of students from participating in the field of mathematics. A Bourdieuian approach enables a richer theorisation of the reproduction of power through school mathematics. Such an approach enables an understanding of why it takes considerably more effort for teachers and learners to enable students from those backgrounds who are typically at risk of failing school mathematics to be successful in their study of this discipline. Drawing on the work of Bourdieu, I elucidate how the practice of school mathematics can be seen as one in which the ways of thinking, acting, talking and working mathematically have become reified through documents to define what is seen as legitimate knowledge. Those students who have this knowledge and can display it appropriately are more likely to be considered successful learners.

[1]The data and examples used in this chapter pre-date my employment at Yulara and in no way should be inferred to be representing the Anungu people.

Bourdieu argues that the habitus is the embodiment of the social, which enables the person to be at one (or not) with the field in which they are located.

> Habitus being the social embodied, it is 'at home' in the field it inhabits, it perceives its [knowledge] immediately as endowed with meaning and interest. (Bourdieu & Wacquant, 1989, p. 128)

The primary habitus is shaped by the family and is brought into the early years of schooling. This habitus in turn shapes, enables, enhances or excludes students from participating in much of the discourse of mathematics.

This comfort, or 'feel' for the game of mathematics, is referred to as 'doxa'. Bourdieu further argues that the feeling for the game, in this case mathematics, is often at the level of the unconscious and, as such, it enables the reproduction of the field (and power).

> The earlier a player enters the game and the less he is aware of the associated learning, the greater his ignorance of all that is tacitly granted through his investment in the field and his interest in it's very existence and perpetuation and in everything that is played for it, and his unawareness of the unthought presuppositions the game produces and endlessly reproduces, thereby reproducing the conditions of its own existence. (Bourdieu, 1990, p. 67)

For example, consider the young child who comes to school knowing how to count, classify and articulate the names of geometric shapes and other mathematical objects. The preschool familial practices have enabled the child to acquire particular forms of knowledge that are valued within the practices of school mathematics. That is, the child has embodied features of the culture (namely numbers) so that these are now part of the habitus that is now brought into the school context. The child is able to display this knowledge (or habitus) in ways prescribed by the field. When the child displays, for example, particular counting skills, then the teacher ascribes the child to the category of being an 'effective counter'.

This habitus now acts as a medium through which the child displays and acquires new forms of knowledge. This habitus is now a form of capital within the field that is converted to symbolic power, such as status of the learner. Conversely, in many remote Indigenous communities, there is little exposure to numbers—the houses are not numbered, few homes (if any) have a telephone, the local store may not display prices, some people do not have their birthdates recorded—so the experience of numbers is considerably different from that of urban/city learners, who are immersed in numbers. For remote Indigenous learners, the potential for developing a habitus rich in numbers is very limited, restricting the transition into the early years curriculum. Thus, the 'learning difficulties' prescribed to many Indigenous students are not some inherent deficits but a difference in the habitus valued within the field. The unawareness of how numbers are a taken-for-granted form of knowledge in Western epistemology enables the exclusion of cultures that bring to school different ways of knowing; that is a different habitus, that is not valued. Knowing how to count becomes an 'at oneness' with the field and hence is not questioned and, as such, supports the knowledge structures of some fields while excluding others.

For Indigenous learners, the home environment offers different potential for meaning making and language development. As has been shown by Watson and

Chambers (1989), Yolgnu[2] people have a rich repertoire of spatial knowledge that is configured in ways that are very different from that represented in school mathematics. History, people and events figure strongly in the ways of mapping the land. The learners bring to school mathematics a rich language and conceptualisation of spatial representation, but this is not recognised or given attention in the curriculum, pedagogy or assessment practices. As such, the spatial habitus of Yolgnu learners does not convert to scholastic capital with the formal school system. Rather, the lack of particular Western constructs works against their potential for success in school, since the curriculum and/or assessment does not recognise their cultural and linguistic knowledge. In the field of Indigenous communities, the spatial knowledge they develop, and the habitus in which this is embedded, is valued and hence scholastic merit is applied to the learner in this field, but it is a field different from that of school mathematics. It is here that the tautological relationships of Bourdieu's constructs become apparent. It is not possible to consider field, habitus and capital independent of each other.

Structured failing: the field of school mathematics

By seeing mathematics as a discipline that represents objective facts, the discourses of teaching and knowledge remain unchallenged. As such, the unconscious actions of educators often support the teaching of mainstream mathematics when working with Indigenous students whose cultures and languages are not integrated within the practice of instruction/teaching. When assuming that mathematics is acultural, educators are at risk of not recognising the strong ways in which mathematics represents a particular culture or, more specifically, cultural practices that may be different from those practices that the students bring to school.

The teaching of mathematics is culturally laden. Consider the example of the teaching of number facts. In this process, number facts are typically seen as the precursor to addition; if students do not recognise the number 4, then it is assumed that addition of two 4s will not be possible. However, when teaching is considered a cultural act whereby particular elements of that culture are embedded within the act of teaching, such assumptions are challenged. These may be evident in the language being used, the ways of interacting, the knowledge being represented, but always remaining invisible to the participants—both teachers and students. It is more so the case in mathematics than any other curriculum area. For example, when teaching number, the approach often requires students to count objects using one-to-one correspondence. Indeed, curriculum documents suggest that one-to-one correspondence supports early counting. Challenging the hegemony of Western counting sequences, Willis (2000) found that some Indigenous students subitised very effectively before they entered formal schooling; that is recognised groups of animals not

[2] Yolgnu country is the north-eastern corner of the Northern Territory, Australia.

by counting but by recognising the group. Forcing students to count using one-to-one correspondence is counter to the skills that these learners brought to the learning situation and, indeed, subitsing is a more advanced skill than is aspired to in school mathematics. As such, one could argue that the ways of organising curriculum is a particular view of the world and when the students' worldview (in the case of Indigenous students their ability to subitise) their cultural knowledge is not recognised as legitimate—even though later on in the curriculum organisation subitising is seen to be a key skill in group recognition. For many Indigenous students, the early years number curriculum forces them to unlearn their numeracy knowledge (in this case subitising) through the structured counting process. Through this unlearning, they come to learn that for the first 3 years of schooling they do not know very much. This can be internalised so as to constitute a mathematical habitus that is often one of 'I cannot do maths'. For Bourdieu, this pattern of practice is critical to understanding the ways in which the field operates to exclude (or include) particular cultural knowledges.

Language, mathematics and linguistic capital

Part of being successful in school mathematics is being able to speak its language. The words spoken in school mathematics never exist in isolation, but are an integral part of the overall discourse that operates within that context. This discourse is part of the field, but in a dialectic relationship with the habitus of those who participate in that discourse. Bourdieu (1991) defines it thus

> The form and content of a discourse depend on the relation between the habitus (itself a the product of the sanctions of a market of a given tension) and the market (field) defined by a level of tension which is more or less heightened, hence by the severity of the sanctions it inflicts on those who pay insufficient attention to 'correctness' and to the 'imposition of form' which formal usage presupposes. (p. 79)

The language of mathematics represents a particular social language—that of the White, middle class. Consider, for example, the processes of comparisons through the use of binary opposites when comparing groups; the terms 'more' and 'less' are used frequently. In her intensive study of mother–child interactions, Walkerdine and Lucey (1989) reported that middle-class mothers were most likely to use the signifiers 'more' and 'less' in their interactions with their children. In comparison, she also noted that working-class mothers were more likely to use the term 'more' only. Similar differences have been noted in Indigenous languages (Zevenbergen, Mousley, & Sullivan, 2004). In the early years of mathematics there are considerable learnings through comparisons—which number is more, which is less, what number is 3 more than 6, what number is 2 less than 9. Similarly, comparisons in the measurement strand are just as common, but more likely to use terms such as wide/narrow, tall/short. Therefore, middle-class, English-speaking students are more likely to use both terms with almost equal fluency, but the term 'less' is not used with students from other social and cultural groups such as working-class and

Indigenous students. As such, when the teacher embarks on activities and questions such as 'Which group has more?' 'Which group has less?' 'Which number is two more than five?' 'Which number is two less than five?' some students gain greater or lesser access to the mathematics, depending on their home language. The language becomes a tool for relaying concepts to students. Where they have the language that aligns with the classroom language, they have greater access to mathematical ideas and knowledge, whereas the converse is also true.

Recognising that language is the key medium through which learning is facilitated, it becomes important to identify the ways in which language and culture become barriers to learning mathematics for students whose language is not aligned with school mathematics. There are now many resources available to support teachers in recognising how the languages and cultures that Indigenous students bring to school are substantially different from those that are typically part of the teaching repertoire. The depth of knowledge that has been brought to bear on the range of teaching support materials is critical to good teaching when working with Indigenous students, if success is to be had by all students. It is not the case that teaching mainstream knowledge will suffice. There is now a longstanding body of evidence to show that the imposition of Western knowledge and practices is failing Indigenous students. As such, a radical change in teaching practice needs to be adopted that will enable students (Indigenous and other students who have been traditionally excluded from schooling) to become active learners. To do this successfully, legitimating the language and cultures in and through practice becomes essential.

The role of language in mathematics learning

Often, the learning of mathematics is seen to be a process of learning mathematical ideas, concepts and processes. Increasing the ways in which mathematical ideas are taught to students is seen as problematic. The teacher may have an idea in their head as to what is to be taught, but the way in which the students interpret the discussion can be quite idiosyncratic. When students respond to teacher questions, there is considerable scope for interpretation (and misinterpretation) by both teachers and students. Once the teaching of mathematics was not seen to have a strong relationship with language, but this is now changing so that the role of language in the teaching of mathematics increasingly is recognised. Consider the example in which a teacher poses the problem of $2 \times 5 =$ and represents this on a whiteboard. The teacher's explanation goes something like 'So, if I have five groups of two … five times two … How much is that?' The teacher's goal is to convey the concept of multiplication. Students sitting at their desks can make multiple interpretations, such as identifying the key of 'much' and seeing it as a problem about money; identifying the key word of 'time' and seeing it as a problem about 'o'clock'. Thus, the language being used to convey a concept can create other difficulties for learning.

Typically, the teacher has an idea that they need to convey to the students, but students interpret the teacher's talk in particular ways, such as shown above. When teaching is problematised and the impact of language is recognised, the focus is increasingly on the pedagogy, rather than the student. Recognising the potential communication problems inherent in the pedagogic interaction can create another level of complexity. A greater recognition of language enables a richer interpretation of success (and failure) in school mathematics.

It is here that Bourdieu's approach has practical application. It can be seen how the teachers' doxa with the game of mathematics creates a blindness to the problematic nature of the field in terms of its potential to exclude students. The misrecognition of language and concepts works to exclude Indigenous students from the game.

In adopting a language approach to teaching mathematics, it becomes important to recognise the specific language of mathematics. Much like any other language, mathematics has its unique features. When the language of instruction is standard Australian English (SAE), it is recognised that the greater the difference between the home language and the school language, the greater the difficulty in coming to learn school mathematics. This phenomenon has been well-documented across social class backgrounds (Bernstein, 1996; Cooper & Dunne, 1999; Zevenbergen, 2000) and non-English speakers (Leder, Rowley, & Brew, 1995). Similar issues arise for Indigenous speakers whose languages range along a continuum depending on their backgrounds. English may be their first language, but the register is different from that of school. Others, may speak Koori English or Murri English[3] so that there are differences in language structure, whereas Pidgins and Creoles are further distanced from SAE. Finally, there are those students whose first (and other languages) are those of their traditional language. In many groups, the need to travel into other people's country means there may be a need to speak multiple languages. It is not uncommon for Indigenous children to enter school speaking up to four different languages and having to learn SAE as they come in contact with the school system.

Language and worldviews

Language frames how we see and interpret the world, but language is also a representation of the world as it is interpreted. Wittgenstein (1953, 1967) argued that through language games, the world becomes interpreted and constructed. For Wittgenstein (1967) and others (Kanes, 1991; Watson, 1989) who have drawn on his work in mathematics, it is proposed that the mathematical 'facts' that are seen to be part of the discourse of school mathematics—such as 'the sum of the internal angles of a triangle add up to 180°'—are not so much facts, but conventions that have been accepted as norms within particular social groups. For example, consider

---

[3]'Koori' and 'Murri' are terms used by Aboriginal peoples of the eastern regions of Australia in reference to themselves. 'Koori' refers to Aboriginal people from Victoria northwards to approximately halfway through New South Wales (NSW). 'Murri' refers to Aboriginal people from midway through NSW to southern Queensland.

mapping conventions. There are particular ways of representing the land in school mathematics, and often these employ particular conventions (scale, icons, etc.). In contrast, in the Garma project in Northern Territory (Watson & Chambers, 1989), the Yolgnu people were involved in a 'both ways' education programme (Watson, 1988). Through this approach, students were exposed to traditional mapping conventions whereby events (such as burials or meetings) were represented according to culturally significant events. Thus, the two different approaches to mapping can be seen as socially negotiated artifacts that become a legitimate part of the game of school mathematics.

Wittgenstein (1967) argued that what are seen as mathematical facts or knowledge are events that have become an accepted part of the culture that has grown out of language games. The dominant culture fails to recognise that the language games played construct particular forms of knowledge, seeing such knowledge as 'natural' or 'normal'. Many Indigenous students come to school with other language games and forms of life. In their out-of-school experiences, different games are played according to the needs of their cultures. For example, depending on their living circumstances, language games are developed to reflect the needs of the community. In her work with people in Northern Territory, Harris (1990) noted that mothers holding their babies and young children talked about directions (north, south, etc.) so that the young members gained a strong sense of direction. This has often been interpreted by Westerners as 'an inbuilt compass'. For the city child whose 'natural' language has a strong component of shape language, the experience of coming to school is qualitatively different from the Indigenous student whose 'natural' language has developed a keen sense of direction. In the early years of schooling, the experiences of the city student are more strongly aligned with the curriculum than those of the Indigenous student. Directional knowledge is part of the curriculum, but in the upper years. Before an Indigenous student gets to experience this success, some 5–6 years of school practices have positioned them as knowing very little.

In thinking about the links between objects and language, different cultures develop language to reflect their views of the world. Depending on their lived experiences, different conditions create different needs that are, in turn, expressed through language. It can be argued that the rich language of shape in Western language has developed through the need for a language to describe the built environment. In contrast, in cultures where there is not a built environment, there is little need for a complex language of shape. Consider the words that are used to describe four-sided shapes—square, rectangle, oblong, trapezium, parallelogram, convex, concave and so on—and that assume that the four-sideness is seen as a defining characteristic. The complexity of the language is shaped by the demands of activities situated within a particular cultural context.

Language, habitus and capital

In the preceding sections, I have drawn on a range of perspectives that seek to highlight the problematic nature of language in the learning of mathematics. To unite these literatures, Bourdieu's theoretical project offers a coherent framing. The field, in this case school mathematics, has particular practices that valorise particular ways of knowing and doing. Being immersed in the game of mathematics often means that those who are involved are unaware of how the game is being played and the unspoken rules of the game.

The game of mathematics education is lived through various *objective* structuring practices. These include curriculum documents, assessment practices or pedagogical practices (including group work, streaming, text books). Also, at play are *subjective* structuring practices, whereby participants internalise the effects of objective structuring practices so that they come to see themselves as particular types of learners of mathematics. Within the field, particular knowledges and dispositions are seen as more valuable so that those learners who are able to display such characteristics are more likely to be seen as successful learners. But, such dispositions are not neutral and are often brought to the school situation from the out-of-school experiences of learners. For students whose home habitus aligns with the practices of schools, there is greater chance for them to be framed in particular (and positive) ways.

> The relation between habitus and field operates in two ways. On the one side, it is a relation of conditioning: the field structures the habitus, which is the product of the embodiment of immanent necessity of the field (or of a hierarchically intersecting set of fields). On the other side, it is a relation of knowledge or cognitive construction: habitus contributes to constituting the field as a meaningful world, a world endowed with sense and with value, in which it is worth investing one's practice. (Bourdieu & Wacquant, 1989, p. 144)

Within a Bourdieuian framing, young children come to develop a primary habitus in the home, which includes a language component, so that when they enter school that habitus is valued to greater or lesser degree depending upon its synergy with the field of school mathematics. For those whose habitus is more strongly aligned with the practices of the field, the habitus offers considerably more capital than for their peers whose habitus is less aligned with the field. The habitus thus becomes of form of capital that can be exchanged within a given field for other forms of rewards. In the case of school mathematics, this might be in the form of grades and other accolades bestowed upon 'successful' learners of the discipline.

Linguistic capital: implications for learning mathematics

In this section, I draw on one particular aspect of language, prepositions, to highlight the role of language in learning mathematics, and how the linguistic habitus of learners can enhance or hinder their success in the field. Within mathematics, prepositions are important in the study of space strand since they refer to how objects are located in relation to other objects. This is in stark contrast to other areas of the curriculum, particularly in the teaching of reading, where it has been shown that effective readers often miss the small words as they skim through text to read the more significant terms. In mathematics, the reader needs to pay attention to detail since terms, such as prepositions, play a key role in making meaning. The lexical density of mathematics means that every word serves a

Table 15.1 Prepositions commonly used in classrooms

Across	After	Against	Around	Among	Along
Beneath	Between	Before	By	Beside	Below
During	Down	In	Into	Like	Near
From	For	On	Off	Over	Of
Past	Toward	Through	Under	Up	With
Underneath	Without	Within			

purpose, so skimming through word problems can create difficulties. Prepositions are the small words often ignored by readers but which have significant value in mathematics.

Within school mathematics, a range of prepositions is commonly used. Some of these prepositions have greater application and use in mathematics classes than others. Most are commonly used in relation to position, although they are also used in other contexts. For example, the preposition 'over' can be used to denote position in an overt way, but can be used to describe position when working with numbers—'the numerator is over the denominator'. It is difficult to think of teaching mathematics without the use of prepositions. Table 15.1 is a list of commonly used prepositions in mathematics classrooms. It is not an extensive list, but it highlights the amount and types of prepositions that are used.

The prepositions used in mathematics classrooms can be particularly problematic for Indigenous students and for students who do not use the middle-class English register. The multiple terms that can be used to describe a particular relationship creates a situation whereby the meaning of the terms is less accessible to students. For example, the terms 'next', 'beside' and 'near' can all be used to describe the same situation. Furthermore, for students whose first language is not school English, the difficulty is compounded. Where the first language is not that of English, the use of prepositions may not be as evident, thereby creating difficulties in translation for students. For example, if the use of 'off' and 'of' are considered, the words sound very similar, yet 25% off $100 is very different from 25% of $100. In many of the sign languages used by deaf and hearing-impaired students, the use of prepositions is only a minor aspect of language development, as students are taught to use key words and rely on contexts in which the signs are being used in order to glean meanings (Hyde, Power, & Zevenbergen, 1999; Zevenbergen, Hyde, & Power, 2001). Similarly, the very subtle differences in these two terms may be indiscernible for some Indigenous learners who are prone to hearing loss due to ear infections.

For learners who have a strong grasp of language; that is a linguistic habitus that is aligned with the field, there is a greater possibility for success in mathematics. Having a strong grasp of the breadth, meaning and applicataion of prepositions represents a strong linguistic habitus that can, in turn, be exchanged for rewards and accolades within the field, thus making the linguistic habitus a form of capital that bestows rewards within the field of school mathematics.

Prepositions in Kriol languages

In contrast, the linguistic habitus that an Indigenous learner brings to the school is shaped by the language games of the home. The richness of prepositions in the primary habitus facilitate greater or lesser chance of converting the linguistic habitus to scholastic capital. In studying Kriol languages of the Kimberley/Pilbara region,[4] Hudson (1983) identified five prepositions. These are outlined in Table 15.2. As can be seen from this table, the use of prepositions within Kriol languages does not easily translate to school English.

Examples such as this highlight a two-fold difficulty. First, there is little direct translation from one language to another, thus making it difficult for the learner to transpose ideas from the home language into the school language. Second, the number of prepositions is considerably less than those that students encounter in school mathematics. This makes the explanation of the prepositions used in school mathematics more complex due to the poor synergies between the two languages. The examples here suggest that the linguistic habitus that these Indigenous learners bring to school do not transfer easily to school mathematics, thus making for considerable difficulties in coming to learn the acute differences in English prepositions and their applications in school mathematics. This requires considerable reconstitution of the linguistic habitus of the learners.

Table 15.2 Prepositions used in Kriol languages

Langa or shortened to *la*	In, at, on, in, near	Det men ben hittim langa hed The man hit her/him on the head
Blanga or shortened to *bla*	For, because of, about and possessive s	Det wumun bin kukum dempa bla ola kid The woman cooked damper for the children
Fo	Used for talking about the reason for doing something or for indicating possession of something	Wi bin lukum Rufus fo met We saw Rufus' friend
From	Similar to English from	Dei bin lukunat as from kemp They watched us from the camp
Garra (longer version is *garram*)	Used to indicate association with a person or thing. Equivaent to with or using	Det boi bin nakam garra stone The boy hit her with a stone

Source: Adapted from Berry and Hudson (1997, p. 118).

[4]The Kimberley/Pilbara region is in far north Western Australia.

Conclusion

In the preceding sections, I have sought to illustrate some of the challenges that language poses for Indigenous learners in coming to learn mathematics. Where students come to school with the middle-class register of the language of instruction, in this case English, they are more likely to be positioned well as learners. For them, there is a strong synergy between their home habitus and the practices of school mathematics. The learning difficulties of many Indigenous learners can be better understood as systematic failure due to the misrecognition of the habitus of the learner and the unquestioned practices of the field. While the issues are complex, the role of language and culture cannot be ignored. Through this chapter, I have sought to illustrate some of the ways in which language is implicated in the teaching and learning of mathematics and how it poses particular learning difficulties for students whose language, and hence familial habitus, is not aligned with school instruction. Many remote Indigenous students enter school with at least two languages, neither of which is SAE. Coming to learn mathematics is about learning the language of instruction, which is embedded in a particular nuanced relay system heavily laden with cultural values that may not be known to the learners. Ignoring the subtle and coersive ways in which mathematical language is implicated in the failure of students coming to learn school mathematics amounts to symbolic violence. Appreciating and redressing these forms of symbolic violence may be a small step in changing the current educational disadvantage faced by significant numbers of Indigenous learners.

Essential next questions

In writing this chapter, my intention has been to challenge the orthodoxy around 'learning difficulties' associated with Indigenous learners. I have used this cultural group for two reasons. First, they are most at risk of performing poorly on most standardised testing schemes, which engenders some deficit thinking around explaining such performance. My goal in this chapter was to highlight the ways in which the structuring practices of the field of mathematics are highly exclusive in terms of excluding particular groups of people. My second intention was to draw on the most excluded group of people to illustrate the reifying processes adopted within the field of mathematics that exclude learners so that a strong case can be presented. These examples highlight the structuring practices within the field that can be applied to other social and cultural groups. The concepts used in this chapter allow for a rich discussion of how the field of mathematics education creates a mythology of 'learning difficulties' that enables the blame for failure to be placed on the learner rather than the field. By using Bourdieu's concepts, a discussion is possible as to a way forward that enables greater access to the most hegemonic of curriculum areas.

What practices can be developed to enable greater access to school mathematics?

Through the writings of Bourdieu, it becomes possible to challenge the practices of school mathematics in order to open them up for critique. In a small chapter it is difficult to do justice to the corpus of work generated by Bourdieu. Many concepts have not been addressed here and those that have are only marginally included. A wider reading of Bourdieu's work enables a much richer conceptualisation of how practice is implicated in scholastic mortality for marginalised groups. From a theoretical perspective, a richer understanding of marginalisation can be developed. From a practical perspective, his tools enable a shift away from deficit thinking towards developing a much richer practice that enables greater access and participation.

How can we challenge and change the status quo around language barriers in school mathematics?

Breaking down the barriers created by the doxa in the field of mathematics is probably the greatest challenge I see in school mathematics. Members of the field have considerable power not only within the field, but also beyond it, given the high status of mathematics in the wider community. I would contend that much of the power of mathematics has been achieved through its exclusion of learners who have come to accept that the field is one for the 'elite' and, by implication, that they are not capable. As such, it becomes important for those who have power within the field to recognise the structuring practices that exclude participation. By opening up and challenging the taken-for-granted orthodoxies of the field, a more equitable field may be created, with greater success for those traditionally excluded by the field.

What are the implications of this work for teachers and teacher education?

The advantage of using a Bourdieuian approach is that it does not engender a victim-blaming approach of either teachers or learners. Rather, the work enables an understanding of the field and how its practices have come to be rendered invisible in terms of access and exclusion. By focusing on practice, greater chance exists for changing practice. My own work has enabled me to work with many educators, and I see the biggest challenges in two areas. For many primary school teachers, their background in mathematics is weak so there is a tendency to defer to the field and a lack of confidence to challenge the dominant orthodoxies. Using

Bourdieu's work enables many to see how their exclusion from the field positions them as 'others' and, as such, they are able to see the ways in which their exclusion works. In contrast, secondary school teachers have had considerable buy-in into the field and have rendered the practices of exclusion as invisible, but on the basis of their strong capital within the field—after all, they were successful in their study of mathematics. Often, they see others as less successful due to some innate feature or characteristic. This is evident in the practice of ability grouping, which is rampant in mathematics but not in other curriculum areas. Using Bourdieu's approach enables teachers to see the structured failing of their students as being due to the practices of the field rather than some innate characteristic of the student.

Glossary

Capital The accumulation of forms of objectified and/or subjective relations that become forms that can be exchanged for other gains. For example, capital may come in the form of culture—such as language—that can be exchanged for other goods (such as test scores, rewards, accolades) within a field. The field will shape the particular forms of capital that are valued within that field. The language valued in schools is different from the language valued in hip-hop music, but within the two fields the language operates as a form of capital shaped by that field.

Doxa A feel for how particular 'games' are played out in particular contexts. In this chapter, having 'doxa' in schools enables learners to understand the practice of teaching situations so as to be able to participate effectively within that practice.

Field The arena in which an object of study can be undertaken. This would include arenas such as education, work, medicine, sport or art.

Habitus The embodiment of culture that provides a lens for seeing and creating the world.

Subitise To immediately recognise collections of objects by 'how many' are in the collection but without counting each item.

References

Bernstein, B. (1996). *Pedagogy, symbolic control and identity: Theory, research and critique*. London: Taylor & Francis.

Berry, R., & Hudson, J. (1997). *Making the jump: A resource book for teachers of aboriginal students*. Kimberleys: Catholic Education Office Kimberly Region.

Bourdieu, P. (1990). *The logic of practice* (R. Nice, Trans.). Cambridge: Polity Press.

Bourdieu, P. (1991). *Language and symbolic power* (G. A. M. Raymond, Trans.). Cambridge: Polity Press.

Bourdieu, P., & Wacquant, L. (1989). Towards a reflexive sociology: A workshop with Pierre Bourdieu. *Sociological Theory*, *7*(1), 26–63.

Cooper, B., & Dunne, M. (1999). *Assessing children's mathematical knowledge: Social class, sex and problem solving*. London: Open University Press.

FaHCSIA (Department of Families Housing, Community Services and Indigenous Affairs). (2009). *Closing the gap for Indigenous Australians*. Accessed April 15, 2009, http://www.fahcsia.gov.au/indigenous/closing_the_gap/default.htm

Harris, P. (1990). *Mathematics in a cultural context: Aboriginal perspectives on time, space and money*. Geelong: Deakin University Press.

Hudson, T. (1983). Correspondences and numerical differences between disjoint sets. *Child Development, 54*, 84–90.

Hyde, M., Power, D., & Zevenbergen, R. (1999). Deaf students solving of arithmetic word problems. In J. Truran & K. Truran (Eds.), *Making the difference: Proceedings from the 22nd MERGA conference* (pp. 275–282). Adelaide: MERGA.

Kanes, C. (1991). Games and language-games: Towards a socially interactive mode for learning mathematics. In F. Furinghetti (Ed.), *Proceedings of the 15th PME conference* (Vol. 2, pp. 229–236). Assisi, Italy: International Group for the Psychology of Mathematics Education.

Leder, G. C., Rowley, G., & Brew, C. (1995). Second language learners: Help or hindrance for mathematics achievement? In R. P. Hunting, G. E. Fitzsimons, P. C. Clarkson, & A. J. Bishop (Eds.), *Regional collaboration in mathematics education* (pp. 425–434). Melbourne: Monash University.

MCEETYA. (2009). *National assessment program literacy and numeracy: Achievement in reading, writing, language conventions and numeracy*. Canberra: Author.

National Curriculum Board. (2008). *National mathematics curriculum: Framing paper*. Accessed April 14, 2009, http://www.ncb.org.au/verve/_resources/National_Mathematics_Curriculum_-_Framing_Paper.pdf

Walkerdine, V., & Lucey, H. (1989). *Democracy in the kitchen: Regulating mothers and socialising daughters*. London: Virago.

Watson, H. (1988). Language and mathematics education for Aboriginal-Australian children. *Language and Education, 2*(4), 255–273.

Watson, H. (1989). A wittgensteinian view of mathematics: Implications for teachers of mathematics. In N. F. Ellerton & M. A. Clements (Eds.), *School mathematics: The challenge to change* (pp. 18–30). Geelong: Deakin University Press.

Watson, H., & Chambers, D. W. (1989). *Singing the land, signing the land*. Geelong, VIC: Deakin University Press.

Willis, S. (2000). Strengthening numeracy: Reducing risk. *Improving numeracy learning: Research conference 2000: Proceedings* (pp. 3–24). Melbourne: Australian Council for Educational Research.

Wittgenstein, L. (1953). *Philosophical investigations* (G. E. M. Anscombe, Trans., 1974 ed.). Oxford: Basil Blackwell.

Wittgenstein, L. (1967). *Remarks on the foundations of mathematics*. London: Basil Backwell.

Zevenbergen, R. (2000). 'Cracking the code' of mathematics: School success as a function of linguistic, social and cultural background. In J. Boaler (Ed.), *Multiple perspectives on mathematics teaching and learning* (pp. 201–221). New York: JAI/Ablex.

Zevenbergen, R., Hyde, M., & Power, D. (2001). Language, arithmetic word problems and deaf students: Linguistic strategies used by deaf students to solve tasks. *Mathematics Education Research Journal, 13*(3), 204–218.

Zevenbergen, R., Mousley, J., & Sullivan, P. (2004). Disrupting pedagogic relay in mathematics classrooms: Using open-ended tasks with indigenous students. *International Journal of Inclusive Education, 8*(4), 391–415.

Chapter 16
'She's Not in My Head or in My Body': Developing Identities of Exclusion and Inclusion in Whole-Class Discussions

Laura Black

Introduction

Whole-class discussions have long been the focus of much research. They are a dominant practice in many classrooms and are, potentially, key sites for fostering children's learning (Barnes, 1976; Cazden, 2001; Edwards & Mercer, 1987; Wells, 1999). In England, we have seen the endorsement of whole-class teaching within educational policy (Department for Education and Employment (DfEE), 1998, 1999, 2001), not only as a pedagogic tool for standards-based educational reform (Reynolds & Muijs, 1999), but also as a means to achieving inclusivity within an increasingly diverse classroom environment. For example, the National Literacy Strategy Framework for Teaching introduced the notion of *interactive whole class teaching*, stating that it

> benefits children who need help to access the curriculum for example low attainers, those with special needs or limited fluency in English, by introducing them to challenging work, teaching them skills in a highly supportive context and raising their expectations and self esteem. (DfEE, 1998, p. 96)

In achieving such inclusivity, teachers have been directed to differentiate their communicative behaviour in accordance with student ability. For instance, the National Strategy for Key Stage 3[1] states that teachers should use 'open and closed questions which should be adjusted and targeted at pupils according to ability so that all pupils can contribute' (DfEE, 2001). Additionally, the aforementioned framework for teaching from the National Literacy Strategy stated the following about its precursor pilot study, the National Literacy Project

L. Black (✉)
University of Manchester, Manchester, UK
e-mail: laura.black@manchester.ac.uk

[1] In England, compulsory education is divided into four key stages: Key stage 1—age 5–7 years, Key stage 2—7–11 years, Key stage 3—11–14 years and Key stage 4—14–16-years. For the majority, Key stages 1 and 2 are covered at primary school, while Key stages 3 and 4 are covered at secondary school.

C. Wyatt-Smith et al. (eds.), *Multiple Perspectives on Difficulties in Learning Literacy and Numeracy*, DOI 10.1007/978-1-4020-8864-3_16,

> By tailoring questions and inviting contributions at more challenging levels, teachers in the National Literacy Project expected [pupils] to be able to explain and justify their ideas, to make generalisations, generate hypotheses, and offer critical comments. All this, in a class situation, challenged the able whilst keeping within contexts that lower attainers could understand and in which they could participate. (DfEE, 1998, p. 96)

While these guidelines may be appealing in a context in which teachers are required to deliver a highly paced curriculum to a diverse group of students over a short period of time, they also offer familiar deficit assumptions surrounding those with learning difficulties or positioned as 'needing help to access the curriculum'. They present a binary division between those who are labelled 'the able' and are to be challenged, and those who are merely able to remain party to the discussion and participate. Thus, the latter are viewed in terms of how they can be included in the activities of their normative, more successful peers, rather than as students who can be challenged and subsequently can play a role in shaping discussions in their own terms (Benjamin, 2002). Additionally, the extracts above also conflate notions of 'needing help to access the curriculum' with the construct of 'ability'—a feat that Benjamin (2002) argues positions students intellectually in subordination to dominant versions of success legitimised by the education system. Obviously, such assumptions are not only made by educational policy makers; for instance, Ainscow, Booth, and Dyson (2004) note how teachers' beliefs regarding student 'deficit' or disadvantage permeate their behaviour and present a barrier to pedagogic practices such as whole-class teaching, which are intended to promote inclusivity. However, these guidelines do raise several questions regarding the nature of whole-class teaching and how teachers are to orchestrate such discussions in an environment in which the construct of 'ability' and the drive for individualised notions of success (that is through performance in examinations) dominate. Consequently, we might ask: what are we to make of this notion of participation that is deemed appropriate for 'lower attainers' or 'those who need help accessing the curriculum'? How does it differ from the kinds of participation experienced by 'the able'? And what may be the long-term consequences for these different groups of students?

In this chapter, I aim to address these questions using the case of Jason, a student who was identified to me by his teacher as a student who 'has his problems' (more detail on this provided shortly). I have selected Jason as a focus for this chapter for two reasons. First, the label the teacher ascribed to him suggested that his institutional identity (Gee, 2001) was as a student who needed help accessing the curriculum, and thus he was a student who, according to the government guidelines referred to above, would benefit from whole-class teaching. Second, his experience of whole-class discussions (or my re-storying of it) highlights the complexity of participation and non-participation in the classroom practices that constitute learning a given subject (in this case, learning school mathematics).

In presenting Jason's case, I draw on sociocultural theory (Lave & Wenger, 1991; Wenger, 1998), to understand how the communicative behaviour of teachers and students impacts upon student participation and non-participation in whole-class discussions. This approach sees learning as the product of reciprocal relations between persons and practice (Solomon, 1998) involving the transition of one's learner identity from novice/new member to expert (Lave & Wenger, 1991; Wenger, 1998) through a process of participation; a process of ontological transformation rather than epistemological effect (Hodges, 1998). Key to participation in practice is the use of common tools and artefacts that are used to enable new members to adopt the goals, belief systems, ground rules and cultural norms of a specific practice (Wenger, 1998) and, over time, to generate common knowledge required for the smooth running of everyday activities (Renshaw, 2003). In many classrooms, whole-class discussions are positioned 'ideally' as a key-mediating tool that allow teachers and students to build a shared foundation of experience from which to negotiate the various concepts, representations and methods associated with the specific subject discipline (Mercer, 2000, 1995). Thus, I argue that whole-class discussions are a key site for the transformation of students into 'successful' learners of the relevant knowledge domain (Boaler, 2000; Lerman, 2001).

Methodology

My re-storying of Jason's experience comes from a wider ethnographic study that explored how student access to productive talk (for learning) in whole-class discussions related to cultural background (that is class, gender and ethnicity). The study involved a 5-month period of participant observation in a single classroom (Class 5W), during which time maths lessons were observed and video recorded two mornings a week—a joint decision made between me and the teacher. Observing whole-class discussions over this period of time was important because it enabled me to analyse how student participation developed as the children became familiar with the teacher, each other and the classroom's micro-environment. Two months after the observation period had ended, the teacher and five selected students were interviewed about their perceptions of teacher–student talk in whole-class discussions and learning mathematics.

The data were analysed using a multi-layered framework that examined individual teacher–student interactions within whole-class discussions, both as moments of participation/non-participation in the practices associated with learning school mathematics and cumulatively, as constructing particular learning trajectories for each student over time (Solomon, 2008). In addition, the teacher's interview data concerning her pedagogic intentions was used to develop an understanding of the cultural models (Gee, 1996) she drew upon in describing her practice, and to analyse how these informed her behaviour within whole-class discussions (see below for further explanation). The ethnographic nature of the research was a key aspect of the process of analysis, since it provided the researcher with 'insider' contextual knowledge, drawn upon to interpret the data. I have written about this multi-layered analysis framework elsewhere (Black, 2007) so here I provide a brief overview of how coding was developed to help orientate the analysis of Jason's story, which is to come.

The data were assigned codes at three different stages:

1. Micro-analysis of teacher–student interactions—Edwards and Mercer's (1987) analysis of classroom talk was used as a basis to identify whether the interaction was active or passive. Active interactions were those in which the student played a dialogic role (Barnes, 1976) and where they appeared to 'tune in' to the teacher's intended meaning. Passive interactions were those in which the teacher used high-controlling behaviour (for example, 'closed questioning') and the student played a limited, monosyllabic role. The aim of this analysis was to provide a picture of student participation in whole-class discussions in this classroom—that is who experienced active or passive interactions and when.
2. Cumulative analysis of teacher–student interactions—a series of analytical steps that quantified the qualitative micro-analysis, with the aim of tracking the continuity of students' experiences of 'active' or 'passive' teacher–student interactions across time. This enabled me to consider the communicative roles and learner identities students took on in class and the nature of the learning trajectory each student was on (Solomon, 2008). This analysis also corroborated earlier interpretations of the data, by representing each teacher–student interaction as one of a repeated number of cases.
3. Cultural models analysis—this stage focused on the cultural models drawn upon by the teacher in her post-observation interview since, given her institutional identity[2] as teacher (Gee, 2001), she was deemed to have control over when and where whole-class discussions took place, who took part in them and in what way. Gee (1996) refers to 'cultural models' as everyday theories of action that are situated in social and cultural experiences. For example, the teacher may draw upon and be informed by cultural models that relate to pedagogy (for example a belief in learning as a social act) or to the specific subject she is teaching (for example the concept of mathematics as a 'hard' subject). Additionally, Gee (1996) states that cultural models are aligned or disaligned with the wider macro-ideologies of the institution—ideologies that assign symbolic value to certain forms of behaviour (for example a child who is keen to participate and is enthusiastic is classed as 'able', whereas a child who is more reticent to contribute is viewed as 'less able'). In most cases, we would expect the cultural models adopted by the teacher to align with those of the institution, since it is the teacher's role to act as representative of the institution in the classroom. A series of cultural model codes were developed, based on the teacher's interview. I then re-read each teacher–student interaction, looking for evidence where such cultural models appeared to inform the behaviour of the teacher. This stage of the analysis enabled me to explore why the teacher engaged in the specific

[2]Gee (2001) describes institutional identities as those that refer to one's position within an institution—for example the status of a teacher or a student is authorised or legitimated by the school system.

forms of behaviour observed, what beliefs and values informed their actions and where such beliefs originated. This chapter focuses specifically on one particular cultural model, which was dominant in both the teacher's interview and in her classroom communication—the individualised learner.

The analysis was conducted chronologically, each stage building on previous findings. This enabled me to understand how moments of participation and non-participation, and the longer-term roles or identities that such moments formed over time, were framed through the teacher's intentions and actions by dominant cultural models (Gee, 1996). The analysis of Jason's experience, below, will demonstrate this analysis process further.

The school

The school was situated in a large town in the north west of England, and was one of two in the area that had a diverse intake according to class and ethnicity due to its geographical location and reputation. The school was viewed by many parents as an alternative option for those who did not get into the neighbouring, more successful school. It was located within a Local Education Authority, which still operated a grammar-school system in which students had to pass an '11-plus' entrance exam in order to attend the two most 'successful' secondary schools in the district. Only 'high attainers' were encouraged by their teachers to sit the exam, and were subsequently supported in head-teacher references. As this exam occurred shortly after the students entered year 6 (aged 10–11 years), the judgment of the year 5 (aged 9–10 years) teacher was crucial in determining which students were capable of passing the exam. The presence of the grammar-school system provided an unusual flavour to the classroom observed, and there was a strong emphasis on sorting and selecting students on the basis of ability, both within and between classrooms.

The classroom observed was a year 5, top-set maths class (out of two) consisting of 12 girls and 17 boys; four students were from minority ethnic groups. Their highly experienced teacher—Mrs Williams—was from the local area and had been teaching for 30 years. The children's opportunity to engage in learning school mathematics came in the form of traditional classroom activities (for example whole-class discussions, one-to-one interactions with the teacher, working independently with the textbook) and whole-class discussions were a key tool in enabling children to acquire the appropriate tools and representations (Boaler, 2000). Whole-class discussions typically took place at the beginning of each lesson and, in organising and controlling these discussions, the teacher had several purposes 'to give them confidence ... to go over the language ... to introduce the idea to them' [teacher interview].

Jason's positioning in year 5W

As already mentioned, Jason was identified to me by his teacher as a child with problems; he had difficulties with reading, described by the teacher as 'very poor with his language', and such difficulties would often infiltrate his experience in maths lessons. For example, his teacher would frequently ask other students to assist him in reading the questions in the textbook. Furthermore, although Jason was a regular participant in class (involved in 28 interactions during the observation period, which was above the class average of 20), the role he played in whole-class discussions was often fairly limited, as the following extract exemplifies:

In this lesson, the children were learning to calculate fractions of a pound and then decimalise these amounts. This was a textbook task, and the questions they were discussing were as follows: 4) 50p = £0.50 and 5) 60p = £0.60.

1. **T**: Alright, so let's just move on then cos we've just really, briefly gone
2. through question one, two and three there, haven't we? Err ten pence of a
3. pound, ten pence piece is ten p or nought dot one nought. Twenty pence is
4. nought, dot two nought. So how would you write a fifty pence? (*Tim puts*
5. *hand up*)
6. **Tim**: err nought dot five nought
7. **T:** nought dot five nought. (*writes it on the board*) Sixty p? (*pause*) Jason?
8. **Jason**: (*no response*) (*pause*)
9. **T:** We're on question five. How would you write sixty pence as part of a
10. pound?*(pause)* Sixty pence. Pretend that is a six. (*pointing to the five in '0.50'*
11. *written on the board*) *(pause)* Alright, so what would you do with it here?
12. *(pointing to '60p' written in the textbook)*
13. **Jason**: *(no response) (pause)*
14. **T:** Do you genuinely not know? *(sounding fustrated)*
15. **Jason**: Change that.
16. **T**: I knew you knew it. Sixty p, course it is, just change that don't you? Write
17. Nought.

Fig. 16.1 Extract from lesson on 'money'

Here, we see Jason playing out a passive role—he does not respond to several of the teacher's questions (lines 8 and 13), and his final input provides an ambiguous suggestion (line 15 'Change that'), which has been heavily cued by the teacher's line of questions (lines 7, 9–12, 14) and her use of a visual clue on the blackboard. The question is: why does Jason behave in this way? And what are the implications for Jason in terms of developing his identity as a learner of school mathematics?

A vast body of research has highlighted that, due to the institutional identities assigned to teachers and students (Gee, 2001), classroom interactions inevitably involve a power imbalance that places the teacher in control of who says what and when (Barnes, 1976; Edwards & Mercer, 1987; Wells, 1992). Thus, it is their behaviour that plays a dominant role in shaping student input. Edwards and Mercer (1987) identify the kind of repeated questioning we see above as 'cued elicitation' which, they argue, evokes a kind of 'guessing game' whereby the student ends up trying to guess the teacher's pre-determined, monosyllabic answer. The pre-determined nature of Jason's required answer is evident at the end of the interaction in Fig. 16.1, where the teacher exclaims 'I knew you knew it' (line 16). As such, we might argue that Jason's reticent behaviour is shaped, in part at least, by the teacher's heavy use of questioning to publicly evaluate his understanding of the task. An act that ultimately offers him a risky position of either providing a wrong answer or admitting that he does not understand (in response to line 14 'Do you genuinely not know?').

Analysis at the cumulative stage revealed that this kind of interaction was not an isolated experience for Jason. Of the interactions he was involved in, 71% appeared to take a similar form, in that they involved highly controlled episodes of cued elicitation, whereby the teacher asked a line of questions with a pre-determined answer in mind. Opportunities for Jason to engage in dialogic forms of interaction, which I outlined earlier as being beneficial to children's learning, were rare. As such, the consistency of his involvement in 'cued elicitation' meant that Jason was repeatedly positioned through the teacher's behaviour as a student who needed to be checked and evaluated on a regular basis.

In order to understand this process of positioning further, I now provide an example of an interaction involving another boy, James, who was one of a small number of boys in the class who regularly experienced interactions of a very different nature (Fig. 16.2).

This extract is from a revision lesson held just prior to when the children took an assessment known as the Richmond Test[3]. Here, the teacher introduces the concept of 'factors' which although, not covered on the National Curriculum[4] at this stage, appear on the test.

1. **T:** We're going to be looking at some new, different types of words, for
2. example, does anybody know what a factor is? If I said, give me a factor of
3. fifteen, can anyone give me one? *(pause)* …
4. **James:** I … I kind of understand.
5. **T:** Go on then, you tell me what a factor of fifteen is.
6. **James:** Is it … is it a number that'll add up ... erm times or add up to ...
7. **T** *(interrupts)*: Times.
8. **James:** Up to fifty?
9. **T:** Up to anything. Up to any number. So what's a factor ... give me a factor of
10. fifteen then. *(Sean puts hand up)* So what number will divide equally into …
11. Fifteen? *(she nods at Sean)*
12. **Sean:** Three.
13. **T:** Three. That's a factor of fifteen.
14. **James:** *(calls out)* And five.
15. **T:** Think of another ... five. Think of another one? *(nods to Chris)*
16. **Chris:** One.
17. **T:** And another one? *(Pause) (Sean has hand up)*
18. **Sean:** Fifteen.
19. **T:** Fifteen. Those are the factors of fifteen.

Fig. 16.2 Extract from revision lesson

Here, James's hesitant use of 'I kind of understand' (line 4) suggests that he is beginning to behave as a 'learner'—a novice who is attempting to understand mathematical concepts and representations but is aware that they are not fully formed yet. Although the teacher still controls the interaction (she asks the questions) and James's understanding of 'a factor' is in fact incomplete (that is in line 8 he believes a factor must make up the number 'fifty'), the interaction is far more dialogic than that which we see with Jason. He sustains his involvement in the dialogue despite his incorrect answer—even breaking the ground rules of classroom discourse by calling out (line 14) (Edwards & Mercer, 1987)—and is instrumental in constructing the definition of a factor.

Thus, unlike Jason, James is able to take on a more active role in constructing new knowledge with the teacher and other pupils. However, further analysis suggests he is able to act in this way because the teacher offers him the space to do so (Wood, 1992). First, she allows him to play teacher and take responsibility for defining specialised terminology (line 5—open directive) and second, she validates his definition despite it being incomplete, by incorporating it into her model for calculating a factor (lines 9–19). Therefore, on this occasion, the teacher sends out a message to the rest of the class that James, as someone who 'kind of understands', is able to make a credible contribution to the local practice of learning school mathematics by defining key concepts such as 'factors'.

Analysis at the cumulative stage revealed that James did not always play such a positive role in whole-class discussions. During the initial weeks of my observation, James was involved in very few interactions and these were predominantly passive. However, by the final 6 weeks of the observation period, he was involved in a total of 23 interactions, 61% of which were identified as active, with James playing the kind of dialogic role we see above. I have discussed elsewhere the reasons behind James's transformation during the school year, suggesting that the teacher may have changed her communicative behaviour in response to several episodes in which James publicly displayed cultural capital (for example James told the class how his father had helped him learn the rule for adding fractions—Black, 2002; see also Solomon, 2008 for further discussion of this same data). As such, it is not my intention to discuss James's experience at length here but rather point out that, in contrast to Jason, during the course of the year James was able to use certain cultural resources to take on the identity of someone who can act like a learner of school mathematics (see Black, 2004a, 2004b for further examples of such positive learner identities).

The question we might ask then is why did these two students have such different experiences of whole-class discussions in Class 5W? Why did the teacher behave differently towards these two pupils? Here, I turn to the cultural models analysis of the teacher interview data in order to unpack the teacher's perceived motivations/intentions and to consider how these might inform our understanding of the observation data.

When asked to describe Jason's participation in class discussions, his teacher reported the following

T: . . . he clams up doesn't he? But that's because I think he gets embarrassed. I don't think he believes he can do it really, I think that's just his nervousness. I mean he sometimes just won't say anything at all *even if* you ask him to justify something or explain something. He might have the right answer or he may well have worked it out totally happily on his own but when you come to ask him 'how', 'why' he clams up . . . he won't explain it even though he's done it right.

Here, the teacher juxtaposes Jason's typical behaviour of reticence and minimal input with what she deems to be good-quality classroom talk involving justification and explanation—the latter are both characteristics of quality talk for learning as outlined in both policy (DfEE, 1998, 1999, 2001) and research on classroom communication referred to earlier (Barnes, 1976; Cazden, 2001; Edwards & Mercer, 1987; Mercer, 1995; Wells, 1999). However, in explaining why Jason played this limited role, the teacher draws on a cultural model of individualism. Through her use of 'even if' in relation to Jason's typical behaviour, the teacher suggests that regardless of how she behaves, Jason is unwilling to act as a participant due to intrinsic traits such as 'nervousness' and 'embarrassment'; a move that locates the responsibility for Jason's failure to benefit from their discussions firmly with Jason rather than herself. A cultural model of individualism views learning as an individual act whereby students are personally responsible for their own success and failures—Boler (1999) calls this a meritocratic fiction that decontextualises students' behaviour and experiences from the social and cultural contexts that shape them. The following extract shows the teacher communicating this belief directly to Jason and thereby clearly informing him of his position in the class.

T: Mr Collingwood *(another Year 5 teacher)* and I have had chats about you. You're losing . . . you're losing that little extra something you had at the beginning of the year, you're not trying as hard and that worries us because you were beginning to really start to make . . . you know, make a little bit of effort . . . you were putting extra effort into working. And I feel at the moment . . . I know there's lots of words but you've had lots of help, haven't you? . . . but you're slowing up again, so you're falling behind. And that's sad. Now come one, let's try and do it.

Once again we see the teacher emphasise to Jason his own responsibility for his success or failure. In doing so, she draws on notions of 'ability' and 'pacing' to prompt Jason into working harder—she emphasises his need to put 'extra effort into working', which is presumably required because 'there's lots of words' and he needs extra help. Furthermore, the reference to 'falling behind' suggests she is comparing his performance with other 'faster' or 'more able' students in the class and feels the need to prevent students such as Jason from 'slowing up' (in terms of progression through the curriculum).

Thus, it appears that, despite her initial subscription to beliefs regarding quality talk (involving justification and explanation), beliefs that may relate to some kind of socio-constructivist pedagogy, the teacher ultimately subscribed to individualism and as such appeared to differentiate her behaviour according to what she perceived to be the individual needs of each pupil. Clearly, such beliefs are not uniquely held by Mrs Williams alone (Ainscow et al., 2004)—a cultural model based around individual learning permeates the dominant policy discourse of performativity (Solomon, 2008) with its emphasis on labelling individual children based on their performance in various forms of summative assessment. Indeed, Dadds (2001) has highlighted how teachers often feel conflicted between a pedagogy for delivery, with its emphasis on maintaining a fast pace through the curriculum, and a pedagogy for learning. However, the data I have presented here highlights how such beliefs can act as cultural models that inform the teacher's communicative behaviour in this classroom and thus offer different spaces for students to take on learner identities.

Student agency in shaping learner identities

I have so far focused on the role of the teacher's behaviour and her beliefs in positioning Jason as a non-participant in terms of learning school mathematics. However, as we can see in the extract above (and, indeed, that involving James) this role was not solely constructed by means of the teacher's communicative behaviour—Jason also contributed through his own behaviour (for example his lack of response to the teacher's questions in the above extract meant that he did not behave in the 'appropriate' manner as expected by the teacher). The following extract reveals Jason further contributing to his non-participatory role by articulating himself as someone who does not share the teacher's view of the world (Fig. 16.3).

Here, Jason and Hasan discuss Mrs Williams' feedback on a test she had marked on fractions. They had been instructed to re-do any incorrect sums.

1. **Jason**: Hasan, what does that say?
2. **Hasan**: 'Corrections are needed, you ... you rushed the last part of this work.'
3. You've rushed it, you did.
4. **Jason**: I never, I never.
5. **Hasan**: You did.
6. **Jason**: I never.
7. **Hasan**: You have to do it all over again. Yeh, it says up there.
8. **Jason**: Yeh, I know but I never rushed.
9. **Hasan**: It says you rushed.
10. **Jason**: I never. She's not in my head or in my body.

Fig. 16.3 Jason and Hasan discuss Jason's work

On this occasion, the teacher has recorded that Jason is someone who rushes his work and Hasan re-asserts this perception by giving voice to the teacher's written comments. However, rather than accept the teacher's judgment (as required to fit the institutionally 'ideal' role of 'the student'), here, Jason's comments suggest he is not willing to take on the passive role that has been assigned to him and simply accept that Mrs Williams has the right to define how he behaved while doing the test. Jason's assertion that 'She's not in my head or in my body' indicates that he is not willing to go along with the teacher's version of events. Thus, he uses the space provided by the context of a conversation with his friend, Hasan, to construct his own identity as a non-participant (although he may well be 'self authoring' himself as a participant in another, non-formal practice-Holland, Lachicotte, Skinner, & Cain, 1998). Thus, we see Jason actively subscribing to his peripheral position and, therefore, injecting some sense of agency into the construction of his identity (Holland et al., 1998). However, due to the hierarchical power structure of the classroom and Jason's limited voice within it, he is unable to do this in a way that would challenge the teacher's expectations and move towards becoming a more active participant, like James. In fact, we might argue that James's adoption of a more positive learner identity only occurs because his behaviour (exchanging his cultural capital) aligned with the dominant behavioural norms and expectations of the institution.

Becoming a participant/non-participant in whole-class discussions

The case of Jason (and, indeed, James), which I have outlined above, indicates how varied students' participation and non-participation in the activities that constitute learning school mathematics can be. But what exactly does such participation/non-participation entail? And what are the implications of such positionings? I have already argued that learning is a process of becoming or 'ontological transformation' (Hodges, 1998) through participation and engagement in social practice (Holland et al., 1998; Lave & Wenger, 1991; Wenger, 1998), and in the case of learning school mathematics, a key feature of participation involves engaging in active, dialogic interactions such as that experienced by James. These provide students with the opportunity to use the tools, specialised register (Zevenbergen, 2000), concepts and methods of the relevant knowledge domain effectively (Barnes & Todd, 1995; Mercer, 2000). However, there are several points to note about the nature of participation in any formal classroom practice, which may help us understand the process by which students such as Jason come to be positioned on the margins of such practices.

First, there are specific power relationships inherent in any institutional practice, which dominate participation in activities one way or another. Within such practices, what counts as appropriate forms of behaviour and participation is non-indigenous, determined largely by the institution and manifested through the teacher's behaviour as the more powerful agent. This is particularly the case in a traditional, didactic classroom, such as Class 5W, where pupils are given very little opportunity to define and re-construct the local practice and have to be willing to accept a fairly passive role in negotiating what signifies 'legitimate membership' (Boaler & Greeno, 2000) (that is the modes of behaviour that suggest one is becoming a successful learner of school mathematics). All the pupils in Class 5W (including James) conformed to such a role, since they were unable to negotiate the curriculum content or format of the lessons, or transform the tools identified as appropriate by the institution (for example use of the mathematical register). In both of the above teacher–student interactions, Mrs Williams firmly sets the agenda through her use of questions, and carefully controls the students' input in an attempt to ensure that it fits with her understanding of the appropriate cultural norms for doing school mathematics. Consequently, although James manages to play an active role in the interaction, he has little say in defining both the discussion topic and the mode of behaviour that is deemed 'appropriate' to that situation. He may behave like a 'learner of school mathematics', but he does not get to determine what such an identity entails.

Second, participation in formal classroom practices also involves being able to 'tune in' to the teacher's perspective and acquire ways of using and engaging with the cultural artifacts of the social practice in question. These include linguistic artifacts (Zevenbergen, 2000) and institutional knowledge in addition to domain-specific objects, tools and norms that are peculiar to the relevant formal classroom practice. For example, in Fig. 16.2 we see James 'tuning in' by testing out and refining ideas in reaction to the teacher's responses and in doing so he also utilises the mathematics register (discursive tool; that is 'times', 'add up to'). Although teacher and student do not ultimately share goals when engaging in dialogue, since both parties approach an interaction with different objectives, there is a sense in which some students display a willingness to go along with the teacher in pursuit of their own educational success. Once this is recognised and the student's status is legitimated by the teacher (for example by having their input validated in some way), such pupils (like James) may come to be regarded as credible contributors to the practice on a local level. By contrast, students like Jason, who are restricted to monosyllabic input in episodes of heavily controlled 'cued elicitation', experience a form of non-participation since they do not have the opportunity to actively contribute to the ideas/concepts under discussion. Boaler (2000) argues that such experiences limit students to the passive reception of knowledge, which is known to have a detrimental impact upon understanding (see Edwards & Mercer's (1987) ritual versus principled understanding).

Furthermore, I argue that the consistency of such passive interactions may also have long-term consequences for Jason's learning career, in that regular episodes of non-participation hinder the learner's transition from novice to active knower and may have a long-term impact on their educational success and the identities they take on in the future. Wenger (1998) argues that one outcome of such passivity is the development of marginalised identities

> Members whose contributions are never adopted develop an identity of non-participation that progressively marginalizes them . . . (p. 203)

Repeated alienation from activities (such as dialogic teacher–student interactions) that comprise the social practice serve to construct a marginalised identity for the student (in relation to practice) on a trajectory of non-participation that gradually makes the prospect of 'belonging' increasingly unattainable (Hodges, 1998). For Jason, this meant repeatedly being positioned in a passive role within whole-class discussions that further perpetuated his identity as a student who 'needed help', and was reticent to participate.

Conclusion

In this chapter, I have argued that teacher–student interactions within the context of whole-class discussions are a key mechanism in regulating student participation in the practices that constitute learning school mathematics, and consequently are significant sites for the co-construction and transformation of student identities from novices into mathematical learners. Within such interactions, students not only attempt to construct an understanding of mathematical concepts and methods using various tools and signs, but they also negotiate their position in relation to the social practice in question and each other. The two cases presented highlight the heterogeneity of this experience for students; some, like James, are able to contribute to the local operation of the practice through active participation in interactions. However, others, like Jason in Fig. 16.1, experience little opportunity to engage in sense-making and problem-solving using the relevant representations, concepts and methods as tools (Boaler & Greeno, 2000).

I have argued that the teacher's behaviour played a major role in positioning these students differently and I have highlighted how a cultural model based on individualism may have informed her behaviour and set up certain norms of participation (evident at the cumulative analysis stage). In class 5W, such norms meant that only a small number of students (mainly boys viewed as 'high ability') had the opportunity and space to genuinely contribute to the ideas under discussion. However, it is not my intention to blame teachers here or place responsibility for Jason's failure to learn with Mrs Williams. Clearly, teachers do not operate in a cultural vacuum and their behaviour is intricately informed and framed by policy and the institutions in which they work—a fact which I have acknowledged by analysing and presenting Mrs Williams' perceptions/beliefs as cultural models. According to Gee (1996), beliefs, values, etc. can only be interpreted as cultural models if they can be connected to everyday commonsense theories about how things work in the world and if they draw on wider institutional or macro-ideologies. Such an interpretation is not difficult to establish with a cultural model of individualism, since many have already noted how this notion has been reproduced within policy and institutional practices throughout history (Chitty, 1989; Thompson, 1980). For example, practices such as selective entry to grammar schools, tracking and de-tracking, streaming, setting, etc. are all predicated on an understanding of learning that places responsibility on the individual and rewards merit to those who align with the values of the institution. More recently in England, we have seen a shift in policy towards personalised learning within the primary sector (Department for Education and Skills (DfES)/Qualifications and Curriculum Authority (QCA), 2003)—the recent National Primary Strategy explicitly directs teachers to focus on students as individuals and to assess their 'needs' and 'abilities'

> Learning must be focused on individual pupils' needs and abilities ... Every teacher knows that truly effective learning focuses on individual children ... The new Primary Strategy will actively support more tailoring of teaching to individuals ... Workforce reform will ... be critical to helping teachers focus on individual children's needs ... Increasing the focus on individual children will serve every child. (p. 39 and paras. 4.1–4.5)

Thus, a cultural model of individualism is now promoted within policy guidelines as essential to 'effective learning', and furthermore as key to an 'inclusive' pedagogy that serves 'every child'. However, this is to be implemented alongside a performance agenda that places pressure on students, teachers and schools to achieve higher and higher grades through summative assessment. The latter further endorses the notion of the individualised learner by requiring teachers to measure students' performance against pre-determined assessment criteria (in England this is manifest in the National Curriculum attainment levels (DfEE, 1999; DfES/QCA, 2003)) and thereby make comparisons of individuals' competence or ability in a given subject. It is this dominant ideology, based on individual performance and merit, which permeates the dialogue analysed in this chapter; while James's 'appropriate behaviour' is rewarded with space to participate (and thereby increase his status), the case of Jason highlights the limited spaces available to students whose behaviour do not fit with the institutional ideal.

I noted at the beginning of this chapter that whole-class teaching has been endorsed within educational policy, in England at least, as a pedagogic practice that meets the inclusive agenda I have mentioned above. Whole-class teaching is presented as offering 'high levels of involvement for all pupils, particularly the least able, many of whom quickly gain confidence' (DfEE, 1998, p. 10). Furthermore, whole-class teaching is directed to be interactive, with all children encouraged to 'play an active part by ... contributing points to discussions, and explaining and demonstrating their methods and solutions to others in the class' (DfEE, 2001, p. 26). It is my view that the case of Jason highlights how such broad aims are not easily reconciled with a dominant policy agenda of performance and accountability. Although the teacher, Mrs Williams, attempts to include Jason in the dialogue, his failure to fit the institutional ideal (by being an active 'able' student such as James) means his inclusion is heavily controlled, monosyllabic and offers little in terms of developing his understanding of mathematics. Yet, it is not the practice of whole-class teaching per se that is problematic here—for example Alexander (2000) notes how a collective pedagogy operating in Russian classrooms means that teachers spend *more* 'talk time' with less-developed students in whole-class discussions because they deserve to achieve no less than their peers. Rather, it is the contradiction between an agenda of inclusivity and the dominant agenda of individualism (Benjamin, 2002) within policy that places pressure on teachers to make regular sacrifices (Dadds, 2001). The use of whole-class teaching to deliver a fast-paced curriculum in a context in which teachers are accountable for the individual performance of their students means that students such as Jason cannot be given the 'talk time' needed to develop a dialogic communicative role.

Thus, I argue that there is a need to move away from a performance agenda that encourages explanations of student behaviour based on fixed, intrinsic and individualised notions such as 'ability'. Such socially constructed concepts mask wider sociological processes at work (for example social background) since they determine educational success by presenting student differences as 'natural' and consequently, difficult to address (Bourne, 1994; Solomon, 2008). In class 5W, the strong ideological emphasis on individual ability or competence (possibly accentuated by the schools' institutional context as a grammar-school feeder) motivated the teacher towards using highly controlled forms of communication with 'low ability' students such as Jason. The consequences of this may be far reaching—when teachers and students interact they contribute to the co-construction of student identities, which may have long-lasting effects on future participation in educational practices (Wenger, 1998).

Essential next questions

In summary, given the deeply entrenched nature of the processes I have hopefully illustrated in this chapter, there remains a number of questions that we might address as a means of challenging pervasive dominant ideological beliefs regarding the individualised learner and the ensuing contradictions that emerge between this and notions of 'inclusive education'. These are

How might we support teachers in challenging the 'deficit assumptions' regarding student competence that are so prevalent in many classrooms and policy documents?

I have already indicated how notions regarding individualism and deficit assumptions operate as cultural models—values pertaining to particular discourses within the education that permeate at every level (Gee, 1996). As such, while we see such models evident in policy documents, we are also aware that they cannot shape, inform or affect what happens in the classroom without teachers and practitioners who, like Mrs Williams, are under increasing pressure (through mechanisms of accountability) to align with such dominant ideologies. This does not suggest we hold teachers accountable, but highlights the need to recognise their role in the legitimisation and constant reification of such beliefs (Ainscow et al., 2004). Thus, we are led to consider ways of enabling teachers to challenge or disalign. Others have suggested ways of encouraging teachers to theorise their practice through reflection (Argyris & Schon, 1974) and action research (Wells & Ball, 2008). What is essential here is that future research continues to empower teachers so that they are able to find the space to challenge what is accepted as 'normal' or natural.

How might we enable students to have a 'voice' in defining classroom activities, lesson content and the behaviour(s) that come to be viewed as legitimate participation?

In England, the recent introduction of the National Primary Strategy (DfES/QCA, 2003) has specified time in the national curriculum for teachers and students to investigate, apply and reflect on effective modes of communication under the heading of *Speaking & Listening*. However, while such opportunities for reflection among students is a step in the right direction, it falls far short of enabling them to have a say in what they learn and the types of behaviour or social norms that come to be legitimated by the teacher as evidence of competence or 'ability'. Indeed, the main focus of such guidelines appear to direct student attention towards how to behave in small-group situations where the teacher is not present (DfES/QCA, 2003). In the United States, several studies have attempted to give students more agency in both the content and nature of their learning (Cazden, 2001; Michaels, O'Connor, Hall, & Resnick, 2002). For example, Wells and Ball (2008) outline an approach known as 'dialogic inquiry' which, they argue, fosters more exploratory modes of talk because students are encouraged to focus on what genuinely interests them, to contribute their own opinions in discussion and they are given the right to share control and evaluate claims with those deemed as having expertise (for example the teacher). What appears to be crucial here is the opportunity such interventions provide to allow students to critically reflect on the nature of the power relationships they experience in the classroom. Future research should continue to find ways of allowing students to do this (in conjunction with teachers) so that they are able to recognise and define those modes of behaviour that represent 'competence' or 'ability' and, consequently, gain access to the expertise that both teachers and students (and others) bring to the classroom.

Glossary

Deficit assumptions Where one assumes that a difference in student behaviour/performance is the result of some kind of cognitive deficit (for example a lack of ability) with resulting limited expectations of what the student can do.

Dominant behavioural norms The modes of behaviour that have become normative in a given classroom because they align with the goals of the education system.

Individualism In the context of education, the notion that individual students are responsible for their own failure/success in learning or achievement and are consequently rewarded or punished on this basis.

Institutional practices The common activities and modes of behaviour that are peculiar to a given institution, such as the education system.

National Literacy Strategy (NLS) A government policy that revamped literacy teaching in England at the primary level (ages 5–11) in 1998. This directed teachers on lesson content, structure (the literacy hour) and delivery. It was not compulsory for schools to follow the NLS, but schools needed to have an alternative framework in place if they were to meet the standard of teaching expected by the government's school inspection body (OFSTED).

National Numeracy Strategy (NNS) similar to the NLS in purpose and direction, this strategy was introduced a year after the NLS, with a particular focus on raising standards in numeracy at the primary level.

National Primary Strategy Introduced in 2003 in primary schools, not as a replacement for the NLS and NNS, but as an 'umbrella' document which incorporated the key aspects of both. This brought new additions to primary teaching in England, including an increased focus on personalised learning and a stronger emphasis on formative assessment (assessment for learning).

National Strategy for Key Stage 3 The practice of directing teachers on lesson content, structure and mode of delivery in literacy and numeracy was extended to secondary schools in 2001. This covered the 11–14 years age range and focused on literacy and numeracy in the one document.

References

Ainscow, M., Booth, T., & Dyson, A. (2004). Understanding and developing inclusive practices in schools: A collaborative action research network. *International Journal of Inclusive Education, 8*(2), 125–140.

Alexander, R. (2000). *Culture and pedagogy: International comparisons in primary education.* Oxford: Blackwell Publishers.

Argyris, C., & Schon, D. A. (1974). *Theory in practice: Increasing professional effectiveness.* San Francisco: Jossey-Bass.

Barnes, D. (1976). *From communication to curriculum.* London: Penguin Books Ltd.

Barnes, D., & Todd, F. (1995). *Communication and learning revisited: Making meaning through talk.* Portsmouth, NH: Boynton/Cook Publishers Inc.

Benjamin, S. (2002). *The micropolitics of inclusive education: An ethnography.* Buckingham: Open University Press.

Black, L. (2002). *The guided construction of educational inequality: How socially disadvantaged children are marginalised in classroom interactions.* Unpublished PhD thesis, Lancaster University.

Black, L. (2004a). Teacher–pupil talk in whole class discussions and processes of social positioning within the primary school classroom. *Language and Education, 18*(5), 347–414.

Black, L. (2004b). Differential participation in whole class discussions and the construction of marginalised identities. *Journal of Educational Enquiry, 5*(1), 34–54.

Black, L. (2007). Analysing cultural models in socio-cultural discourse analysis. *International Journal of Educational Research, 46*(1–2), 20–30.

Boaler, J. (2000). Mathematics from another world: Traditional communities and the alienation of learners. *Journal of Mathematical Behavior, 18*(4), 1–19.

Boaler, J., & Greeno, J. (2000). Identity, agency and knowing in mathematics worlds. In J. Boaler (Ed.), *Multiple perspectives on mathematics teaching and learning.* Westport: Ablex Publishing.

Boler, M. (1999). *Feeling power: Emotions and education.* London: Routledge.

Bourne, J. (1994). A question of ability. In J. Bourne (Ed.), *Thinking through primary practice* (pp. 212–222). London: Routledge.

Cazden, C. (2001). *Classroom discourse: The language of teaching and learning.* Portsmouth, NH: Greenwood Press.

Chitty, C. (1989). *Towards a new education system: The victory of the new right?* London: Routledge.

Dadds, M. (2001). The politics of pedagogy. *Teachers and Teaching: Theory and Practice, 7*(1), 43–58.
Department for Education and Employment (DfEE). (1998). *The national literacy strategy: A framework for teaching*. London: HMSO.
Department for Education and Employment (DfEE). (1999). *The national numeracy strategy: Framework for teaching mathematics from reception to year 6*. London: HMSO.
Department for Education and Employment (DfEE). (2001). *The framework for teaching mathematics: Years 7, 8 and 9*. Nottingham: DfEE Publications.
Department for Education and Employment (DfEE). (2003). *Excellence and enjoyment: A strategy for primary schools*. London: Department for Education and Skills.
DfES/QCA. (2003). *Speaking, listening, learning: Working with children in key stages 1 and 2*. London: Crown Copyright.
Edwards, D., & Mercer, N. (1987). *Common knowledge: The development of understanding in the classroom*. London: Routledge.
Gee, J. (1996). *Social linguistics and literacies: Ideology in discourses*. New York: Routledge Falmer.
Gee, J. (2001). Identity as an analytic lens for research in education. *Review of Research in Education, 25*, 99–125.
Hodges, D. (1998). Participation as dis-identification with/in a community of practice. *Mind, Culture and Activity, 5*, 272–290.
Holland, D., Lachicotte, W., Skinner, D., & Cain, C. (1998). *Identity and agency in cultural worlds*. London: Harvard University Press.
Lave, J., & Wenger, E. (1991). *Situated learning: Legitimate peripheral participation*. Cambridge: Cambridge University Press.
Lerman, S. (2001). Cultural, discursive psychology: A socio-cultural approach to studying the teaching and learning of mathematics. *Educational Studies in Mathematics, 46*, 87–113.
Mercer, N. (1995). *The guided construction of knowledge*. Clevedon: Multilingual Matters.
Mercer, N. (2000). *Words and minds: How we use language to think together*. London: Routledge.
Michaels, S., O'Connor, M., Hall, M., & Resnick, L. (2002). *Accountable talk: Classroom conversation that works* (E-book from CD-ROM Series). Pittsburgh: University of Pittsburgh.
Renshaw, P. (2003). Community and learning: Contradictions, dilemmas and prospects. *Discourse: Studies in the Cultural Politics of Education, 24*(3), 355–370.
Reynolds, D., & Muijs, D. (1999). The effective teaching of mathematics: A review of research. *School Leadership and Management, 19*(3), 273–288.
Solomon, Y. (1998). Teaching mathematics: Ritual, principle and practice. *Journal of Philosophy of Education, 32*, 377–390.
Solomon, Y. (2008). *Mathematical literacy: Developing identities of inclusion*. London: Routledge.
Thompson, E. P. (1980). *The making of the english working class*. New York: Penguin Books.
Wells, G. (1992). The centrality of talk in education. In K. Norman (Ed.), *Thinking voices: The work of the National Oracy Project*. London: Hodder & Stoughton.
Wells, G. (1999). *Dialogic inquiry: Towards a socio-cultural practice and theory of education*. Cambridge: Cambridge University Press.
Wells, G., & Ball, T. (2008). Exploratory talk and dialogic inquiry. In N. Mercer & S. Hodgkinson (Eds.), *Exploring talk in school*. London: Sage.
Wenger, E. (1998). *Communities of practice: Learning, meaning and identity*. Cambridge: Cambridge University Press.
Wood, D. (1992). Teaching talk. In K. Norman (Ed.), *Thinking voices: The work of the National Oracy Project*. London: Hodder & Stoughton.
Zevenbergen, R. (2000). Cracking the code of mathematics classrooms: School success as a function of linguistic, social and cultural background. In J. Boaler (Ed.), *Multiple perspectives on mathematics teaching and learning*. Westport: Ablex Publishing.

Chapter 17
Breaking Down the Silos: The Search for an Evidentiary Base

John Elkins and Claire Wyatt-Smith

A key challenge for educators and researchers in the 21st century lies with the matter of evidence and evidence-based practice. A major question is, What are the desirable characteristics of the research evidentiary base to inform efforts to improve learning and learning outcomes for all students? A related challenge, and one that is perhaps more important, is to better understand different theoretical, methodological and conceptual arguments about the nature of learning itself, and to consider the complementarity and differences among them. Our starting proposition for this book, as discussed in Chapters 1 and 3, was that research into learning difficulties involves intensely complex, interdisciplinary work. This book shows a diversity of interdisciplinary, theoretical work in the field of learning difficulties. It is now our understanding that it is time to go well beyond simply recognising different relationships between theory and method, if we are to be serious about socially just and equitable education for all. It is time to break down the theoretical and methodological 'silos' in studying learning and learning difficulties, and to bring theory–method relationships to centre stage. It is time to connect research-based claims to different research methods and their related underpinning theoretical arguments and conceptualisations (for example of learning, of teaching, of assessment).

This chapter aims to consider the complexities of these challenges and how they can be addressed. In order to reach this aim, first we identify several overarching themes emerging from the chapters in this book. While no singular theme was present in every chapter, the suite of themes across the boundaries of literacy, numeracy and learning difficulties constitute an opportunity to witness points of convergence across fields, perspectives and methods.

Following discussion of these themes, we present a number of reflections that range more widely than the invited chapters, and draw upon our own activities that were the inspiration for the book, and also the fields of literacy, numeracy and learning difficulties as they continue to develop.

C. Wyatt-Smith (✉)
Griffith University, Brisbane, QLD, Australia
e-mail: c.wyatt-smith@griffith.edu.au

C. Wyatt-Smith et al. (eds.), *Multiple Perspectives on Difficulties in Learning Literacy and Numeracy*, DOI 10.1007/978-1-4020-8864-3_17,

Finally, we ask how we might prepare classroom teachers and researchers for thinking in a more integrated fashion about the learning of students in core curriculum areas. We consider issues relating to multiple research perspectives and operating across traditional domains: literacy, numeracy and learning difficulties. We also speculate about how we can synthesise the results of research that will support evidence-based practice but that can utilise the findings from a wide range of research methodologies and theoretical perspectives.

Overarching themes

A clear theme across the chapters was *matching the instruction and the learning needs* of the child at the point-in-time (as distinct from official or mandated curriculum specifications). While it is widely recognised that students may well engage in different ways with instructional programs, some authors make the strong point about the need to address who the students are who are expected to benefit from an instructional program or practice. From this vantage point, the selection of a program based on reported benefits is of itself insufficient. Several writers highlight the need to have the classroom program utilise ideas that come from more targeted research. Related to this, authors identify the need to understand the characteristics of those students who participated in the studies in which a conclusion is made of program or practice effectiveness. Overall, the authors suggest that this is imperative in ensuring a suitable match with the needs of the student. There is a call to ensure that decisions about learning support be based on suitably considered empirical evidence. This is the case even though the authors themselves do not extend to how evidence itself can be 'read' and 're-read' relative to different theoretical methods and underpinning conceptualisations of learning. Understandably, this was not their task.

However, this observation opens the space for consideration of issues relating to *contexts and conditions* and how these are important for optimal student results. A complete description of the setting and conditions under which interventions are implemented in research studies is critical to understanding where, when and how interventions are found to be effective. Whether an instructional program is effective in other settings, such as the general education classroom, and how that setting may affect the performance of students with learning difficulties as well as other students, are important concerns that need to be investigated.

The *relational dimension of teaching and learning* was a strong theme throughout the book. Within this dimension, several chapters presented discussions of how learning opportunities in the classroom might be broadened through the use of technologies. Supporting this stance is the recognition that online reading comprehension instruction is important for all students, including struggling readers. Castek and colleagues (Chapter 4) make the strong point that

> In today's world, all students must become efficient information managers and reflective thinkers who can collaborate and communicate effectively in new and complex online contexts that are constantly changing. Will providing more Internet activities help students with learning difficulties improve their academic achievement? How can we ensure this occurs?

Several writers, including Castek et al. and Harrison (Chapter 5), indicate the need for sustained research on how to structure classroom activities to enable all students to become active participants and contributors to online reading comprehension and strategy instruction, in particular. Further, the potential of technology to provide opportunities for developing students' creativity and higher-order thinking was identified, with a clear need for new research into ways of working with online knowledge. Parallel to this was the recognition that digital records that include live video of literacy and numeracy activities are a rich source of data for studying students' learning *in situ*.

Across the fields and methods, the writers indicate that there is a need to know more about classroom contexts and the types of interactions, including talk, action and interaction that occur in those contexts. Collectively, the chapters show how the everyday routine of school classrooms and what Brown in Chapter 13 (drawing on Cobb, Confrey, diSessa, Lehrer, & Schauble, 2003) refers to as the 'learning ecology' are much more than sites for explicit transmission of curricular knowledge. Instead, they show how learning is interactionally accomplished and, in many ways, is reliant upon complex patterns of action and interaction among students and between the teacher and students. Also, striking across the chapters is that the role of the teacher is central in explicit instruction, modelling, providing feedback, scaffolding and extending learning opportunities for students, individually and in whole-class settings.

Extending the focus on context and the relational dimension of learning and teaching, some chapters point to the need for increased time and heightened motivation through recruiting the community–home–learning partnership. Wearmouth and Berryman (Chapter 8), for example, argue that:

> We need to learn more about how schools can most effectively and sensitively support home-based literacy programs for students where parents' and/or carers' own experience of difficulties in acquiring literacy makes them particularly vulnerable to criticism or comment by others, unaware of family circumstances or cultural background.

A main point of convergence across the chapters was the powerful recognition that, while barriers to success in school learning play out in the classroom, some are traceable to wider cultural and socioeconomic factors. It is at this point that *contexts and the search for educational equity* come together. Kramer-Dahl and Kwek (Chapter 7) present the challenge

> What 'deficiencies' matter to educators? ... We acknowledged earlier in the chapter that there are material realities that students encounter daily that impact on their educational experiences, yet there is a need to ask exactly which 'deficiencies' in the home are important to educators for the specific purpose of *redressing* educational inequity in schools. Following through this question requires opening up issues of class and cultural differences, ethnicity and socioeconomic disparities, with the recognition that educational achievement is currently defined by the dominant (middle to upper) classes.

The observation that *teaching, learning and curriculum are inherently cultural* is also evident in several chapters, as is the observation that classrooms are sites in which powerful assumptions come to be played out and deficit beliefs persist. A contribution of several chapters is to show how, as students learn literacy and numeracy, they take on academic identities. These are as much about the actions that they undertake as they are about the collective talk, actions and interactions of those with whom they are learning (or resisting learning). Jorgensen's discussion (Chapter 15) of language learning and mathematics and her call for culturally relevant learning resources in Indigenous classrooms is a telling example, as is Wheldall and Beaman's call (Chapter 12) for intervening early in promoting receptive language development for Indigenous students.

Also, apparent across the chapters was the need for greater student agency and participatory action in learning in the classroom. Brown (Chapter 13) and Black (Chapter 16), for example, both address the issue of how students might be enabled to have a 'voice' in defining classroom activities, lesson content and the behaviours that come to be viewed as legitimate participation.

Reflections

Once there is a shift in focus from literacy and numeracy to the difficulties that students experience in learning these key skills, the diversity of approaches that can be used to investigate such difficulties tends to narrow considerably—leaving the field largely to the traditional dominance of psychologically oriented studies of the causes of difficulties and the treatment of atypical learners. Overall, the chapters in this book reflect a broad range of theoretical and methodological perspectives, even though there are others that might have found a place in the book. In this section we reflect upon a number of issues that we believe deserve attention.

While some researchers offer highly explicit theoretical foundations for their research, many expect readers to have a tacit understanding of the logic of studies. While the former should lead to trustworthy findings, it may blinker the researcher and readers to other ways of understanding research questions. The latter does not help readers unfamiliar with a particular research paradigm to appreciate the strengths and weaknesses of the work.

We appeal for wide training in theory and research technique, even if only to help researchers better choose problems and match research designs. The benefits for interpreting findings may be even more important, since awareness of other ways of approaching research problems may open up richer and more insightful accounts.

Of particular value may be the ways in which complementary and contradictory outcomes of studies can be addressed. If studies that use different theory and method lead to similar conclusions, this complementarity is reassuring. If results conflict, then being able to relate different theories and methods may be crucial in understanding results and in planning further studies. Further, such conflict may be productive in shedding light on the chosen theories and methods, giving insight to their application and sufficiency in particular studies.

There is always room for new theoretical formulations. When the first author of this chapter (Elkins) was introduced to educational research, many theories and methods, widely used today were not available to be applied to the study of difficulties in learning literacy and numeracy. Unfortunately, traditional approaches still dominate and the potential of this wide variety of theoretical orientations and methodologies has not been realised to date.

The three dominant fields of research that are concerned with learning difficulties in literacy and numeracy (literacy education, numeracy education and Special Education) overlap in many ways. Research mostly reflects its home literature, and researchers seem most comfortable when following the norms of their parent field. Most evident is that a Special Education heritage predisposes researchers to identifying particular 'aberrant' students to be studied, whereas a background in literacy or numeracy research increases the likelihood that the focus will be the events and conditions that result in students experiencing difficulties.

It is likely in the 21st century that the main responsibility for identifying and supporting students who experience difficulties in learning will lie increasingly with the classroom teacher. This is particularly true in schools with high proportions of students from indigenous, immigrant/refugee and low socioeconomic backgrounds. This is because a referral to an expert model is inapplicable, if most of the students need long-term, continuing support. Classroom teachers in such schools cannot expect to have low-achieving students supported by learning support specialists working outside the classroom, unless teachers have provided the best classroom learning opportunities. Thus, research needs to address literacy and numeracy instruction in classroom contexts where most students are low achieving. Research of this type is essential to secure the best outcomes for all students.

Even in schools in which most students respond to classroom opportunities to learn, there will be some students who experience difficulties. Here, the challenge remains for classroom teachers to adapt programs and provide additional support whether or not specialist assistance is available. Research on how this can be done efficiently is still needed, despite our wealth of knowledge about exemplary whole-class instruction for typical classrooms and about effective tutoring techniques in those settings. For example, comparison of the information provided by external achievement tests with that obtained by classroom teacher monitoring of student learning may help to reduce the time spent on assessment and thus increase learning time. Also, we need to know more about why some students do not prosper in classrooms that provide effective outcomes for their peers.

Thus, far we have considered literacy and numeracy separately. However, they need to be considered together since these key domains are the concern of the same teachers in at least the first 6, and sometimes 8, years of schooling. As a minimum, researchers with expertise in both fields might work together, since there may be efficiencies from co-researching learning activities that address both literacy and numeracy.

While response to intervention (RTI) has become widely supported as a way of addressing literacy difficulties, and to a lesser extent numeracy difficulties, it has not yet been widely or extensively studied as it has been applied in schools. Rather, it is a plausible hypothesis arising from the failure of previous definitional approaches to identifying and 'treating' learning disabilities. It seems worthwhile looking at RTI over a substantial timeframe that permits study of the transition from acquisition to application of literacy and numeracy. Criticisms include its imprecision when used in schools (Reynolds & Shaywitz, 2009), lack of research support for groups such as limited English proficient and ethnic groups, and lack of clarity about how to decide on referral for more intensive support (Burns, Jacob, & Wagner, 2008).

Research that takes a particular element of literacy or numeracy difficulty may prove relatively tidy to apply, but provides a challenge when implications are sought. For example, although phonemic awareness is a crucial component of early reading success, it is unwise to conclude that students who score low on tests of phonemic awareness have not been taught phonics. This is a reasonable hypothesis, but classroom-based observations may be needed to demonstrate that phonics has been neglected. It is equally possible that these students have failed to profit from phonics instruction and may have developed negative attitudes towards word recognition and reading, more generally. Further, we need to consider what the effect might be of excessive emphasis on phonics teaching for students who are well in command of the alphabetic principle.

These observations point to the need for caution in drawing conclusions and making recommendations for classroom literacy and numeracy instruction based on studies of a small element of these domains. In contrast, research that is broad in scope may fail to reveal the role of various components. Indeed, many 'programs' in literacy or numeracy are so multi-faceted that it may be difficult to do more that intuit the elements that are crucial to their effectiveness.

The chief beneficiaries of research are students, teachers and administrators. However, researchers are often at a remove from the classroom, with their need to publish and meet the pressures of funding bodies and editorial boards. Researchers may be more convinced of the benefits that ensue from their work than consumers such as teachers and administrators. Students are rarely informed about the results, let alone enabled to contribute to the research process. Studies that involve students actively in the research process may yield richer understanding as they are best placed to give 'insider accounts' of learning. Having professional researchers and teachers more collaboratively engaged in planning, carrying out and interpreting research deserves to be used more often. This would position teachers well to convert the results of research into changed practices that improve student learning. It would help teachers to move away from judging the utility of research findings in terms of their ease of implementation and acceptance by students. It would also go some way to redressing our concern that research in literacy and numeracy difficulties is often too narrowly focused on the practical, and therefore may miss opportunities to enrich theory.

While the research–policy–practice nexus is widely recognised, policy makers themselves tend to be constrained by what they believe to be political imperatives and, in turn, they influence the research projects that come to be funded. For example, testing has assumed a major role in education, with large-scale, standardised achievement tests being ubiquitous in many countries. How this influences teaching and learning is widely debated. If we consider research that is carried out in hospitals by clinicians, it is distinctive for their intimate understanding of the topic and their need to improve treatments. Schools could profitably be funded to carry out research, for example, by using external achievement test data together with classroom data, both assessment and instructional information. One way forward would be for schools to invest in teacher researcher positions to lead research in local contexts. One important consideration for teacher-led research is quality control, which is handled for university researchers through the peer review of proposals and reports. Quality assurance could be achieved through a process of panels in which teacher researchers and university researchers participate.

Most chapters in this book have studied difficulties in learning literacy and numeracy without particular attention as to whether methods and results may have differed if support had been organised differently. Usually differences in support are considered as differences in intervention techniques in a particular setting such as the whole class or the resource room. However, analysing the comparative effect of individual or small-group tutoring, modified classroom instruction or the contribution of home support is rarely undertaken systematically. Similarly, the interactions between organisational and instructional variables are seldom examined. The writing in this book points to how research studies on similar topics using different theoretical frameworks may also be fruitfully explored.

Conclusion

In conclusion, we emphasise the need to bring to centre stage theory–method relationships related to educational issues of our time and ask: How can classroom teachers and researchers be prepared to think in a more integrated fashion about the learning of students in core curriculum areas? This question necessitates consideration of issues relating to multiple research perspectives and operating across traditional domains: literacy, numeracy and learning difficulties. Increasingly, these domains need to be studied together, since they coexist with all areas of the school curriculum, and even in the formative years prior to formal education. Initial teacher preparation typically pays little attention to research, and there is a major challenge to bring the results of research to bear upon the pre-service experience and to present them in ways that can readily be incorporated into the growing repertoire of teaching skills and curriculum insights that are the main focus of teacher education.

Ways forward must include the breaking down of the traditional ‘silos’ such as have operated between the domains of literacy, numeracy and learning disabilities, and the related theoretical and research traditions of these domains. Most chapters in this book show that the influence of the separate domains of research is still strong. There is opportunity now to synthesise the results of research that will support evidence-based practice but which can utilise the findings from a wide range of research methodologies and theoretical perspectives. We need a way to accumulate the results of multiple studies that go beyond quantitative meta-analyses to encompass a wider range of theoretical perspectives and methodologies. The complexity of this task is hard to underestimate. The types of research used in literacy and numeracy research, even when restricted to learning difficulties, have grown steadily. Some research focuses on the efficacy of teaching, while other studies enrich our understanding of the conditions of learning. Furthermore, larger investigations increasingly resort to mixed methods, which has become one way in which the conflict between quantitative/empiricist versus qualitative/constructivist theoretical positions has been handled.

Further, with the rapid advances of new technologies and digital storage facilities, we are able to accumulate systematic data banks of quantitative, narrative and multimodal information that could be accessed by researchers for analysis in contexts beyond those for which the data were originally collected. This idea goes well beyond the notion of cross-national validation and offers promise of multi-faceted investigation of learning difficulties, and perhaps even towards a meta-theory. Further, it also offers promise of how longitudinal databases could be revisited and re-examined for the further development of practice-centred theories of teaching and assessment.

We consider that there has been a considerable shift in education wrought by the widespread use of national achievement testing programs. In particular, the focus is less on identifying particular students and more on improving the performance of students at all levels. This, in turn, could lead to an integration of the efforts of classroom and specialist teachers, and the interests of both the majority who are learning well and the minority who struggle are served in a coordinated fashion. Teachers will need greater familiarity with the collection and analysis of classroom learning data.

This leads into consideration of how a growing body of research has indicated that a substantial proportion of school effectiveness can be attributed to teachers, with teacher effects being cumulative and additive. Given this, teacher knowledge about the history of a field, theoretical perspectives and varying approaches and beliefs about learning are seen to be at the heart of any effort to improve education. Further, the importance of the teacher in the provision of high-quality instruction and the amelioration of learning difficulties has been routinely mentioned in several major studies. However, often theory development and research represent a top-down process from academic researchers to practitioners. Due to restraints on teacher time, robust mechanisms that routinely allow practicing teachers to be genuine partners in the research process have not been developed. Ideally, collaboration should exist between researchers and teachers in order to advance the profession, with mechanisms routinely established to ensure active teacher participation.

The challenge facing research and practice communities alike is to find ways of accumulating the results of multiple studies that go beyond quantitative meta-analyses to encompass a wider range of theoretical perspectives and methodologies. While multitheoretical and cross-disciplinary studies have been undertaken previously (see Cumming & Wyatt-Smith, 2001; Wyatt-Smith & Cumming, 2001), new technologies provide the means for digital data records to be readily stored and retrieved. This permits systematic and sustained analysis of data through different lenses, the goal being to generate new insights not otherwise possible. We look forward to a richer, more inclusive era in educational research that gives researchers wider vision as they plan and interpret their work, and that better informs teachers as they seek to support learning.

References

Burns, M. K., Jacob, S., & Wagner, A. (2008). Ethical and legal issues associated with using response-to-intervention to assess learning disabilities. *Journal of School Psychology*, *46*, 263–279.

Cobb, P., Confrey, J., diSessa, A., Lehrer, R., & Schauble, L. (2003). Design experiments in educational research. *Educational Researcher*, *32*(1), 9–13.

Cumming, J., & Wyatt-Smith, C. (Eds.). (2001). *Literacy and the curriculum: Success in senior secondary schooling*. Melbourne, VIC: ACER Press.

Reynolds, C. R., & Shaywitz, S. E. (2009). Response to intervention: Ready or not? Or, from wait-to-fail to watch-them-fail. *School Psychology Quarterly*, *24*(2), 130–145.

Wyatt-Smith, C. M., & Cumming, J. J. (Special Issue Eds.). (2001). Examining the literacy–curriculum connection. *Linguistics and Education*, *11*(4), 295–312.

Author Index

C. Wyatt-Smith et al. (eds.), *Multiple Perspectives on Difficulties in Learning Literacy and Numeracy*, DOI 10.1007/978-1-4020-8864-3,

Subject Index

Kolibri 360 GmbH
Ferdinand-Jühlke-Straße 7
99095 Erfurt, Deutschland
produktsicherheit@kolibri360.de